Advances in Plant-Sourced Natural Compounds as Anticancer Agents

Advances in Plant-Sourced Natural Compounds as Anticancer Agents

Guest Editors

Violeta Popovici
Emma Adriana Ozon
Cerasela Elena Gîrd
Dumitru Lupuliasa

MDPI

Basel • Beijing • Wuhan • Barcelona • Belgrade • Novi Sad • Cluj • Manchester

Guest Editors
Violeta Popovici
Department of Integrated Mountain Territorial Development
Romanian Academy
Vatra Dornei
Romania

Emma Adriana Ozon
Department of Pharmaceutical Technology and Biopharmacy
"Carol Davila" University of Medicine and Pharmacy
Bucharest
Romania

Cerasela Elena Gîrd
Department of Pharmacognosy, Phytochemistry and Phytotherapy
"Carol Davila" University of Medicine and Pharmacy
Bucharest
Romania

Dumitru Lupuliasa
Department of Pharmaceutical Technology and Biopharmacy
"Carol Davila" University of Medicine and Pharmacy
Bucharest
Romania

Editorial Office
MDPI AG
Grosspeteranlage 5
4052 Basel, Switzerland

This is a reprint of the Special Issue, published open access by the journal *Molecules* (ISSN 1420-3049), freely accessible at: https://www.mdpi.com/journal/molecules/special_issues/Q6661R9V2K.

For citation purposes, cite each article independently as indicated on the article page online and as indicated below:

Lastname, A.A.; Lastname, B.B. Article Title. *Journal Name* **Year**, *Volume Number*, Page Range.

ISBN 978-3-7258-6450-8 (Hbk)
ISBN 978-3-7258-6451-5 (PDF)
https://doi.org/10.3390/books978-3-7258-6451-5

Contents

About the Editors

Violeta Popovici

Violeta Popovici is a Senior Clinical Pharmacist, Ph.D. in Dental Medicine, and Senior Researcher III at the Mountain Economy Center (CE-MONT), Vatra-Dornei. She finished her Post-Doc Studies at Ovidius University of Constanta. Her interdisciplinary research over 6 years focused on the lichen Usnea barbata in the Călimani Mountains, Romania. The phytochemical profile and pharmacological properties of U. barbata extract in different solvents were explored. By incorporating the most active U. barbata extracts into mucoadhesive oral films, promising effects against oral cancer and various oral infections caused by opportunistic pathogens have been reported. These extensive studies were published in reputable WOS-indexed Journals (Q1 and Q2), and the oral films loaded with U. barbata extracts remain unique worldwide. She currently collaborates with interdisciplinary research teams across the medical field. Her research interests include plant/lichen secondary metabolites—identification and pharmacological potential; antibiotics; antimicrobial activity; in vitro anticancer activity; oral cancer; ROS; oxidative stress; antioxidants/pro-oxidants; food; diets; and case studies.

Emma Adriana Ozon

Emma Adriana Ozon is a Doctor of Pharmacy and Professor in the Department of Pharmaceutical Technology and Biopharmacy at Carol Davila University of Medicine and Pharmacy in Bucharest, Romania. Her work focuses on pharmaceutical formulation, preformulation studies, and physicochemical characterization of various dosage forms. She has significant experience enhancing drug solubility and bioavailability, particularly for poorly soluble drugs, using methods such as cyclodextrin inclusion complexes and lyophilization. Key research areas include the development of oral solid dosage forms, mouth-dissolving films, topical gel-creams, and cosmetic preparations, as well as innovative methods such as 3D printing and hot-melt extrusion in drug manufacturing. Currently, an interdisciplinary research team conducted by Professor Ozon investigates other pharmaceutical formulations loaded with U. barbata extracts, with other different pharmacological properties.

Cerasela Elena Gîrd

Cerasela Elena Gîrd is Professor at the Faculty of Pharmacy of the "Carol Davila" University of Medicine and Pharmacy in Bucharest. She is the head of the Discipline of Pharmacognosy, Phytochemistry and Phytotherapy, and a prominent figure in Romania's pharmaceutical sector, recognized for her extensive expertise in pharmacognosy, phytochemistry, phytotherapy, apitherapy, and aromatherapy. An elected member of the Council of the Second Department and the Professorial Council of the Faculty of Pharmacy, she is actively involved in medicinal plant research, aiming to develop pharmaceutical formulations for diverse therapeutic applications. She has numerous scientific publications and participation in national and international scientific events. She has published many essential books and university courses in the field with reputable publishers. She is a member of several professional organizations, including the Society of Pharmaceutical Sciences of Romania and the Romanian Association of Phytotherapists.

Dumitru Lupuleasa

Dumitru Lupuleasa, Emeritus Professor at the "Carol Davila" University of Medicine and Pharmacy in Bucharest, is a recognized authority in Pharmaceutical Technology and Biopharmacy. He has contributed significantly to the scientific literature in the pharmaceutical field through numerous research papers and award-winning books, addressing topics including pharmaceutical technology, bio-pharmacy, and pharmaceutical practice. He is Doctor Honoris Causa of Ovidius University of Constanta and of the University of Oradea. He is a member of the Academy of Medical Sciences and the scientific council of the National Medicines Agency. He is also president of the Romanian Society of Pharmaceutical Sciences and the College of Pharmacists of Romania. He is an essential voice in the regulation and modernization of the pharmacy profession at the national level.

Preface

Cancer remains one of the leading causes of death worldwide, particularly in countries with higher life expectancy, education levels, and standards of living. Despite significant advances in early detection and treatment over recent decades, cancer continues to pose a considerable challenge to global health.

The primary goal of current cancer therapy is to selectively target and eliminate malignant cells in early-stage neoplasms or those remaining after surgical intervention, with minimal harm to normal physiological processes. Additionally, new therapies seek to trigger an immune response against residual tumor cells to prevent recurrence. However, even with the advent of targeted and immunotherapies for invasive neoplasms, these approaches often yield only modest improvements in patient survival, mainly due to rapid drug resistance and tumor adaptation.

As a result, there is growing interest in exploring alternative and complementary strategies, such as using natural compounds as adjuvant therapies, to overcome these limitations and improve patient outcomes.

The plant world is a rich source of bioactive molecules for the discovery and development of new anticancer drugs. Several plant metabolites have served as lead structures for the development of FDA-approved chemotherapeutics. Research into medicinal plants continues to grow because novel therapeutics derived from them offer the promise of more effective, less toxic treatment options for patients, particularly when conventional therapies are limited by resistance or intolerable side effects. They may enhance cancer cells' sensitivity to existing treatments. Therefore, this Special Issue Reprint includes advanced studies on the anticancer activity of various plant extracts (*Laurus nobilis, Berberis vulgaris*), essential oils (from *Juniperus excelsa, Lavandula vera, Salvia libanotica*, and *Lantana camara*), and plant-derived compounds (fucoxanthin, curcumenol, cordycepin derivatives, and tocopherols). Their intrinsic mechanisms involve key signaling pathways implicated in tumor growth, metastasis, and immune evasion, and these findings suggest their potential applications in the prophylaxis and treatment of various neoplasms.

Violeta Popovici, Emma Adriana Ozon, Cerasela Elena Gîrd, and Dumitru Lupuliasa
Guest Editors

molecules

MDPI

Article

The Therapeutic Potential of *Laurus nobilis* L. Leaves Ethanolic Extract in Cancer Therapy

Farah Al-Mammori [1], Ashraf M. A. Qasem [1], Deniz Al-Tawalbeh [2], Duaa Abuarqoub [3,4] and Ali Hmedat [5,*]

[1] Department of Pharmaceutical Sciences, Faculty of Pharmacy, Zarqa University, Zarqa 13110, Jordan; fmamoori@zu.edu.jo (F.A.-M.); aqasem@zu.edu.jo (A.M.A.Q.)
[2] Department of Medicinal Chemistry and Pharmacognosy, Faculty of Pharmacy, Yarmouk University, Irbid 21163, Jordan; deniz.altawalbeh@yu.edu.jo
[3] Department of Pharmacology and Biomedical Sciences, Faculty of Pharmacy and Medical Sciences, University of Petra, Amman 11196, Jordan; duaa.abuarqoub@uop.edu.jo
[4] Cell Therapy Center, The University of Jordan, Amman 11942, Jordan
[5] Department of Pharmaceutical Technology and Pharmaceutics, Faculty of Pharmacy, Yarmouk University, Irbid 21163, Jordan
* Correspondence: ali.hmedat@yu.edu.jo; Tel.: +962-797875512

Abstract

This study explores the anticancer, antioxidant, and phytochemical activities of *Laurus nobilis* L. ethanolic leaf extract. The extract demonstrated selective cytotoxicity against four human cancer cell lines, showing strong cytotoxic effect against ovarian (ES2), head and neck (SAS), and colorectal (HT-29) cancer cells, with IC_{50} values ranging from 3.8 ± 0.3 to 4.4 ± 0.6 µg/mL. Notably, it exhibited only moderate inhibition of the MDA-MB-231 breast cancer cell line (IC_{50} = 10.5 ± 0.8 µg/mL), possibly reflecting intrinsic differences in cell line sensitivity. Importantly, the extract showed low toxicity toward normal human fibroblasts (HDF), with an IC_{50} value exceeding 100 µg/mL, indicating a favorable selectivity profile. The flow cytometry analysis showed that the extract caused cell death and stopped the cell cycle in both SAS and ES2 cancer cell lines. In SAS cells, extract treatment significantly increased apoptotic cells (21.1% ± 0.3%) compared to the control (6.3% ± 0.4%), along with G2 phase accumulation, indicating G2 arrest. Similarly, in ES2 cells, apoptosis increased (16.2% ± 1.3% vs. control 8.1% ± 1.0%), and a significant cell accumulation in the S phase was observed, suggesting disruption of cell cycle progression. Antioxidant screenings showed impressive dose-dependent DPPH radical scavenging activity (25–2000 µg/mL), although less potent than ascorbic acid (2.6 µg/mL). UPLC-QTOF/MS phytochemical analysis revealed various phenolic constituents, such as flavonoids and phenolic acids, and an inferred association with the recorded bioactivities. This preliminary work indicates that *L. nobilis* extracts may act as natural anticancer and antioxidant agents; however, it was limited to *in vitro* testing with non-standardized samples, underscoring the need for further research to validate and extend these findings for future applications.

Keywords: *Laurus nobilis*; anticancer; antioxidant; breast cancer; ovarian cancer; colorectal cancer; head and neck cancer; apoptosis

1. Introduction

Cancer is the second-leading cause of death worldwide, and the global cancer burden is predicted to grow over the next two decades [1]. Conventional cancer treatments, such as chemotherapy, can impose significant physical and psychological burdens, prompting increased interest in plant-based remedies as safer, adjunctive options [2]. Throughout history, people have turned to nature as a source of medicine, a tradition that persists in modern healthcare. The global popularity of herbal medicine has grown significantly, particularly as complementary and alternative therapies gain attention for their potential role in managing diseases, including cancer [3,4].

Laurus nobilis (family Lauraceae), commonly known as bay leaf, is a well-known medicinal plant widely used in Mediterranean regions, including Jordan. Its leaves and fruits are valued for their aromatic properties and have traditionally been used to treat gastrointestinal issues such as bloating and indigestion [5]. The essential oil extracted from the leaves has been shown to have antibacterial, antioxidant, and anti-inflammatory properties. Additional leaves of *L. nobilis* have been used to cure rheumatism, neuralgia, and scabies [6]. The studies suggest that *L. nobilis* possesses phenolic compounds, including flavonoids, phenolic acids, and tannins (proanthocyanidins) [7], with antioxidant properties, potentially contributing to its anticancer effects [5]. Because oxidative stress is a well-documented phenomenon in cancer, it is reasonable that antioxidants can greatly reduce cancer incidence and development [8]. However, further research is required to validate its therapeutic applications and understand its mechanisms of action. In Jordan, breast cancer is the most common cancer among women and is often diagnosed at advanced stages and younger ages than in Western countries [9]. Ovarian cancer ranks as the sixth leading cause of cancer-related deaths among women globally [10], while colorectal cancer is the most prevalent cancer in Jordanian men and the second most common in women [8]. Additionally, head and neck cancers, the seventh most common malignancies worldwide, contribute significantly to male cancer mortality [11]. Despite the growing global interest in herbal medicine and preliminary reports highlighting the antioxidant and potential anticancer properties of *L. nobilis*, evidence for its efficacy against specific cancer types remains limited, particularly those prevalent in Jordan. This study is novel in demonstrating its selective activity against multiple cancer cell lines, including SAS head and neck cancer cells, while confirming safety on normal fibroblasts and revealing mechanisms such as apoptosis and cell cycle arrest. Therefore, the present work aims to evaluate the anticancer and antioxidant activities of the ethanolic leaf extracts of *L. nobilis* and to identify their main phytochemical constituents using UPLC-QTOF/MS. The study is preliminary and seeks to establish the potential use of *L. nobilis* extract components in the future.

2. Results and Discussion

2.1. Cytotoxic Activity

The MTT assay was performed to evaluate the anticancer activity of the extract against MDA-MB-231, SAS, HT-29, and ES2 cell lines. As shown in Figure 1, treatment with increasing concentrations of the extract (0.03–100 μg/mL) resulted in a dose-dependent decrease in cell viability, indicating its potential cytotoxic effect across all tested cell lines.

To quantify this effect, IC_{50} values were determined from dose–response curves, including normal human fibroblasts, to evaluate the extract's selectivity toward cancer cells. As summarized in Table 1, the extract exhibited potent inhibition on ES2 (IC_{50} = 4.2 ± 0.2 μg/mL), SAS (IC_{50} = 3.8 ± 0.2 μg/mL), and HT-29 (IC_{50} = 4.4 ± 0.6 μg/mL) cell lines, while demonstrating moderate inhibition against MDA-MB-231 cells (IC_{50} = 18.5 ± 0.8 μg/mL). The noticeably higher IC_{50} value observed in MDA-MB-231 cells (18.5 μg/mL) compared to the values obtained for ES2, HT-29, and SAS cells (3.8–4.4 μg/mL) may reflect intrinsic differences in cell line sensitivity. MDA-MB-231 is a triple-negative breast cancer cell line known for its aggressive phenotype, high metastatic potential, and resistance to a wide range of chemotherapeutic agents, which may contribute to the reduced sensitivity observed in our assay [12]. This highlights the potential cell type-specific activity of the extract, which may be more effective against certain tumor types.

Notably, the extract displayed low toxicity toward normal fibroblast cells (IC_{50} > 100 μg/mL), indicating selectivity for cancer cells.

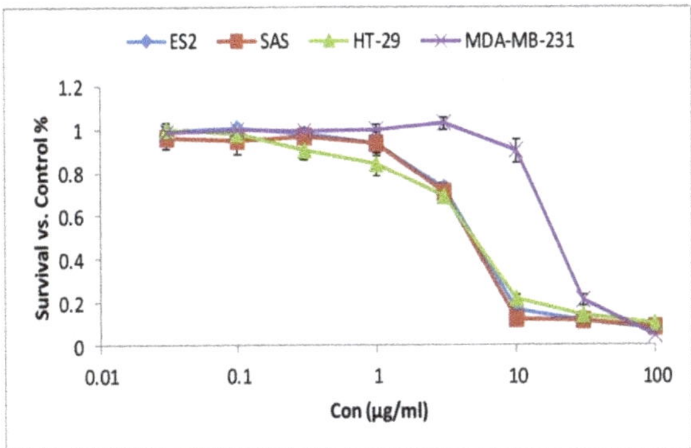

Figure 1. Cytotoxic activity of *L. nobilis* leaves ethanolic extract, as determined by the MTT assay. Dose–response curves show a concentration-dependent decrease in cell viability. Data are presented as mean ± SD from three independent experiments.

Table 1. IC_{50} values of the *L. nobilis* extract and cisplatin against cancer cell lines and normal fibroblasts (μg/mL).

Cell Line	*L. nobilis* Extract	Cisplatin
ES2	4.2 ± 0.2	3.6 ± 0.3
SAS	3.8 ± 0.3	2.9 ± 0.5
HT-29	4.4 ± 0.6	10.7 ± 2.1
MDA MB-231	18.5 ± 0.8	16.7 ± 1.1
HDF	>100	>100

To benchmark the anticancer efficacy of *L. nobilis* ethanolic extract, its IC_{50} values were compared with those of cisplatin, a widely used chemotherapeutic agent. As shown in Table 1, the extract exhibited comparable cytotoxicity to cisplatin in ES2 and SAS cells, with IC_{50} values of 4.2 ± 0.2 and 3.8 ± 0.2 μg/mL, respectively, versus 3.6 ± 0.3 and 2.9 ± 0.5 μg/mL for cisplatin. Notably, the extract demonstrated superior cytotoxicity in HT-29 colorectal cancer cells (4.4 ± 0.6 μg/mL) compared to cisplatin (10.7 ± 2.1 μg/mL), underscoring its potential therapeutic relevance in colorectal cancer. For MDA-MB-231, a triple-negative breast cancer cell line known for its chemoresistance, both agents showed relatively higher IC_{50} values (18.5 ± 0.8 μg/mL for the extract vs. 16.7 ± 1.1 μg/mL for cisplatin), reflecting lower susceptibility of this phenotype.

These findings suggest a broad-spectrum anticancer activity of the extract, as it effectively inhibited the growth of multiple cancer cell lines from different tissue origins, including head and neck, colorectal, ovarian, and breast cancer [13]. The significant cytotoxic effects observed in ES2, SAS, and HT-29 cells, along with the moderate effect on MDA-MB-231, indicate that the extract may interfere with key oncogenic pathways shared across various cancer types. However, further investigation is needed to identify the specific molecular targets and mechanisms underlying its anticancer effects, which could provide valuable insights for future therapeutic development. Additionally, its low toxicity toward normal HDF highlights its potential selectivity for malignant cells, making it a promising candidate for further preclinical development as an anticancer agent [14]. This selective cytotoxicity may be explained by inherent differences between cancerous and normal cells. The selective cytotoxicity of the *L. nobilis* ethanolic extract toward malignant cells over normal human fibroblasts may be attributed to several cancer-specific biological characteristics.

One key factor is the significantly higher proliferation rate of cancer cells compared to normal fibroblasts [15]. Rapidly dividing cells are more susceptible to agents that interfere with DNA replication, cell cycle progression, and oxidative stress—all of which are known mechanisms of action for phenolic compounds such as quercetin, kaempferol, and gallic acid identified in our extract [16]. Moreover, cancer cells typically exhibit elevated basal levels of reactive oxygen species and a disrupted redox balance [17], rendering them more vulnerable to additional oxidative stress induced by phenolic compounds [18].

2.2. Apoptosis Evaluation and Cell Cycle Analysis

To further investigate the mechanisms underlying the reduced cell viability, the induction of apoptosis in SAS and ES2 cells following extract treatment was assessed using Annexin V/PI double staining, as described in Section 3 [19]. The results, presented in Figure 2A,B, reveal a significant increase in Annexin V-positive apoptotic cells after treatment with the extract (SAS: control 6.3% ± 0.4% vs. extract 21.1% ± 0.3%, n = 3) and (ES2: control 8.1% ± 1.0% vs. extract 16.2% ± 1.3%, n = 3). These findings suggest that the extract may promote cell death through apoptosis, consistent with previous observations of plant extract-induced apoptosis in cancer cells [20]. However, further investigations are required to confirm this apoptotic effect by examining caspase activation, mitochondrial membrane potential, and ROS generation, which are key markers of intrinsic apoptotic pathways [21].

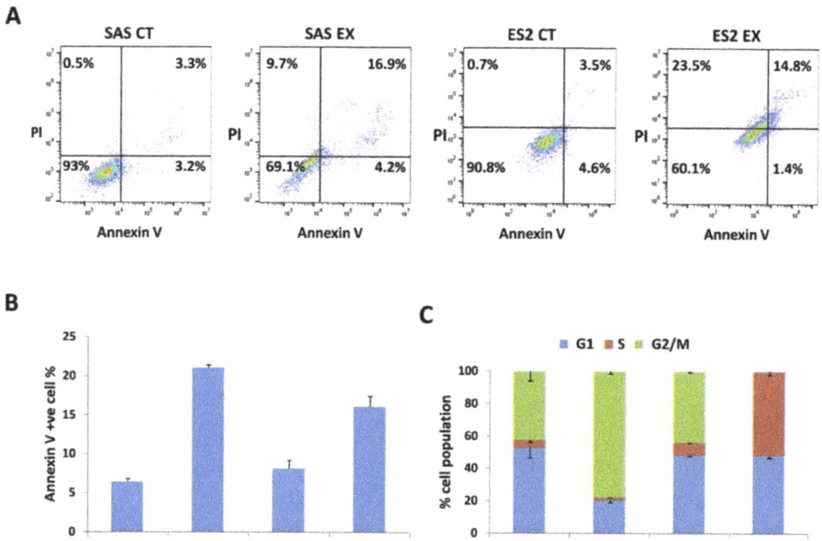

Figure 2. The extract induces apoptosis and cell cycle arrest in SAS and ES2 cells. Cells were treated with DMSO (CT) or the *L. nobilis* leaves ethanolic extract (EX) for 72 h. (**A**) Representative fluorescence images of different treatment groups. The x-axis represents Annexin V fluorescence intensity, while the y-axis indicates PI fluorescence intensity. (**B**) Numerical representation of Annexin +ve cells. (**C**) Cell cycle distribution of SAS and ES2 cells analyzed by flow cytometry following PI staining. The proportions of cells in G1, S, and G2/M phases are shown as percentages of the total cell population. Results in (**B**,**C**) are expressed as means and standard deviation taken from 3 independent experiments.

To further investigate the mechanism of action of the *L. nobilis* extract, we examined its impact on cell cycle progression in SAS and ES2 cells. As shown in Figure 2C, treatment with the extract resulted in a significant accumulation of cells in the G2 phase (SAS: control 42.5% ± 6.1% vs. extract 77.9% ± 1.7%, n = 3). This was accompanied by a marked reduction in the proportion of cells in the G1 and S phases, indicating G2 phase arrest (G1 phase: control 52.9% ± 6.5% vs. extract 20.7% ± 1.9%, $n = 3$; S phase: control 4.6% ± 1.4% vs. extract 1.4% ± 0.2%, $n = 3$). In contrast, ES2 cells exhibited a distinct response, with extract treatment leading to a marked accumulation in the S phase (control 7.4% ± 0.3% vs. extract 51.2% ± 1.9%, $n = 3$) and a substantial reduction in the G2 phase population (control 43.9% ± 0.7% vs. extract 20.7% ± 1.9%). These findings indicate that *L. nobilis* extract induces cell cycle arrest in a cell type-dependent manner, consistent with the phase-specific responses seen in many plant-derived compounds [22,23]. The observed disruption in normal cell cycle progression, particularly G2 arrest in SAS and S phase accumulation in ES2, may prevent cancer cells from entering mitosis, thus inhibiting their proliferation and contributing to the extract's anticancer potential [24]. This finding indicates that the extract may exert its anticancer effects by inducing cell cycle arrest at a critical checkpoint, thereby inhibiting tumor cell proliferation [25]. Further studies are needed to elucidate the molecular mechanisms underlying this effect, including the potential involvement of key cell cycle regulators such as CDK1, Cyclin B1, and WEE1 kinase. Cell cycle arrest, notably at the G2 and S phases, is often associated with the activation of apoptotic pathways, especially when cells are unable to repair damage or resume progression [26]. Here, the extract's ability to halt cell cycle progression may also trigger apoptosis, thereby enhancing its anticancer potential. These findings suggest a dual mechanism of action, involving both cell cycle disruption and induction of programmed cell death [27].

Moreover, previous studies have evaluated various *L. nobilis* extracts for their cytotoxic potential across different cancer cell lines. Essential oils derived from *L. nobilis* leaves and seeds demonstrated weak to moderate cytotoxicity, with IC_{50} values exceeding 500 µg/mL against SH-SY5Y neuroblastoma cells [28] and 75–95 µg/mL against K562 leukemia cells [29]. In stark contrast, our phenolic-rich ethanolic leaf extract exhibited markedly stronger anticancer activity, with IC_{50} values ranging from 3.8 to 18.5 µg/mL against the investigated cell lines. This represents more than a 20-fold increase in potency compared to essential oil extracts. These findings underscore the importance of non-volatile, polyphenolic constituents—such as quercetin, kaempferol, and gallic acid—identified in our extract via UPLC-QTOF/MS and known for their pro-apoptotic and antiproliferative effects [30].

Similarly, chloroform extracts and their fractions showed lower cytotoxic activity against SK-N-BE (2)-C and SH-SY5Y neuroblastoma cell lines, with IC_{50} values ranging from 19 to 250 µg/mL [23]. A previous ethanolic extract also exhibited an IC_{50} of 24.49 µg/mL against the MCF-7 breast cancer cell line [31] This aligns with our findings against the MDA-MB-231 breast cancer cell line (IC_{50} = 18.5 ± 0.8 µg/mL), although our extract demonstrated even greater potency against SAS, ES2, and HT-29 cells. Collectively, these comparisons highlight the potential of phenolic-rich ethanolic extracts as promising sources for the development of selective anticancer agents from *L. nobilis*.

Although the *L. nobilis* extract is slightly less potent than standard chemotherapeutic agents such as cisplatin, which typically exhibit IC_{50} values of 1.6–4 µg/mL against the investigated cell lines [32], its cytotoxic activity remains remarkable for a crude plant extract. Notably, these values are substantially lower than those reported for many other botanical extracts [33].

2.3. *Antioxidant Activity Extracts by DPPH Assay*

The antioxidant activity of the *L. nobilis* leaves extract was determined by DPPH radical scavenging assay. The scavenging of DPPH radicals caused by hydrogen donation from antioxidants result in decrease DPPH radical absorbance at 517 nm [34]. As shown in Table 2, the extract showed significant concentration-dependent free radical scavenging activity from (25–2000 μg/mL), with a IC_{50} at a concentration of 713 μg/mL as compared with the standard ascorbic acid (2.6 μg/mL). Low IC_{50} values correspond to high antioxidant activity [34].

Table 2. DPPH scavenging activity of *L. nobilis* leaves ethanolic extract.

Concentration (μg/mL)	DPPH Scavenging Activity %
25	4.2 ± 0.002
50	8.9 ± 0.001
100	15.7 ± 0.0006
200	27.2 ± 0.002
500	41.9 ± 0.002
1000	55.5 ± 0.0006
2000	77.5 ± 0.0006

The results could hint to considerable quenching actions of phenolic compounds like flavonoids found in laurel leaves, which have been detected by LC-MS/MS, against DPPH radicals. As excessive free radicals contribute to many chronic clinical conditions, the extract may help to minimize the radicals damage [34]. This antioxidant capability is especially important in the context of oxidative stress and disease development. Antioxidants are widely characterized as "substances that delay, prevent, or remove oxidative damage to specific target molecules by reacting with oxidants," such as ROS. Antioxidants and oxidants in equilibrium maintain redox homeostasis, while an imbalance in favor of oxidants can cause oxidative stress, which has been associated with cancer etiology. Cancer tissues are thought to produce more ROS because to altered metabolism, inflammation, a hypoxic environment, and oncogene-induced activation of ROS-generating enzymes, highlighting the role of antioxidants in buffering excess ROS and potentially slowing cancer growth [35]. Besides from redox regulation, antioxidants may also induce intrinsic cellular responses that contribute to cancer cell death. one of the most potent cytotoxic processes activated by these natural antioxidant is apoptosis. This approach can cause planned cell death and work alongside with necrosis. Because there are distinct indicators for determining which apoptotic pathway has happened, the action mechanism of each medication may be studied more precisely in future investigations [36].

2.4. Bioactive Compounds in L. nobilis Leaves Ethanolic Extract

In order to provide insight into the polyphenolic composition of the *L. nobilis* leaves ethanolic extract, UPLC-MS/MS analysis was performed. A number of phenolic compounds, including flavonoids, and phenolic acids, were identified in the extract Table 3.

Phenolic compounds have attracted great attention on the strength of their capacity to enhance the anticancer activity of conventional chemotherapeutic agents. There is growing evidence that suggests such bioactive compounds have the potential to exert synergistic anticancer effects in various cancer cell lines and therefore represent potential adjuncts to anticancer therapy [37].

Flavonoids are promising cytotoxic anticancer compounds due to their biological potential and minimal side effects. Flavonoids exert their anticancer effect by various mechanisms like ROS scavenging activity, suppress cancer cells proliferation, and participating in arresting the cell cycle [38]. Several key flavonoids have been identified in the *L. nobilis* leaves ethanolic extract (Table 3) like quercetin, kaempferol, naringenin, and rutin where previous studies have shown their potential role against various cancer cell lines [39] and therefore they contribute to the observed cytotoxicity of *L. nobilis* leaves extract against ES2, SAS, and HT-29 cancer cells.

Table 3. Phenolic compounds and their phytochemical class identification in *L. nobilis* leaves ethanolic extract using the LC-MS/MS.

N	RT (min)	Identified Compound Name	Phytochemical Class	Chemical Formula	Molecular Weight (Da)	*m*/*z*	Ion Mode
1	3.39	Gallic acid	Phenolic acid	$C_7H_6O_5$	170.0216	171.0289	$[M+H]^+$
2	5.29	Quercetin	Flavonoid	$C_{15}H_{10}O_7$	302.04226	303.04954	$[M+H]^+$
3	5.35	Isorhamnetin	Flavonoid	$C_{16}H_{12}O_7$	316.05796	317.06524	$[M+H]^+$
4	5.8	Kaempferol	Flavonoid	$C_{15}H_{10}O_6$	286.04752	287.0548	$[M+H]^+$
5	6.24	Resveratrol	Stilbenoid	$C_{14}H_{12}O_3$	228.07855	229.08582	$[M+H]^+$
6	11.15	4-hydroxycoumarin	Phenolic acid	$C_9H_6O_3$	162.0317	163.03894	$[M+H]^+$
7	1.52	Anisic acid	Phenolic acid	$C_8H_8O_3$	152.0474	153.05472	$[M+H]^+$
8	1.89	Gallocatechin	Flavonoid	$C_{15}H_{14}O_7$	306.07196	307.07923	$[M+H]^+$
9	3.07	Caffeic Acid	Phenolic acid	$C_9H_8O_4$	180.0399	181.04718	$[M+H]^+$
10	3.17	Chlorogenic acid	Phenolic acid	$C_{16}H_{18}O_9$	354.09439	355.10166	$[M+H]^+$
11	3.98	o-Coumaric acid	Phenolic acid	$C_9H_8O_3$	164.04732	165.0546	$[M+H]^+$
12	4.24	Catechin	Flavonoid	$C_{15}H_{14}O_6$	290.07941	291.08652	$[M+H]^+$
13	4.25	Naringenin	Flavonoid	$C_{15}H_{12}O_5$	272.06789	273.07516	$[M+H]^+$
14	4.51	Vanillic acid	Phenolic acid	$C_8H_8O_4$	168.04219	169.04947	$[M+H]^+$
15	4.74	Scopoletin	Coumarin	$C_{10}H_8O_4$	192.04185	193.04913	$[M+H]^+$
16	4.82	Vitexin	Flavonoid	$C_{21}H_{20}O_{10}$	432.10515	433.11243	$[M+H]^+$
17	4.91	Rutin	Flavonoid	$C_{27}H_{30}O_{16}$	610.15234	611.1598	$[M+H]^+$
18	4.96	Hyperoside	Flavonoid	$C_{21}H_{20}O_{12}$	464.0949	465.10217	$[M+H]^+$
19	5.28	Saponarin	Flavonoid	$C_{27}H_{30}O_{15}$	594.1577	595.16512	$[M+H]^+$
20	6.51	Cinnamic acid	Phenolic acid	$C_9H_8O_2$	148.05233	149.05961	$[M+H]^+$
21	3.82	ProcyanidinB2	Flavonoid	$C_{30}H_{26}O_{12}$	578.14289	577.13561	$[M-H]^-$

The identified chemicals were compared with some existing literature on anticancer mechanisms. The findings are elaborated in Supplementary Table S1.

Naringin, a flavonoid polyphenol found primarily in grapefruit and certain citrus fruits, has been reported to exhibit chemosensitizing activity, particularly in combination with paclitaxel. Naringin has been shown by research to enhance the cytotoxicity of paclitaxel in both androgen-independent and androgen-dependent human prostate cancer cells [40]. The enhancement procedure was performed through the mediation of intrinsic apoptosis pathway activation and the G1 cell cycle arrest. Naringin also had the ability to enhance PTEN expression, a negative regulator of the PI3K/Akt signaling cascade. In parallel, it prevents cancer cell invasion and migration through the downregulation of NF-κB, Snail, Twist, and c-Myc mRNA levels, all of which are implicated in metastasis and cancer development [41].

Similarly, quercetin, a well-studied phenolic compound, was also identified to be a potentiator of temozolomide, an oral alkylating agent employed in glioblastoma treatment. In vitro testing on U87 and U251 glioblastoma cell lines revealed that quercetin sensitize the cells to temozolomide by blocking heat-shock proteins, which are typically at the center of chemoresistance mechanisms [41].

Furthermore, in a phase I clinical trial, resveratrol dramatically decreased the levels of circulating cancer biomarkers such as Insulin-like growth factor 1 and Insulin-like growth factor-binding protein 3 [42]. Another phase I, double-blind, randomized clinical investigation found that resveratrol raised the levels of cleaved caspase-3 in malignant liver tissue [43].

Phenolic acids, a large group of non-flavonoid phenolics of plant origin, exist either in the free form (aglycones) or as glycosides. They are distributed extensively in all food sources like oilseeds, cereals, legumes, fruits and vegetables, herbs, and an enormous variety of beverages. These phenolic acids have attracted increasing attention due to their versatile biological activities, with particular interest in their potential anticancer activity [44]. The UPLC-MS/MS analysis of the *L. nobilis* leaves ethanolic extract indicated the presence of biologically important phenolic acids like gallic acid, caffeic acid, cinnamic acid, and vanillic acid. It was previously reported that phenolic acids show their anticancer effect through antioxidant activity, cell cycle arrest, and induction of apoptosis [45] which may explain the observed antioxidant and cytotoxicity of *L. nobilis* leaves extract.

Gallic acid, which is well-characterized phenolic acid, showed potential anticancer activity in vitro. Its activity was recently studied by Ko et al. against the A549 NSCLC cell line. Cell viability was studied after 48 h through a dose-dependent cell proliferation inhibition assay, suggesting its potential as a cytotoxic agent against NSCLC cells [46]. Similarly, caffeic acid has shown activity against estrogen receptor-positive (ER+) and estrogen receptor-negative (ER−) breast cancer cell lines. Rosendahl et al. established that caffeic acid is an anti-estrogenic mimic and disrupts significant signaling pathways, including the IGF-IR/pAkt and ER/cyclin D1 axes. Disruption results in cell cycle arrest and reduced proliferation, and the findings suggest that caffeic acid sensitizes tumor cells to endocrine treatment such as tamoxifen, enhancing the treatment response [47]. Vanillic acid has also been reported to have anticancer properties. In an experiment that tested its effect on breast and prostate cancer cell lines, it was determined that vanillic acid level is associated with reduced viability of cells. These results suggest that vanillic acid may be among the major compounds responsible for the antiproliferative effects in these cancer models, thus a suitable candidate as a bioactive compound to be involved in dietary cancer prevention [48].

The present study highlights a key strength compared to previous studies in that it evaluated the selective anticancer activity of the crude ethanolic extract of *L. nobilis* against different cancer cell lines, including some (e.g., SAS head and neck cancer cells) that have not been studied before, while simultaneously assessing its safety on normal fibroblasts. Unlike many earlier studies that focused primarily on general antioxidant or cytotoxic properties, this work offers evidence of both selectivity and underlying mechanisms, including apoptosis induction and cell cycle arrest. Collectively, these findings underscore the therapeutic promise of phenolic compounds in overcoming drug resistance and enhancing the anticancer efficacy of standard chemotherapeutic agents. Though the current study is limited to in vitro assays and crude extract analysis, and the samples examined were not standardized—making the reproducibility of the findings dependent on the stability of the extract's composition—which highlights the need for further in vivo and mechanistic studies to validate and extend these results.

3. Materials and Methods

3.1. Plant Material and Extraction

The leaves of *L. nobilis* were purchased from the local traditional herbal market in Amman, Jordan. The plant was taxonomically identified with the assistance of Dr. Al-Gharaibeh, Faculty of Agriculture, Jordan Department of Plant Production, University of Science and Technology. The voucher specimens (No. Bay-2023) were deposited at the Department of Pharmaceutical Sciences, Faculty of Pharmacy, Zarqa University, Zaqa, Jordan (Pharmacognosy Lab). The ground leaves of *L. nobilis* leaves (50 g) were extracted with 70% ethanol at 60 °C in a Soxhlet apparatus until the refluxed solvent became colorless. The extract was separately evaporated to dryness in a rotary vacuum evaporator, then weighted. The ethanolic extract was stored at 4 °C until subjected to analysis and in vitro tests.

3.2. Cell Culture

Prof. Stephan Feller, Institute of Molecular Medicine, Martin Luther University, Germany, kindly provided the human cancer cell lines MDA-MB-231 (breast cancer cells), SAS (head and neck cancer cells), HT-29 (colorectal cancer cells), and ES2 (ovarian cancer cells). The human dermal fibroblast (HDF) cell line was obtained as a kind gift from Prof. Lina Dahabiyeh (Jordan University, Amman, Jordan). The choice of cancer cell lines was based on their clinical and epidemiological significance, given that breast, colorectal, ovarian, and head and neck cancers rank among the most prevalent malignancies in Jordan. HDFs were included as a non-malignant control to evaluate the extract's selectivity toward cancer cells.

Cancer cell lines were cultured in Dulbecco's Modified Eagle's Medium (DMEM) (Sigma-Aldrich, Darmstadt, Germany) at 37 °C in a humidified incubator with 5% CO_2. Human dermal fibroblasts were maintained under the same conditions using Iscove's Modified Dulbecco's Medium (IMDM (Sigma-Aldrich, Darmstadt, Germany). The media were supplemented with 10% fetal bovine serum (FBS), 100 µg/mL streptomycin, 100 U/mL penicillin, and 1% non-essential amino acids. Cell cultures were grown to confluence, with the medium replaced twice weekly.

3.3. *In Vitro Cytotoxicity Assay*

The MTT (3-[4,5-dimethylthiazol-2-yl]-2,5-diphenyl tetrazolium bromide) assay was used to examine cell viability and to calculate the half-maximal inhibitory concentration (IC_{50}) values of the *L. nobilis* extract against various cancer cells and the HDF cell line. A stock solution of the extract was prepared at 100 mg/mL in DMSO, where it dissolved completely. Cells were seeded in 96-well plates at a density of 6000 cells/well for MDA-MB-231, HT-29, and HDF cells and 4000 cells/well for SAS and ES2 cells. After 24 h of incubation, cells were treated with serial concentrations of the extract or cisplatin (100, 30, 10, 3, 1, 0.3, 0.1, and 0.03 µg/mL) for 72 h. At the end of the treatment period, the supernatant was aspirated, and 50 µL of fresh DMEM containing MTT (0.5 mg/mL) was added to each well. Cells were then incubated at 37 °C for 2–4 h. The medium was then completely removed, and 50 µL of DMSO was added to dissolve the precipitated formazan crystals formed by viable cells [49].

The Synergy HTX Multimode Reader (BioTek, Shoreline, WA, USA) was used to measure cell viability, as represented by the absorbance of the formazan solution, at 570 nm. For each concentration, the relative cell viability was calculated as follows:

$$\text{Relative cell viability} = (\text{mean treatment absorbance}/\text{mean control absorbance}) \times 100\%.$$

Assays were performed in duplicate on three independent experiments. IC_{50} values were determined utilizing the four-parameter logistic function within the Sigmaplot 12.5 software and are expressed as the mean ± SD derived from three independent experiments.

3.4. *Cell Cycle Analysis*

SAS and ES2 cells were seeded in 6-well plates at a density of 1×10^5 cells/well and treated for 72 h with the IC_{50} concentration of the extract or DMSO as a vehicle control. Following treatment, both the floating dead cells and attached cells were collected and fixed in cold ethanol at −20 °C for 30 min. Cells were then stained with a propidium iodide (PI) working solution (1 µg/mL PI and 10 µg/mL RNase A in PBS) for 30 min at 37 °C in the dark, followed by PBS washing [50]. Fluorescence was analyzed using a FACS Accuri C6 Plus flow cytometer (BD Biosciences, San Jose, CA, USA). Histograms representing cell counts versus fluorescence intensity were generated, and cell cycle distribution was determined based on mean fluorescence intensity values. A total of 10,000 events were measured in each experiment and analyzed using FlowJo software (version 10). All flow cytometry experiments were performed in three independent biological replicates. Results are reported as mean ± standard deviation (SD) to reflect reproducibility across independent experiments.

3.5. Apoptosis Evaluation by Annexin V-FITC/PI Staining

SAS and ES2 cells were seeded in 6-well plates and treated with the extract or DMSO as previously described. Following treatment, both floating dead cells and attached cells were collected, washed with PBS, and resuspended in 100 μL of Annexin V solution for 30 min, following the manufacturer's instructions (Miltenyi Biotec, Bergisch Gladbach, Germany) #130-092-052) [51]. After Annexin V staining, cells were briefly stained with PI to exclude dead cells. Apoptosis quantification was performed using a FACS Accuri C6 Plus flow cytometer (BD Biosciences, San Jose, CA, USA), with 10,000 events recorded and analyzed using FlowJo software (version 10).

3.6. Antioxidant Assay

The free radical scavenging activity of *L. nobilis* leaves extract was determined by measuring the decrease in absorbance of DPPH at 517 nm as described by Gulcin et al. (2003) [52]. Briefly, 1 mL of 0.1 mM DPPH in methanol was added to 3 mL of bay extract solution in methanol at various concentrations (25–2000 μg/mL). The resulting mixture was incubated in the dark for 30 min, and then absorbance was measured at 517 nm. The control sample was prepared by adding 1 mL of DPPH to 3 mL of methanol. Vitamin C was used as a reference standard.

The ability of the extract to scavenge DPPH radical was calculated according to the following equation:

$$\text{DPPH Scavenging Activity} = [(Ao - A1/Ao) \times 100]$$

where Ao is the absorbance of the control reaction and A1 is the absorbance in the presence of the sample of *L. nobilis* extract. The results were expressed as mean value ± standard deviation (SD).

3.7. Identification of Bioactive Compounds in L. nobilis Leaves Ethanolic Extract

Ultra-performance liquid chromatography (UPLC) equipped with quadrupole time-of-flight mass spectrometry (Q-TOF/MS) was utilized to characterize bioactive chemicals. The UPLC chromatographic conditions are detailed in Table 4. The Apollo II ion funnel electrospray source was employed for ionization in both positive and negative modes. The mass spectrometry conditions were as follows:

- Capillary voltage: 2500 volts (VT);
- Nebulizer gas pressure: 2.0 bar;
- Dry gas (nitrogen) flow rate: 8 L/min;
- Dry gas temperature: 200 °C.

The instrument provided a mass resolution of 50,000 full sensitivity resolution (FSR) and a mass accuracy of 1 ppm. The time-of-flight (TOF) repetition rate reached up to 20 kHz. Data analysis was conducted using the Data Analysis 4.2 software (Bruker Daltonics, Bremen, Germany). Compound identification was carried out by LC-MS/MS using two complementary approaches: (i) direct comparison with authentic standards when available, and (ii) tentative assignment based on accurate mass and fragmentation patterns with spectral library matching.

Table 4. Summary of chromatographic conditions.

UPLC Conditions	Injection Volume	Flow Rate	Autosampler Temperature	Column Oven Temperature	Total Run Time
	3 μL	0.51 mL/min	8 °C	40 °C	35 min.
Chromatography	Mobile phase	Solvents: (A) Water with 0.1% methanol (B) Methanol 0–3 min, 5% B; 3–23 min, 60% B; 23–28 min, 95% B; 28–31 min, 95% B; 31.0–31.1 min, 5% B; 31.1–35 min 5% B;			
	Column type	Bruker Daltonik (Berlin, Germany) C-18 column (100 × 2.1 mm × 2 μm) (120 A°)			

4. Conclusions

L. nobilis leaf extract exhibited selective anticancer activity against ovarian (ES2), head and neck (SAS), and colorectal (HT-29) cancer cell lines, while showing minimal cytotoxicity toward normal fibroblasts. Its effects included the induction of apoptosis and interruption of cancer cell proliferation by causing G2 phase arrest in SAS cells and S phase accumulation in ES2 cells. The extract was also found to have antioxidant activity, though less than ascorbic acid. These effects can be attributed to the known phenolic compounds such as flavonoids and phenolic acids.

Notably, the crude ethanolic extract (10% yield from dried leaves) was directly assessed without additional purification, which may pose challenges in pinpointing the exact bioactive molecules. Nevertheless, the LC-MS chromatogram of the crude extract supports the presence of diverse phenolic constituents, likely contributing synergistically to the observed effects. Future studies involving fractionation, purification, and detailed profiling are warranted to improve reproducibility and to detect the compounds most responsible for the anticancer and antioxidant activities.

Based on this study, *L. nobilis* may be deemed to possess excellent potential as a source of natural development of selective anticancer and antioxidant drugs and deserves further research.

Supplementary Materials: The following supporting information can be downloaded at: https://www.mdpi.com/article/10.3390/molecules30194012/s1, Table S1: Some of previously reported anticancer activities of identified phenolic compounds in *Laurus nobilis* ethanolic leaf extract; Figure S1: Chromatograms of compounds identified in *L. noblis* extract, see [53–111].

Author Contributions: Conceptualization, F.A.-M. and A.M.A.Q.; methodology, F.A.-M., A.M.A.Q. and A.H.; validation, investigation and data curation F.A.-M., A.M.A.Q., D.A.-T., D.A. and A.H.; writing—review and editing, F.A.-M., D.A.-T. and A.H.; supervision, F.A.-M.; project administration, F.A.-M. All authors have read and agreed to the published version of the manuscript.

Funding: This research is partially funded by Zarqa University-Jordan.

Institutional Review Board Statement: Not applicable.

Informed Consent Statement: Not applicable.

Data Availability Statement: The original contributions presented in this study are included in the article. Further inquiries can be directed to the corresponding author.

Acknowledgments: The authors express their sincere gratitude to Stephan Feller, Institute of Molecular Medicine, Martin Luther University, Germany, for generously offering the human cancer cell lines used in the current study and for Lina Dahabiyeh for generously providing the Human Dermal Fibroblast (HDF) cell line. They also extend their sincere appreciation to Naba Hikma Industrial and Testing Services for their technical assistance and support. Appreciation is expressed to Al-Gharaibeh, Faculty of Agriculture, Department of Plant Production, Jordan University of Science and Technology, for providing assistance in the taxonomic identification of *Laurus nobilis*.

Conflicts of Interest: The authors declare no conflicts of interest.

Abbreviations

The following abbreviations are used in this manuscript:

DMSO	Dimethyl Sulfoxide
DMEM	Dulbecco's Modified Eagle's Medium
FBS	Fetal Bovine Serum
FSR	Full Sensitivity Resolution
IC_{50}	Half Maximal Inhibitory Concentration
HDF	Human Dermal Fibroblast
IGF-IR	Insulin-like Growth Factor-1 Receptor
IMDM	Iscove's Modified Medium
NSCLC	Non-Small Cell Lung Cancer
NF-κB	Nuclear Factor KappaB
PTEN	Phosphate and TENsin Homolog
PBS	Phosphate-Buffer Saline
PI3K	Phosphoinositide 3-Kinases
PI	Propidium Iodide
Q-TOF/MS	Quadrupole Time-of-Flight Mass Spectrometry
ROS	Reactive Oxygen Species
TOF	Time-of-Flight
UPLC	Ultra Performance Liquid Chromatography

References

1. Kocarnik, J.M.; Compton, K.; Dean, F.E.; Fu, W.; Gaw, B.L.; Harvey, J.D.; Henrikson, H.J.; Lu, D.; Pennini, A.; Xu, R.; et al. Cancer incidence, mortality, years of life lost, years lived with disability, and disability-adjusted life years for 29 cancer groups from 2010 to 2019: A systematic analysis for the Global Burden of Disease Study 2019. *JAMA Oncol.* **2022**, *8*, 420–444. [CrossRef] [PubMed]
2. Wani, M.; Parakh, S.; Dehghan, M.; Polshettiwar, S.; Chopade, V.; Chepurwar, S. Herbal medicine and its standardization. *Pharm. Rev.* **2007**, *5*, 411038–411049.
3. Ochwang'i, D.O.; Kimwele, C.N.; Oduma, J.A.; Gathumbi, P.K.; Mbaria, J.M.; Kiama, S.G. Medicinal plants used in treatment and management of cancer in Kakamega County, Kenya. *J. Ethnopharmacol.* **2014**, *151*, 1040–1055. [CrossRef] [PubMed]
4. Greenwell, M.; Rahman, P. Medicinal plants: Their use in anticancer treatment. *Int. J. Pharm. Sci. Res.* **2015**, *6*, 4103–4112. [CrossRef]
5. Abu-Dahab, R.; Kasabri, V.; Afifi, F.U. Evaluation of the Volatile Oil Composition and Antiproliferative Activity of *Laurus nobilis* L. (Lauraceae) on Breast Cancer Cell Line Models. *Rec. Nat. Prod.* **2014**, *8*, 136–147.
6. Mohammed, R.R.; Omer, A.K.; Yener, Z.; Uyar, A.; Ahmed, A.K. Biomedical effects of *Laurus nobilis* L. leaf extract on vital organs in streptozotocin-induced diabetic rats: Experimental research. *Ann. Med. Surg.* **2021**, *61*, 188–197. [CrossRef]
7. Khodja, Y.K.; Bachir-Bey, M.; Belmouhoub, M.; Ladjouzi, R.; Dahmoune, F.; Khettal, B. The botanical study, phytochemical composition, and biological activities of *Laurus nobilis* L. leaves: A review. *Int. J. Second. Metab.* **2023**, *10*, 269–296. [CrossRef]
8. Luo, M.; Zhou, L.; Huang, Z.; Li, B.; Nice, E.C.; Xu, J.; Huang, C. Antioxidant therapy in cancer: Rationale and progress. *Antioxidants* **2022**, *11*, 1128. [CrossRef]
9. Abdel-Razeq, H.; Mansour, A.; Jaddan, D. Breast cancer care in Jordan. *JCO Glob. Oncol.* **2020**, *6*, 260–268. [CrossRef]
10. Azar, J.; Kaddoura, T.; Timonian, M.A.; Karam, E.S.; Abou-Kheir, W.; Daoud, G. Ovarian cancer in the Arab world: An updated review. *Gene Rep.* **2024**, *35*, 102025. [CrossRef]
11. Alotaibi, T.N.; Alsheshani, T.M.; Mohammed, K. Systematic Review of Head and Neck Cancer in Jordan. *Int. J. Med. Sci. Clin. Res. Rev.* **2024**, *7*, 1010–1020.

12. Błaszczak, E.; Miziak, P.; Odrzywolski, A.; Baran, M.; Gumbarewicz, E.; Stepulak, A. Triple-Negative Breast Cancer Progression and Drug Resistance in the Context of Epithelial–Mesenchymal Transition. *Cancers* **2025**, *17*, 228. [CrossRef] [PubMed]
13. Chandra, S.; Gahlot, M.; Choudhary, A.N.; Palai, S.; de Almeida, R.S.; de Vasconcelos, J.E.L.; dos Santos, F.A.V.; de Farias, P.A.M.; Coutinho, H.D.M. Scientific evidences of anticancer potential of medicinal plants. *Food Chem. Adv.* **2023**, *2*, 100239. [CrossRef]
14. Nalkiran, I.; Sevim Nalkiran, H. Phytochemical Profile and Anticancer Potential of *Helichrysum arenarium* Extracts on Glioblastoma, Bladder Cancer, and Breast Cancer Cells. *Pharmaceuticals* **2025**, *18*, 144. [CrossRef]
15. Hu, D.; Zhuo, W.; Gong, P.; Ji, F.; Zhang, X.; Chen, Y.; Mao, M.; Ju, S.; Pan, Y.; Shen, J. Biological differences between normal and cancer-associated fibroblasts in breast cancer. *Heliyon* **2023**, *9*, e19140. [CrossRef]
16. Yang, P.; Li, X.; Wen, Q.; Zhao, X. Quercetin attenuates the proliferation of arsenic-related lung cancer cells via a caspase-dependent DNA damage signaling. *Mol. Carcinog.* **2022**, *61*, 655–663. [CrossRef]
17. Attique, I.; Haider, Z.; Khan, M.; Hassan, S.; Soliman, M.M.; Ibrahim, W.N.; Anjum, S. Reactive Oxygen Species: From Tumorigenesis to Therapeutic Strategies in Cancer. *Cancer Med.* **2025**, *14*, e70947. [CrossRef]
18. Li, L.; Jin, P.; Guan, Y.; Luo, M.; Wang, Y.; He, B.; Li, B.; He, K.; Cao, J.; Huang, C.; et al. Exploiting Polyphenol-Mediated Redox Reorientation in Cancer Therapy. *Pharmaceuticals* **2022**, *15*, 1540. [CrossRef]
19. Hmedat, A.N.; Morejón, M.C.; Rivera, D.G.; Pantelić, N.Đ.; Wessjohann, L.A.; Kaluđerović, G.N. Cyclic Lipopeptides as Selective Anticancer Agents: In Vitro Efficacy on B16F10 Mouse Melanoma Cells. *Anti-Cancer Agents Med. Chem.* **2025**, *25*, 873–882. [CrossRef]
20. Alaouna, M.; Molefi, T.; Khanyile, R.; Chauke-Malinga, N.; Chatziioannou, A.; Luvhengo, T.E.; Raletsena, M.; Penny, C.; Hull, R.; Dlamini, Z. The Potential of the South African Plant *Tulbaghia violacea* Harv for the Treatment of Triple-Negative Breast Cancer. *Sci. Rep.* **2025**, *15*, 5737. [CrossRef]
21. Wani, A.K.; Akhtar, N.; Mir, T.u.G.; Singh, R.; Jha, P.K.; Mallik, S.K.; Sinha, S.; Tripathi, S.K.; Jain, A.; Jha, A.; et al. Targeting Apoptotic Pathway of Cancer Cells with Phytochemicals and Plant-Based Nanomaterials. *Biomolecules* **2023**, *13*, 194. [CrossRef]
22. Bian, Y.; Yang, L.; Sheng, W.; Li, Z.; Xu, Y.; Li, W.; Zeng, L. Ligustrazine Induces the Colorectal Cancer Cells Apoptosis via p53-Dependent Mitochondrial Pathway and Cell Cycle Arrest at the G0/G1 Phase. *Ann. Palliat. Med.* **2021**, *10*, 1578–1588. [CrossRef]
23. Li, W.M.; Liu, H.T.; Li, X.Y.; Wu, J.Y.; Xu, G.; Teng, Y.Z.; Ding, S.T.; Yu, C. The Effect of Tetramethylpyrazine on Hydrogen Peroxide Induced Oxidative Damage in Human Umbilical Vein Endothelial Cells. *Basic Clin. Pharmacol. Toxicol.* **2010**, *106*, 45–52. [CrossRef]
24. Pacifico, S.; Gallicchio, M.; Lorenz, P.; Potenza, N.; Galasso, S.; Marciano, S.; Fiorentino, A.; Stintzing, F.C.; Monaco, P. Apolar *Laurus nobilis* Leaf Extracts Induce Cytotoxicity and Apoptosis towards Three Nervous System Cell Lines. *Food Chem. Toxicol.* **2013**, *62*, 628–637. [CrossRef] [PubMed]
25. El-Huneidi, W.; Shehab, N.G.; Bajbouj, K.; Vinod, A.; El-Serafi, A.; Shafarin, J.; Bou Malhab, L.J.; Abdel-Rahman, W.M.; Abu-Gharbieh, E. *Micromeria fruticosa* Induces Cell Cycle Arrest and Apoptosis in Breast and Colorectal Cancer Cells. *Pharmaceuticals* **2020**, *13*, 115. [CrossRef] [PubMed]
26. Rattanapornsompong, K.; Khattiya, J.; Phannasil, P.; Phaonakrop, N.; Roytrakul, S.; Jitrapakdee, S.; Akekawatchai, C. Impaired G2/M Cell Cycle Arrest Induces Apoptosis in Pyruvate Carboxylase Knockdown MDA-MB-231 Cells. *Biochem. Biophys. Rep.* **2021**, *25*, 100903. [CrossRef] [PubMed]
27. Chaudhry, G.-e.-S.; Md Akim, A.; Sung, Y.Y.; Sifzizul, T.M.T. Cancer and apoptosis: The apoptotic activity of plant and marine natural products and their potential as targeted cancer therapeutics. *Front. Pharmacol.* **2022**, *13*, 842376. [CrossRef]
28. Caputo, L.; Nazzaro, F.; Souza, L.F.; Aliberti, L.; De Martino, L.; Fratianni, F.; Coppola, R.; De Feo, V. *Laurus nobilis*: Composition of essential oil and its biological activities. *Molecules* **2017**, *22*, 930. [CrossRef]
29. Saab, A.M.; Tundis, R.; Loizzo, M.R.; Lampronti, I.; Borgatti, M.; Gambari, R.; Menichini, F.; Esseily, F.; Menichini, F. Antioxidant and antiproliferative activity of *Laurus nobilis* L. (Lauraceae) leaves and seeds essential oils against K562 human chronic myelogenous leukaemia cells. *Nat. Prod. Res.* **2012**, *26*, 1741–1745. [CrossRef]
30. Nguyen, L.T.; Lee, Y.H.; Sharma, A.R.; Park, J.B.; Jagga, S.; Sharma, G.; Lee, S.S.; Nam, J.S. Quercetin induces apoptosis and cell cycle arrest in triple-negative breast cancer cells through modulation of Foxo3a activity. *Korean J. Physiol. Pharmacol.* **2017**, *21*, 205–213. [CrossRef]
31. Al-Kalaldeh, J.Z.; Abu-Dahab, R.; Afifi, F.U. Volatile oil composition and antiproliferative activity of *Laurus nobilis*, *Origanum syriacum*, *Origanum vulgare*, and *Salvia triloba* against human breast adenocarcinoma cells. *Nutr. Res.* **2010**, *30*, 271–278. [CrossRef]
32. Ota, T.; Jono, H.; Ota, K.; Shinriki, S.; Ueda, M.; Sueyoshi, T.; Nakatani, K.; Hiraishi, Y.; Wada, T.; Fujita, S.; et al. Downregulation of midkine induces cisplatin resistance in human oral squamous cell carcinoma. *Oncol. Rep.* **2012**, *27*, 1674–1680. [CrossRef] [PubMed]
33. Bagheri, E.; Hajiaghaalipour, F.; Nyamathulla, S.; Salehen, N.A. The apoptotic effects of *Brucea javanica* fruit extract against HT29 cells associated with p53 upregulation and inhibition of NF-κB translocation. *Drug Des. Devel. Ther.* **2018**, *12*, 657–671. [CrossRef] [PubMed]

34. Elmastaş, M.; Gülçin, İ.; Işildak, Ö.; Küfrevioğlu, Ö.; İbaoğlu, K.; Aboul-Enein, H. Radical scavenging activity and antioxidant capacity of bay leaf extracts. *J. Iran. Chem. Soc.* **2006**, *3*, 258–266. [CrossRef]
35. Hecht, F.; Zocchi, M.; Alimohammadi, F.; Harris, I.S. Regulation of antioxidants in cancer. *Mol. Cell* **2024**, *84*, 23–33. [CrossRef]
36. Oliveira, C.; Nicolau, A.; Teixeira, J.A.; Domingues, L. Cytotoxic effects of native and recombinant frutalin, a plant galactose-binding lectin, on HeLa cervical cancer cells. *J. Biomed. Biotechnol.* **2011**, *2011*, 568932. [CrossRef]
37. Lewandowska, U.; Gorlach, S.; Owczarek, K.; Hrabec, E.; Szewczyk, K. Synergistic Interactions Between Anticancer Chemotherapeutics and Phenolic Compounds and Anticancer Synergy Between Polyphenols. *Postepy Hig. Med. Dosw.* **2014**, *68*, 528–540. [CrossRef]
38. de Luna, F.C.F.; Ferreira, W.A.S.; Casseb, S.M.M.; de Oliveira, E.H.C. Anticancer potential of flavonoids: An overview with an emphasis on tangeretin. *Pharmaceuticals* **2023**, *16*, 1229. [CrossRef]
39. Kopustinskiene, D.M.; Jakstas, V.; Savickas, A.; Bernatoniene, J. Flavonoids as anticancer agents. *Nutrients* **2020**, *12*, 457. [CrossRef]
40. Erdogan, S.; Doganlar, O.; Doganlar, Z.B.; Turkekul, K. Naringin sensitizes human prostate cancer cells to paclitaxel therapy. *Prostate Int.* **2018**, *6*, 126–135. [CrossRef]
41. Kikuchi, H.; Yuan, B.; Hu, X.; Okazaki, M. Chemopreventive and anticancer activity of flavonoids and their potential for clinical use with conventional chemotherapeutic agents. *Am. J. Cancer Res.* **2019**, *9*, 1517–1535. [PubMed]
42. Brown, V.A.; Patel, K.R.; Viskaduraki, M.; Crowell, J.A.; Perloff, M.; Booth, T.D.; Vasilinin, G.; Sen, A.; Schinas, A.M.; Piccirilli, G.; et al. Repeat dose study of the cancer chemopreventive agent resveratrol in healthy volunteers: Safety, pharmacokinetics, and effect on the insulin-like growth factor axis. *Cancer Res.* **2010**, *70*, 9003–9011. [CrossRef] [PubMed]
43. Howells, L.M.; Berry, D.P.; Elliott, P.J.; Jacobson, E.W.; Hoffmann, E.; Hegarty, B.; Brown, K.; Steward, W.P.; Gescher, A.J. Phase I randomized, double-blind pilot study of micronized resveratrol (SRT501) in patients with hepatic metastases: Safety, pharmacokinetics, and pharmacodynamics. *Cancer Prev. Res.* **2011**, *4*, 1419–1425. [CrossRef] [PubMed]
44. Bento, C.; Gonçalves, A.C.; Jesus, F.; Simões, M.; Silva, L.R. Phenolic compounds: Sources, properties and applications. In *Bioactive Compounds: Sources, Properties and Applications*; Porter, R., Parker, N., Eds.; Nova Science Publishers, Inc.: Hauppauge, NY, USA, 2017; pp. 271–299.
45. Abotaleb, M.; Liskova, A.; Kubatka, P.; Büsselberg, D. Therapeutic potential of plant phenolic acids in the treatment of cancer. *Biomolecules* **2020**, *10*, 221. [CrossRef] [PubMed]
46. Ko, E.B.; Jang, Y.G.; Kim, C.W.; Go, R.E.; Lee, H.K.; Choi, K.C. Gallic acid hinders lung cancer progression by inducing cell cycle arrest and apoptosis in A549 lung cancer cells via PI3K/Akt pathway. *Biomol. Ther.* **2022**, *30*, 151–161. [CrossRef]
47. Rosendahl, A.H.; Perks, C.M.; Zeng, L.; Markkula, A.; Simonsson, M.; Rose, C.; Ingvar, C.; Holly, J.M.; Jernström, H. Caffeine and caffeic acid inhibit growth and modify estrogen receptor and insulin-like growth factor I receptor levels in human breast cancer. *Clin. Cancer Res.* **2015**, *21*, 1877–1887. [CrossRef]
48. Kumar, P.; Ammani, K.; Mahammad, A.; Gosala, J. Vanillic acid induces oxidative stress and apoptosis in non-small cell lung cancer cell line. *Int. J. Recent Sci. Res.* **2013**, *4*, 1077–1083.
49. Hmedat, A.N.; Morejón, M.C.; Rivera, D.G.; Pantelić, N.Đ.; Wessjohann, L.A.; Kaluđerović, G.N. *In vitro* anti-cancer studies of a small library of cyclic lipopeptides against the human cervix adenocarcinoma HeLa cells. *J. Serbian Chem. Soc.* **2024**, *89*, 471–484. [CrossRef]
50. Rahmé, R. Assaying cell cycle status using flow cytometry. In *Cell Cycle Checkpoints: Methods and Protocols*; Humana: New York, NY, USA, 2021; pp. 165–179. [CrossRef]
51. Lakshmanan, I.; Batra, S.K. Protocol for apoptosis assay by flow cytometry using annexin V staining method. *Bio-Protocol* **2013**, *3*, e374. [CrossRef]
52. Gulcin, I.; Oktay, M.; Kirecci, E.; Kufrevioglu, I.O. Screening of antioxidant and antimicrobial activities of anise (*Pimpinella anisum* L.) seed extracts. *Food Chem.* **2003**, *83*, 371–382. [CrossRef]
53. Zhang, T.; Ma, L.; Wu, P.; Li, W.; Li, T.; Gu, R.; Dan, X.; Li, Z.; Fan, X.; Xiao, Z. Gallic acid has anticancer activity and enhances the anticancer effects of cisplatin in non-small cell lung cancer A549 cells via the JAK/STAT3 signaling pathway. *Oncol. Rep.* **2019**, *41*, 1779–1788. [CrossRef]
54. Jiang, Y.; Pei, J.; Zheng, Y.; Miao, Y.-J.; Duan, B.-Z.; Huang, L.-F. Gallic acid: A potential anti-cancer agent. *Chin. J. Integr. Med.* **2022**, *28*, 661–671. [CrossRef]
55. Aborehab, N.M.; Osama, N. Effect of Gallic acid in potentiating chemotherapeutic effect of Paclitaxel in HeLa cervical cancer cells. *Cancer Cell Int.* **2019**, *19*, 154. [CrossRef]
56. Tang, S.-M.; Deng, X.-T.; Zhou, J.; Li, Q.-P.; Ge, X.-X.; Miao, L. Pharmacological basis and new insights of quercetin action in respect to its anti-cancer effects. *Biomed. Pharmacother.* **2020**, *121*, 109604. [CrossRef]
57. Reyes-Farias, M.; Carrasco-Pozo, C. The anti-cancer effect of quercetin: Molecular implications in cancer metabolism. *Int. J. Mol. Sci.* **2019**, *20*, 3177. [CrossRef] [PubMed]

58. Hashemzaei, M.; Delarami Far, A.; Yari, A.; Heravi, R.E.; Tabrizian, K.; Taghdisi, S.M.; Sadegh, S.E.; Tsarouhas, K.; Kouretas, D.; Tzanakakis, G.; et al. Anticancer and apoptosis-inducing effects of quercetin in vitro and in vivo. *Oncol. Rep.* **2017**, *38*, 819–828. [CrossRef] [PubMed]
59. Bhatiya, M.; Pathak, S.; Jothimani, G.; Duttaroy, A.K.; Banerjee, A. A comprehensive study on the anti-cancer effects of quercetin and its epigenetic modifications in arresting progression of colon cancer cell proliferation. *Arch. Immunol. Ther. Exp.* **2023**, *71*, 6. [CrossRef] [PubMed]
60. Biswas, P.; Kaium, M.A.; Tareq, M.M.I.; Tauhida, S.J.; Hossain, M.R.; Siam, L.S.; Parvez, A.; Bibi, S.; Hasan, M.H.; Rahman, M.M.; et al. The experimental significance of isorhamnetin as an effective therapeutic option for cancer: A comprehensive analysis. *Biomed. Pharmacother.* **2024**, *176*, 116860. [CrossRef] [PubMed]
61. Mei, C.; Liu, Y.; Lyu, X.; Jiang, Z.; Liu, Z.; Zhi, Y.; Xu, X.; Wang, H. Advances in Isorhamnetin Treatment of Malignant Tumors: Mechanisms and Applications. *Nutrients* **2025**, *17*, 1853. [CrossRef]
62. Khaled, R. Biological activities of isorhamnetin: A review. *Plantae Sci.* **2020**, *3*, 78–81. [CrossRef]
63. Park, C.; Cha, H.-J.; Choi, E.O.; Lee, H.; Hwang-Bo, H.; Ji, S.Y.; Kim, M.Y.; Kim, S.Y.; Hong, S.H.; Cheong, J.; et al. Isorhamnetin induces cell cycle arrest and apoptosis via reactive oxygen species-mediated AMP-activated protein kinase signaling pathway activation in human bladder cancer cells. *Cancers* **2019**, *11*, 1494. [CrossRef]
64. Wang, X.; Yang, Y.; An, Y.; Fang, G. The mechanism of anticancer action and potential clinical use of kaempferol in the treatment of breast cancer. *Biomed. Pharmacother.* **2019**, *117*, 109086. [CrossRef] [PubMed]
65. Imran, M.; Salehi, B.; Sharifi-Rad, J.; Gondal, T.A.; Saeed, F.; Imran, A.; Shahbaz, M.; Fokou, P.V.T.; Arshad, M.U.; Khan, H.; et al. Kaempferol: A key emphasis to its anticancer potential. *Molecules* **2019**, *24*, 2277. [CrossRef] [PubMed]
66. Kaur, S.; Mendonca, P.; Soliman, K.F. The Anticancer Effects and Therapeutic Potential of Kaempferol in Triple-Negative Breast Cancer. *Nutrients* **2024**, *16*, 2392. [CrossRef]
67. Alyami, N.M.; Alyami, H.M.; Almeer, R. Using green biosynthesized kaempferol-coated sliver nanoparticles to inhibit cancer cells growth: An in vitro study using hepatocellular carcinoma (HepG2). *Cancer Nanotechnol.* **2022**, *13*, 26. [CrossRef]
68. Huminiecki, L.; Horbańczuk, J. The functional genomic studies of resveratrol in respect to its anti-cancer effects. *Biotechnol. Adv.* **2018**, *36*, 1699–1708. [CrossRef]
69. Varoni, E.M.; Lo Faro, A.F.; Sharifi-Rad, J.; Iriti, M. Anticancer molecular mechanisms of resveratrol. *Front. Nutr.* **2016**, *3*, 8. [CrossRef]
70. Öner, Ç.; Soyergin, D.; Özyurt, A.; Çolak, E. 4-Hydroxycoumarin effects on both cellular and genetic characteristics of hepatocellular carcinoma cells. *Cytol. Genet.* **2022**, *56*, 292–300. [CrossRef]
71. Dimić, D.S.; Kaluđerović, G.N.; Avdović, E.H.; Milenković, D.A.; Živanović, M.N.; Potočňák, I.; Samoľová, E.; Dimitrijević, M.S.; Saso, L.; Marković, Z.S.; et al. Synthesis, Crystallographic, quantum chemical, antitumor, and molecular docking/dynamic studies of 4-hydroxycoumarin-neurotransmitter derivatives. *Int. J. Mol. Sci.* **2022**, *23*, 1001.
72. Czarnecka, M.; Świtalska, M.; Wietrzyk, J.; Maciejewska, G.; Gliszczyńska, A. Synthesis, characterization, and in vitro cancer cell growth inhibition evaluation of novel phosphatidylcholines with anisic and veratric acids. *Molecules* **2018**, *23*, 2022. [CrossRef]
73. Palko-Łabuz, A.; Gliszczyńska, A.; Skonieczna, M.; Poła, A.; Wesołowska, O.; Środa-Pomianek, K. Conjugation with phospholipids as a modification increasing anticancer activity of phenolic acids in metastatic melanoma—In vitro and in silico studies. *Int. J. Mol. Sci.* **2021**, *22*, 8397. [CrossRef]
74. Kanimozhi, G.; Prasad, N. Anticancer effect of caffeic acid on human cervical cancer cells. In *Coffee in Health and Disease Prevention*; Elsevier: Amsterdam, The Netherlands, 2015; pp. 655–661.
75. Espíndola, K.M.M.; Ferreira, R.G.; Narvaez, L.E.M.; Rosario, A.C.R.S.; da Silva, A.H.M.; Silva, A.G.B.; Vieira, A.P.O.; Monteiro, M.C. Chemical and pharmacological aspects of caffeic acid and its activity in hepatocarcinoma. *Front. Oncol.* **2019**, *9*, 541. [CrossRef] [PubMed]
76. Alam, M.; Ahmed, S.; Elasbali, A.M.; Adnan, M.; Alam, S.; Hassan, I.; Pasupuleti, V.R. Therapeutic implications of caffeic acid in cancer and neurological diseases. *Front. Oncol.* **2022**, *12*, 860508. [CrossRef] [PubMed]
77. Pelinson, L.P.; Assmann, C.E.; Palma, T.V.; da Cruz, I.; Pillat, M.M.; Mânica, A.; Stefanello, N.; Weis, G.C.C.; de Oliveira Alves, A.; De Andrade, C.M.; et al. Antiproliferative and apoptotic effects of caffeic acid on SK-Mel-28 human melanoma cancer cells. *Mol. Biol. Rep.* **2019**, *46*, 2085–2092. [CrossRef] [PubMed]
78. Huang, S.; Wang, L.-L.; Xue, N.-N.; Li, C.; Guo, H.-H.; Ren, T.-K.; Zhan, Y.; Li, W.-B.; Zhang, J.; Chen, X.-G.; et al. Chlorogenic acid effectively treats cancers through induction of cancer cell differentiation. *Theranostics* **2019**, *9*, 6745. [CrossRef]
79. Yamagata, K.; Izawa, Y.; Onodera, D.; Tagami, M. Chlorogenic acid regulates apoptosis and stem cell marker-related gene expression in A549 human lung cancer cells. *Mol. Cell. Biochem.* **2018**, *441*, 9–19. [CrossRef]
80. Bender, O.; Atalay, A. Polyphenol chlorogenic acid, antioxidant profile, and breast cancer. In *Cancer*; Elsevier: Amsterdam, The Netherlands, 2021; pp. 311–321.
81. Wang, L.; Pan, X.; Jiang, L.; Chu, Y.; Gao, S.; Jiang, X.; Zhang, Y.; Chen, Y.; Luo, S.; Peng, C. The biological activity mechanism of chlorogenic acid and its applications in food industry: A review. *Front. Nutr.* **2022**, *9*, 943911. [CrossRef]

82. Mutlu, D. Cytotoxic activity, anti-migration and in silico study of o-coumaric acid on H1975 non-small cell lung cancer cells. *Cumhur. Sci. J.* **2025**, *46*, 35–40. [CrossRef]
83. Mercado, Y.G.; Díaz, J.M.; Hernández, D.O.; Gonzalez, F.L.; Zaldivar, E.R.; Sapiens, M.H.; Pinedo, U.G.; Estrada, R.; Carballo, M.M.; Aguirre, A.C. Ortho-coumaric acid derivatives with therapeutic potential in a three-dimensional culture of the immortalised U-138 MG glioblastoma multiforme cell line. *Neurol. Perspect.* **2022**, *2*, S19–S30. [CrossRef]
84. Burhanoğlu, T.; Halbutoğulları, Z.S.; Turhal, G.; Demiroglu-Zergeroglu, A. Evaluation of the anticancer effects of hydroxycinnamic acid isomers on breast cancer stem cells. *Med. Oncol.* **2025**, *42*, 73. [CrossRef]
85. Cheng, Z.; Zhang, Z.; Han, Y.; Wang, J.; Wang, Y.; Chen, X.; Shao, Y.; Cheng, Y.; Zhou, W.; Lu, X.; et al. A review on anti-cancer effect of green tea catechins. *J. Funct. Foods* **2020**, *74*, 104172. [CrossRef]
86. Sun, H.; Yin, M.; Hao, D.; Shen, Y. Anti-cancer activity of catechin against A549 lung carcinoma cells by induction of cyclin kinase inhibitor P21 and suppression of cyclin E1 and P–AKT. *Appl. Sci.* **2020**, *10*, 2065. [CrossRef]
87. Silva, C.; Correia-Branco, A.; Andrade, N.; Ferreira, A.C.; Soares, M.L.; Sonveaux, P.; Stephenne, J.; Martel, F. Selective pro-apoptotic and antimigratory effects of polyphenol complex catechin: Lysine 1: 2 in breast, pancreatic and colorectal cancer cell lines. *Eur. J. Pharmacol.* **2019**, *859*, 172533. [CrossRef]
88. Stabrauskiene, J.; Kopustinskiene, D.M.; Lazauskas, R.; Bernatoniene, J. Naringin and naringenin: Their mechanisms of action and the potential anticancer activities. *Biomedicines* **2022**, *10*, 1686. [CrossRef] [PubMed]
89. Rauf, A.; Shariati, M.A.; Imran, M.; Bashir, K.; Khan, S.A.; Mitra, S.; Bin Emran, T.; Badalova, K.; Uddin, S.; Mubarak, M.S.; et al. Comprehensive review on naringenin and naringin polyphenols as a potent anticancer agent. *Environ. Sci. Pollut. Res.* **2022**, *29*, 31025–31041. [CrossRef] [PubMed]
90. Choi, J.; Lee, D.-H.; Jang, H.; Park, S.-Y.; Seol, J.-W. Naringenin exerts anticancer effects by inducing tumor cell death and inhibiting angiogenesis in malignant melanoma. *Int. J. Med. Sci.* **2020**, *17*, 3049. [CrossRef]
91. Venkidasamy, B.; Subramanian, U.; Almoallim, H.S.; Alharbi, S.A.; Lakshmikumar, R.R.C.; Thiruvengadam, M. Vanillic Acid Nanocomposite: Synthesis, Characterization Analysis, Antimicrobial, and Anticancer Potentials. *Molecules* **2024**, *29*, 3098. [CrossRef] [PubMed]
92. Gong, J.; Zhou, S.; Yang, S. Vanillic acid suppresses HIF-1α expression via inhibition of mTOR/p70S6K/4E-BP1 and Raf/MEK/ERK pathways in human colon cancer HCT116 cells. *Int. J. Mol. Sci.* **2019**, *20*, 465. [CrossRef]
93. Kaur, J.; Gulati, M.; Gowthamarajan, K.; Vishwas, S.; Chellappan, D.K.; Gupta, G.; Dua, K.; Pandey, N.K.; Kumar, B.; Singh, S.K. Combination therapy of vanillic acid and oxaliplatin co-loaded in polysaccharide based functionalized polymeric micelles could offer effective treatment for colon cancer: A hypothesis. *Med. Hypotheses* **2021**, *156*, 110679. [CrossRef]
94. Velli, S.K.; Sundaram, J.; Murugan, M.; Balaraman, G.; Thiruvengadam, D. Protective effect of vanillic acid against benzo (a) pyrene induced lung cancer in Swiss albino mice. *J. Biochem. Mol. Toxicol.* **2019**, *33*, e22382. [CrossRef]
95. Tian, Q.; Wang, L.; Sun, X.; Zeng, F.; Pan, Q.; Xue, M. Scopoletin exerts anticancer effects on human cervical cancer cell lines by triggering apoptosis, cell cycle arrest, inhibition of cell invasion and PI3K/AKT signalling pathway. *J BUON* **2019**, *24*, 997–1002. [PubMed]
96. Yuan, C.; Wang, M.-H.; Wang, F.; Chen, P.-Y.; Ke, X.-G.; Yu, B.; Yang, Y.-F.; You, P.-T.; Wu, H.-Z. Network pharmacology and molecular docking reveal the mechanism of Scopoletin against non-small cell lung cancer. *Life Sci.* **2021**, *270*, 119105. [CrossRef] [PubMed]
97. Shi, Z.; Chen, L.; Sun, J. Novel scopoletin derivatives kill cancer cells by inducing mitochondrial depolarization and apoptosis. *Anti-Cancer Agents Med. Chem. (Former. Curr. Med. Chem.-Anti-Cancer Agents)* **2021**, *21*, 1774–1782. [CrossRef] [PubMed]
98. Liu, X.; Jiang, Q.; Liu, H.; Luo, S. Vitexin induces apoptosis through mitochondrial pathway and PI3K/Akt/mTOR signaling in human non-small cell lung cancer A549 cells. *Biol. Res.* **2019**, *52*, 7. [CrossRef]
99. Wang, Q.; Zhang, J.; Ye, J.; Guo, J. Vitexin exerts anti-tumor and anti-angiogensis effects on cervical cancer through VEGFA/VEGFR2 pathway. *Eur. J. Gynaecol. Oncol.* **2022**, *43*, 86.
100. Ghazy, E.; Taghi, H.S. The autophagy-inducing mechanisms of vitexin, cinobufacini, and physalis alkekengi hydroalcoholic extract against breast cancer in vitro and in vivo. *J. Gastrointest. Cancer* **2022**, *53*, 592–596. [CrossRef]
101. Caparica, R.; Júlio, A.; Araújo, M.E.M.; Baby, A.R.; Fonte, P.; Costa, J.G.; Santos de Almeida, T. Anticancer activity of rutin and its combination with ionic liquids on renal cells. *Biomolecules* **2020**, *10*, 233. [CrossRef]
102. Paudel, K.R.; Wadhwa, R.; Tew, X.N.; Lau, N.J.X.; Madheswaran, T.; Panneerselvam, J.; Zeeshan, F.; Kumar, P.; Gupta, G.; Anand, K.; et al. Rutin loaded liquid crystalline nanoparticles inhibit non-small cell lung cancer proliferation and migration in vitro. *Life Sci.* **2021**, *276*, 119436. [CrossRef]
103. Kong, Y.; Sun, W.; Wu, P. Hyperoside exerts potent anticancer activity in skin cancer. *Front. Biosci.-Landmark* **2020**, *25*, 463–479.
104. Qiu, J.; Zhang, T.; Zhu, X.; Yang, C.; Wang, Y.; Zhou, N.; Ju, B.; Zhou, T.; Deng, G.; Qiu, C. Hyperoside induces breast cancer cells apoptosis via ROS-mediated NF-κB signaling pathway. *Int. J. Mol. Sciences* **2019**, *21*, 131. [CrossRef]
105. Hu, Z.; Zhao, P.; Xu, H. Hyperoside exhibits anticancer activity in non-small cell lung cancer cells with T790M mutations by upregulating FoxO1 via CCAT1. *Oncol. Rep.* **2020**, *43*, 617–624. [CrossRef]

106. Hunke, M.; Martinez, W.; Kashyap, A.; Bokoskie, T.; Pattabiraman, M.; Chandra, S. Antineoplastic actions of cinnamic acids and their dimers in breast cancer cells: A comparative study. *Anticancer. Res.* **2018**, *38*, 4469–4474. [CrossRef]
107. Yenigül, M.; Akçok, İ.; Gencer Akçok, E.B. Ethacrynic acid and cinnamic acid combination exhibits selective anticancer effects on K562 chronic myeloid leukemia cells. *Mol. Biol. Rep.* **2022**, *49*, 7521–7530. [CrossRef]
108. Imai, M.; Yokoe, H.; Tsubuki, M.; Takahashi, N. Growth inhibition of human breast and prostate cancer cells by cinnamic acid derivatives and their mechanism of action. *Biol. Pharm. Bull.* **2019**, *42*, 1134–1139. [CrossRef] [PubMed]
109. Zhang, R.; Yu, Q.; Lu, W.; Shen, J.; Zhou, D.; Wang, Y.; Gao, S.; Wang, Z. Grape seed procyanidin B2 promotes the autophagy and apoptosis in colorectal cancer cells via regulating PI3K/Akt signaling pathway. *OncoTargets Ther.* **2019**, *12*, 4109–4118. [CrossRef] [PubMed]
110. Li, Y.; Lu, X.; Tian, P.; Wang, K.; Shi, J. Procyanidin B2 induces apoptosis and autophagy in gastric cancer cells by inhibiting Akt/mTOR signaling pathway. *BMC Complement. Med. Ther.* **2021**, *21*, 76. [CrossRef] [PubMed]
111. Xue, H.-K.; Tan, J.-Q.; Li, Q.; Tang, J.-T. Procyanidin B2-induced apoptosis in MCF-7 cells and its mechanism. *Food Sci.* **2021**, *42*, 91–99.

molecules

MDPI

Article

Cytotoxic Activity of Essential Oils from Middle Eastern Medicinal Plants on Malignant Keratinocytes

Rima Othman [1,†], Vanessa Moarbes [2,†], Muriel Tahtouh Zaatar [3,*], Diane Antonios [4], Rabih Roufayel [5], Marc Beyrouthy [6], Ziad Fajloun [7,8,*], Jean-Marc Sabatier [9,*] and Marc Karam [2]

1 Department of Pulmonary Medicine and Critical Care, Johns Hopkins University, Baltimore, MD 21205, USA; remaothman3@gmail.com
2 Department of Biology, Faculty of Sciences, University of Balamand, Al-Kourah, Tripoli P.O. Box 100, Lebanon; vanessa.moarbes@mail.mcgill.ca (V.M.); marckaram1@gmail.com (M.K.)
3 Department of Biological and Physical Sciences, American University in Dubai, Dubai P.O. Box 28282, United Arab Emirates
4 Laboratory of Toxicology, Faculty of Pharmacy, Saint Joseph University of Beirut, Damascus Road, Beirut P.O. Box 11-5076, Lebanon; diane.antonios@usj.edu.lb
5 College of Engineering and Technology, American University of the Middle East, Egaila 54200, Kuwait; rabih.roufayel@aum.edu.kw
6 Department of Agriculture and Food Engineering, School of Engineering, Holy Spirit University of Kaslik, Jounieh P.O. Box 446, Lebanon; marcelbeyrouthy@usek.edu.lb
7 Department of Cell Culture, Laboratory of Applied Biotechnology (LBA3B), Azm Center for Research in Biotechnology and Its Applications, EDST, Lebanese University, Tripoli 1300, Lebanon
8 Faculty of Sciences 3, Lebanese University, Michel Slayman Tripoli Campus, Ras Maska 1352, Lebanon
9 Inst Neurophysiopathol (INP), CNRS, Aix-Marseille Université, 13385 Marseille, France
* Correspondence: mzaatar@aud.edu (M.T.Z.); ziad.fajloun@ul.edu.lb (Z.F.); sabatier.jm1@gmail.com (J.-M.S.)
† These authors contributed equally to this work.

Abstract

Skin cancer, including melanoma and non-melanoma cancers (basal and squamous cell carcinomas), is the most common type of cancer. UV radiation, family history, and genetic predisposition are the main risk factors. Although surgical excision is the standard treatment, essential oils are attracting growing interest for their anti-cancer effects. This study tested the effects of *Juniperus excelsa* M. Bieb. (Cupressaceae), *Lavandula vera* DC. (Lamiaceae), and *Salvia fruticosa (Mill).* (Lamiaceae) essential oils extracted from Middle Eastern medicinal plants on HaCaT (normal), A5 (benign), and II4 (low-grade malignant) keratinocytes. Essential oils were extracted from *Juniperus excelsa, Lavandula vera*, and *Salvia libanotica* using steam distillation and then were chemically analyzed. The oils were sterilized, dissolved in DMSO, and prepared at concentrations of 0.75, 0.5, and 0.25 mg/mL. Human keratinocyte (HaCaT), benign (A5), and malignant (II4) cell lines were cultured in DMEM and treated with the essential oils for 24 or 48 h. Cell viability was assessed using the Trypan Blue Exclusion Test, while cell proliferation was evaluated using the MTT assay. Statistical analysis was performed using ANOVA with appropriate post hoc tests, considering $p < 0.05$ as significant. The results show that *J. excelsa* is cytotoxic but lacks selectivity, limiting its efficacy. In contrast, *L. vera* and *S. fruticosa* preferentially target malignant cells, particularly at low concentrations, while sparing normal cells. These oils have dose-dependent anticancer effects, with *L. vera* efficacy increasing as the concentration increases. In conclusion, *L. vera* and *S. fruticosa* are promising candidates for the treatment of skin cancer, although further in vivo studies are required.

https://doi.org/10.3390/molecules30132844

Keywords: anti-cancer effect; MTT assay; essential oil; HaCat; A5 cell lines; II4 cell lines; *Juniperus excelsa* M. Bieb. (Cupressaceae); *Lavandula vera* DC. (Lamiaceae); *Salvia fruticosa Mill.* (Lamiaceae)

1. Introduction

Skin cancer is the most prevalent form of cancer globally, representing a diverse and complex group of malignancies that arise from various cellular components within the skin. These cancers can affect individuals of all ethnic backgrounds and age groups, underscoring their widespread impact on public health. Among these, cutaneous carcinomas—or skin cancers—are broadly categorized into two primary groups: melanoma and non-melanoma skin cancers (NMSCs). The latter includes the two most common types, basal cell carcinoma (BCC) and squamous cell carcinoma (SCC), which together account for the majority of skin cancer diagnoses worldwide [1].

Skin cancer is the most commonly diagnosed cancer globally and encompasses a range of malignancies, including melanoma and non-melanoma skin cancers (NMSCs) such as basal cell carcinoma (BCC) and squamous cell carcinoma (SCC) [2]. Ultraviolet (UV) radiation is the leading environmental risk factor, while genetic predisposition also plays a significant role [3]. The skin's structure, particularly the keratinocyte-rich epidermis, is critical to understanding cancer initiation. Keratinocytes mature as they migrate through the epidermal layers and are vulnerable to UV-induced DNA damage, which may lead to tumorigenesis [4,5].

Although surgical excision remains the cornerstone of treatment for most forms of skin cancer, particularly when diagnosed early, there is a growing need for complementary therapies that can reduce recurrence, improve outcomes, and lower treatment-associated morbidity. In recent years, natural compounds derived from plants—especially phytochemicals found in essential oils and extracts—have emerged as promising candidates for cancer prevention and therapy. These bioactive molecules have long been used in traditional medicine to promote skin health and regeneration, and recent studies suggest that they may also possess potent anticancer properties [6].

Our research focuses on exploring the anticancer potential of specific native Middle Eastern plants, including *Juniperus excelsa* M. Bieb. (Cupressaceae) (juniper berries), *Lavandula vera* DC. (Lamiaceae) (lavender flowers), and *Salvia fruticosa* (Mill.). (Lamiaceae) (Lebanese sage). We investigate the cytotoxic effects of their extracts on keratinocytes, aiming to assess their viability as novel therapeutic agents against skin cancer. By combining ethnobotanical knowledge with modern biomedical research, this study seeks to contribute to the growing field of plant-based anticancer treatments, with a particular emphasis on their application in dermatological oncology [7,8].

Essential oils, which are concentrated volatile compounds extracted from aromatic plants, have long been valued in traditional medicine for their antimicrobial, anti-inflammatory, and regenerative properties. In recent decades, research has increasingly focused on their potential anticancer activities, particularly through mechanisms relevant to skin cancer pathology. The bioactive constituents of essential oils, such as terpenes, phenolics, flavonoids, and aldehydes, can interact with multiple cellular pathways involved in carcinogenesis, including those governing oxidative stress, inflammation, apoptosis, and cell cycle regulation [9].

This study primarily focused on evaluating the cytotoxic potential of selected essential oils against keratinocyte cell lines, including benign and malignant variants, without delving into the underlying mechanisms of cell death. By employing two fundamental assays—the Trypan Blue Exclusion Test and the MTT assay—we assessed whether the tested essential oils could induce cell death and quantified the percentage of dead cells relative to the DMSO control. While the mechanistic pathways remain to be elucidated, this work provides preliminary evidence of cytotoxic activity, forming a basis for future mechanistic investigations.

2. Results

2.1. Chemical Composition of Essential Oils

The chemical composition of the essential oils (EOs) was first determined using Gas Chromatography–Mass Spectrometry (GC-MS), Co-injection Gas Chromatography (Co-GC), and Retention Index (Ri) comparisons. The main constituents identified in each oil are presented in Table 1, along with their extraction yields and relative percentages.

Table 1. Different chemical compositions of essential oils.

	Plant Name		*Juniperus excelsa*	*Salvia fruticosa*	*Lavandula vera*
	Yield % (*v*/*w*)		**1.17**	**0.9**	**1.25**
R_i	**Compound ID**	**Identification**			
938	α-Pinene	R_i, MS, CoGC	86.8	0.8	0.1
953	Camphene	R_i, MS, CoGC	-	-	0.1
980	β-Pinene	R_i, MS, CoGC	2.5	3.2	
993	Myrcene	R_i, MS, CoGC	3.2	1.4	0.4
1013	δ-3-Carene	R_i, MS	2.4	2.5	
1030	Limonene	R_i, MS, CoGC	2.2	-	
1034	1,8-Cineole	R_i, MS, CoGC	-	48.7	2.8
1057	γ-Terpinene	R_i, MS, CoGC	0.3	-	
1074	Linalool oxide	R_i, MS			0.5
1098	Linalool	R_i, MS, CoGC			42.5
1105	α-Thujone	R_i, MS	-	0.4	
1138	Menthone	R_i, MS	-	-	
1145	Camphor	R_i, MS, CoGC	trace < 0.05	1.1	10.4
1156	Isoborneol				1.1
1166	Borneol				5.6
1168	Lavandulol	R_i, MS	-	-	0.2
1176	Terpinen-4-ol				10.5
1189	alpha-terpineol	R_i, MS	0.4	-	0.9
1191	Hexyl butyrate				2.5
1217	Verbenone	R_i, MS	0.1	-	
1233	Pulegone	R_i, MS, CoGC	-	-	
1264	Linalyl acetate	8.6			
1289	Lavandulyl acetate				1.9
1329	Piperitone	R_i, MS, CoGC	-	-	
1343	Piperitenone	R_i, MS	-	-	
1365	Neryl acetate				0.3
1369	Piperitenone oxide	R_i, MS			
1383	Geranyl acetate				1.1
1404	(Z)-Caryophyllene				1.1
1415	β-Caryophyllene	R_i, MS, CoGC	-	30.8	
1437	Aromadendrene	R_i, MS	-	3.3	
1452	β-Farnesene				3.1
1455	α-Humulene	R_i, MS	-	2.8	
1463	allo-Aromadendrene	R_i, MS	-	0.3	
1491	Bicyclogermacrene	R_i, MS	-	-	

Table 1. *Cont.*

R_i	Plant Name Yield % (*v*/*w*) Compound ID	Identification	*Juniperus excelsa* 1.17	*Salvia fruticosa* 0.9	*Lavandula vera* 1.25
1500	Lavandulyl isovalerate				1.6
1515	δ-Cadinene	R_i, MS	trace < 0.05	0.1	
1566	α-Nerolidol				0.2
1577	Caryophyllene oxide				0.5
	Total		97.8	95.5	87.0

Extraction yield (% *v*/*w*) and chemical composition (% relative content) of the essential oils from *Juniperus excelsa*, *Salvia fruticosa*, and *Lavandula vera*. Identification was performed based on Retention Index (Ri), Mass Spectrometry (MS), and Co-injection Gas Chromatography (CoGC).

Juniperus excelsa oil was predominantly composed of α-pinene (86.8%), while Salvia libanotica was rich in 1,8-cineole (48.7%) and β-caryophyllene (30.8%). The main components of Lavandula vera oil were linalool (42.5%) and terpinen-4-ol (10.5%). These results align with previous studies [8,10,11] that have identified similar major constituents in these species, although minor variations may result from differences in geographical origin, seasonal timing of harvest, or extraction techniques.

A more detailed presentation of the identified compounds, including their potential biological relevance, is essential as these chemical constituents likely contribute to the observed cytotoxic effects of the oils. For example, α-pinene has known anti-proliferative and pro-apoptotic activities, 1,8-cineole exhibits anti-inflammatory and anticancer properties, and linalool is widely reported for its cytotoxic potential against various cancer cell lines.

Following the chemical characterization, the essential oils were subjected to in vitro cytotoxicity assays to assess their effects on the HaCaT, A5, and II4 cell lines.

2.2. *Assessment of Essential Oils' Effect on Cell Viability*

2.2.1. Effect of *Juniperus excelsa* M. Bieb. (Cupressaceae)

The potential dose-dependent effect of *J. excelsa* essential oil was assessed in each of the cell lines of interest in which the essential oil, in its three studied concentrations, was revealed to significantly promote a dose-independent cell death. Indeed, increasing the concentration of this essential oil from 0.25 mg/mL to 0.5 mg/mL and 0.75 mg/mL induced a roughly constant high cell death percentage. These cell death percentages were demonstrated to comprise an elevated extent of statistical significance upon comparison with the control representative cells of each stage treated with DMSO alone, where the % cell death displayed minimal irrelevant percentages of 8.83% in HaCaT, 4.25% in A5, and 10.07% in II4.

In addition to being shown to be dose independent, *J. excelsa* essential oil's effect on HaCaT, A5, and II4 keratinocyte cell viability was found to be time independent whereby prolonging the treatment exposure period from 24 h to 48 h did not significantly alter the cytotoxic effect against any cell line scrutinized (Figure 1).

Furthermore, *J. excelsa* essential oil's effect on cell viability was shown throughout our investigation to be non-differential since the % cell death induction of each of the tested concentrations, 0.25 mg/mL, 0.5 mg/mL and 0.75 mg/mL, in benign A5 and low-grade malignant II4 keratinocytes demonstrated statistical insignificance when compared to the HaCaT control cell line's % cell death; this indiscriminate effect and insignificant % cell death variation were further conserved upon treatment exposure prolongation to 48 h without any further significant amendment (Figure 1).

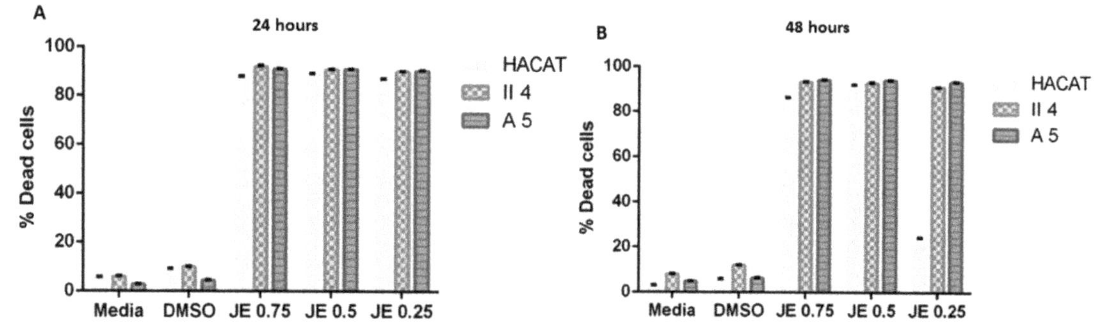

Figure 1. Comparison of *Juniperus excelsa* essential oil's effect on cell viability among keratinocytes representative of different skin cancer stages following 24 and 48 h treatments shown in (**A**,**B**) respectively. Data are presented as mean ± SEM of three experiments, performed in triplicate. Essential oil treatments of tumorigenic A5 and II4 keratinocytes were compared in reference to essential oil treatments of control non-tumorigenic HaCaT keratinocytes. Significance was detected at ($p < 0.5$).

2.2.2. Effect of *Lavandula vera* DC. (Lamiaceae)

The *L. vera* essential oil's dose-associated effect was assessed in each of the previously described target cell lines in which the three investigated essential oil concentrations were capable of significantly instigating a concentration-dependent % cell death. Moreover, except for the effect displayed by the *L. vera* essential oil concentration of 0.25 mg/mL on HaCaT keratinocyte cell death, whose significance was only moderately high, the previously mentioned essential oil-attributed cell death percentages contained a high degree of statistical significance upon comparison with the control representative cells of each stage treated with DMSO alone.

Despite being shown to be dose dependent, *L. vera* essential oil's effect on HaCaT, A5, and II4 keratinocyte cell viability was found to have time-independent features whereby prolonging the treatment exposure period from 24 h to 48 h failed to induce any significant alterations in the cytotoxic effects against any inspected cell line (Figure 2).

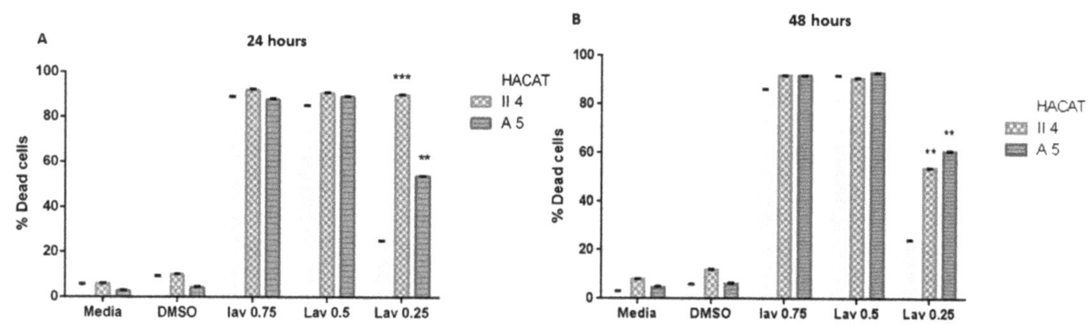

Figure 2. Comparison of *Lavendula vera* essential oil's effect on cell viability among keratinocytes representative of different skin cancer stages following 24 and 48 h treatments shown in (**A**,**B**) respectively. Data are presented as mean ± SEM of three experiments, performed in triplicate. Essential oil treatments of tumorigenic A5 and II4 keratinocytes were compared in reference to essential oil treatments of control non-tumorigenic HaCaT keratinocytes. ** $p < 0.01$; *** $p < 0.001$ compared to control. Significance was detected at ($p < 0.5$).

Of particular interest, *L. vera* essential oil's cytotoxic effect at low concentrations strikingly manifested itself throughout our investigation as highly discriminatory between the investigated cell lines with a preferential toxicity towards cancerous keratinocytes in which the low concentrations, which were shown to be non-toxic to normal cells, exhibited lethal effects on the human skin cancer cells examined. Indeed, 0.25 mg/mL *L. vera* essential oil induced a relatively slight cell death percentage of 24.98% in control non-tumorigenic HaCaT keratinocytes, in contrast to comparatively preeminent cell death percentages of 53.85% cell death in benign A5 keratinocytes and 89.93% cell death in low-grade malignant II4 keratinocytes (Figure 2). Accordingly, at this essential oil concentration, *L. vera*'s formerly stated cytotoxic effect on A5 and II4 keratinocytes was revealed to encompass a prominent grade of statistical significance upon comparison with the % cell death attributed to the HaCaT control cell line. This significance in the discriminant and preferential cell death effect was preserved without any further noteworthy adjustments upon treatment exposure prolongation to 48 h (Figure 2).

2.2.3. Effect of *Salvia fruticosa* (Mill.) (Lamiaceae)

S. fruticosa essential oil's tentative dose-reliant effect was evaluated in each of the previously demarcated control and tumorigenic target cell lines in which the three explored essential oil concentrations were proficient at significantly prompting a concentration contingent cell death. Moreover, excluding the effect presented by the *S. fruticosa* 0.25 mg/mL essential oil concentration on HaCaT keratinocyte cell death, whose significance was only temperately elevated, this essential oil-induced cell death percentages were shown to have an eminent degree of statistical significance when compared to the archetypal control cells of each stage subjected to DMSO treatment.

Notwithstanding the former dose-dependent cell death influence, *S. fruticosa* essential oil's effect on HaCaT, A5, and II4 keratinocyte cell viability was revealed to have time-independent characteristics wherein lengthening the treatment exposure period from 24 h to 48 h proved unsuccessful at causing any significant shifts in the studied cytotoxic effects against any inspected cell line (Figure 3).

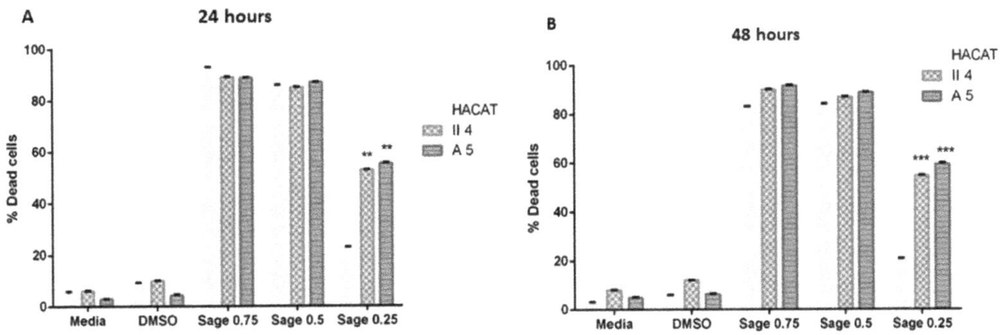

Figure 3. Comparison of *Salvia fruticosa* essential oil's effect on cell viability among keratinocytes representative of different skin cancer stages following 24 and 48 h treatments shown in (**A**,**B**) respectively. Data are presented as mean ± SEM of three experiments, performed in triplicate. Essential oil treatments of tumorigenic A5 and II4 keratinocytes were compared in reference to essential oil treatments of control non-tumorigenic HaCaT keratinocytes. ** $p < 0.01$; *** $p < 0.001$ compared to control. Significance was detected at ($p < 0.5$).

Of particular interest, *S. fruticosa* essential oil's cytotoxic effect at low concentrations strikingly established itself throughout the course of our inquiry as possessing highly discriminant features between the considered cell lines, with toxicity favoring cancerous keratinocytes in which low concentrations, which were found to be non-toxic to normal cells, exhibited lethal effects on the human skin cancer cells examined. Indeed, 0.25 mg/mL *S. fruticosa* essential oil prompted a relatively trivial cell death percentage of 22.88% in non-tumorigenic HaCaT keratinocytes in contrast to a reasonably dominant cell death in benign A5 keratinocytes of 55.39% and in low-grade malignant II4 keratinocytes of 52.89%. Consequently, at this essential oil concentration, *S. fruticosa*'s previously quantified cytotoxic effect on A5 and II4 keratinocytes exposed a conspicuous grade of statistical significance upon comparison with the % cell death ascribed to the HaCaT control cell line. This significance in the discriminant and privileged cell death effect was devoid of any further remarkable variations upon treatment exposure continuation to 48 h (Figure 3).

2.3. *Evaluation of Essential Oils' Effect on Cell Proliferation*

2.3.1. Effect of *Juniperus excelsa* M. Bieb. (Cupressaceae)

Following the described MTT assay, the probable dose-dependent effect of *J. excelsa* essential oil on cell proliferation was quantified and found to have prominent inhibitory effects on HaCaT, A5, and II4 keratinocyte cell proliferation. Indeed, increasing the concentration of this essential oil from 0.25 mg/mL to 0.5 mg/mL and 0.75 mg/mL induced an approximately constant growth suppression displayed as a notable decrease in cell viability: 5.98%, 5.16%, and 4.57%, respectively (Figure 4). These cell viability percentages were demonstrated to comprise an elevated extent of statistical significance upon comparison with the control representative cells of each stage treated with DMSO alone.

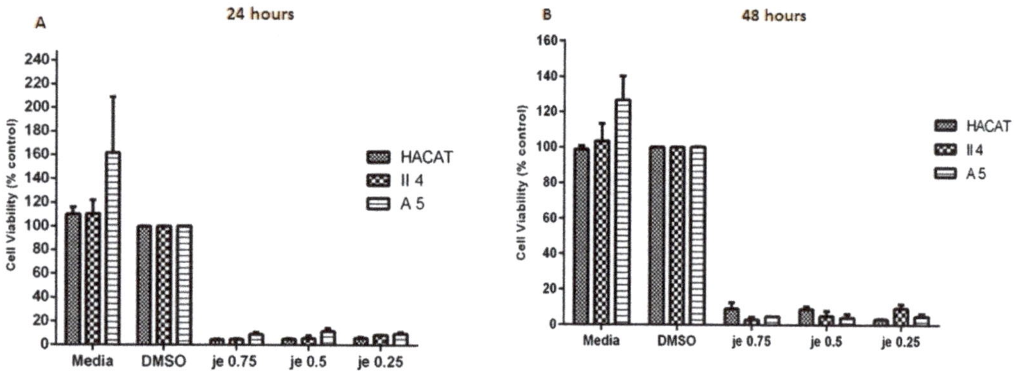

Figure 4. Comparison of *Juniperus excelsa* essential oil's effect on metabolic activity (MTT assay), used as an indicator of cell viability among keratinocytes representative of the different skin cancer stages following 24 and 48 h treatments shown in (**A**,**B**) respectively. Data are presented as mean ± SEM of three experiments, performed in triplicate. Essential oil treatments of tumorigenic A5 and II4 keratinocytes were compared in reference to essential oil treatments of control non-tumorigenic HaCaT keratinocytes. Control represents DMSO-treated cells. Significance was detected at ($p < 0.5$).

In addition to being shown to be dose independent, *J. excelsa* essential oil's effect on HaCaT, A5, and II4 keratinocyte cell proliferation displayed itself as time independent whereby prolonging the treatment exposure period from 24 h to 48 h did not significantly alter the growth inhibitory effects against any cell line scrutinized.

Furthermore, *J. excelsa* essential oil's effect on cell proliferation was shown throughout our investigation to be non-differential since the percentage cell viability of each of the tested concentrations, 0.25 mg/mL, 0.5 mg/mL, and 0.75 mg/mL, in benign A5 and low-grade malignant II4 keratinocytes demonstrated statistical insignificance when compared to the HaCaT control cell line's percentage cell viability; this indiscriminate effect and insignificant percentage cell viability variation were further conserved upon treatment exposure prolongation to 48 h without any further significant amendment (Figure 4).

2.3.2. Effect of *Lavandula vera* DC. (Lamiaceae)

After performing the MTT assay, the dose-dependent effect of *L. vera* essential oil on cell proliferation was calculated in the previously described target cell lines; the three investigated essential oil concentrations were capable of significantly influencing the attributed cell proliferation in a concentration-dependent manner as evidenced from the fluctuations in percentage cell viability. Moreover, except for the effect displayed by the *L. vera* essential oil concentration of 0.25 mg/mL on HaCaT keratinocyte cell proliferation, whose significance was only moderately high, the essential oil-attributed cell viability percentages contained a high degree of statistical significance upon comparison with the control representative cells of each stage treated with DMSO alone in which the % cell viability was revealed as 100% due to its minimal irrelevant growth suppression in the three cell line of interest (Figure 5).

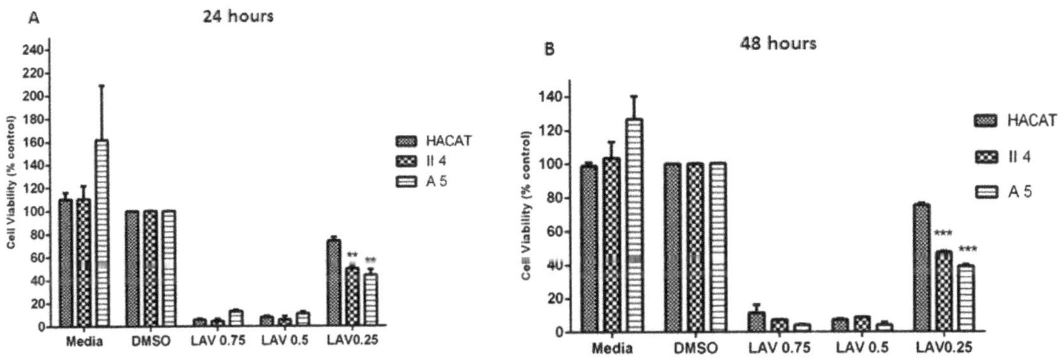

Figure 5. Comparison of *Lavendula vera* essential oil's effect on metabolic activity (MTT assay), used as an indicator of cell viability among keratinocytes representative of different skin cancer stages following 24 and 48 h treatments as shown in (**A**,**B**) respectively. Data are presented as mean ± SEM of three experiments, performed in triplicate. Essential oil treatments of tumorigenic A5 and II4 keratinocytes were compared in reference to essential oil treatments of control non-tumorigenic HaCaT keratinocytes. Control represents DMSO-treated cells. ** $p < 0.01$; *** $p < 0.001$ compared to control. Significance was detected at ($p < 0.5$).

Despite being shown to be dose dependent, the *L. vera* essential oil effect on HaCaT, A5, and II4 keratinocyte cell viability exhibited time-independent features wherein prolonging the treatment exposure period from 24 h to 48 h failed to induce any significant alterations in the growth suppression effects against any inspected cell line (Figure 5).

Of particular interest, at low concentrations, *L. vera* essential oil's effect on cell proliferation strikingly manifested itself throughout our investigation as highly discriminant between the investigated cell lines, with a preferential growth inhibitory trend towards cancerous keratinocytes in which the low concentrations, which were shown to be ineffective in normal cells, exhibited potent growth inhibitory effects on the human skin cancer cells examined. Indeed, 0.25 mg/mL *L. vera* essential oil induced a relatively slight decline in percentage cell viability of 73.94% in control non-tumorigenic HaCaT keratinocytes, in contrast to a comparatively preeminent cell viability reduction in benign A5 keratinocytes of 44.31% and in low-grade malignant II4 keratinocytes of 49.91% (Figure 5). Accordingly, at this essential oil concentration, *L. vera*'s formerly stated growth inhibitory effect on A5 and II4 keratinocytes was revealed to encompass a prominent grade of statistical significance upon comparison with the % cell viability attributed to the HaCaT control cell line This significance in the discriminant and preferential cell death effect was preserved without any further noteworthy adjustments upon treatment exposure prolongation to 48 h (Figure 5).

2.3.3. Effect of *Salvia fruticosa* Mill. (Lamiaceae)

Following the MTT assay, the prospective dose-dependent effect of *S. fruticosa* essential oil on cell proliferation was measured in the target cell lines in which the three considered essential oil concentrations were proficient at significantly influencing the attributed cell proliferations in a concentration-dependent fashion as evidenced by apparent fluctuations in the percentage cell viability. For instance, upon increasing the essential oil concentration, significant percentage cell viability reductions were observed. Furthermore, excluding the effect exhibited by the *S. fruticosa* 0.25 mg/mL essential oil concentration on HaCaT keratinocyte cell proliferation, whose significance was only moderately high, this essential oil's associated cell viability percentages were confirmed to have an elevated grade of statistical significance upon comparison with the control representative cells of each stage treated with DMSO alone, where the % cell viability was considered as 100% due to its minimal irrelevant growth suppression effects in the three cell lines of interest (Figure 6).

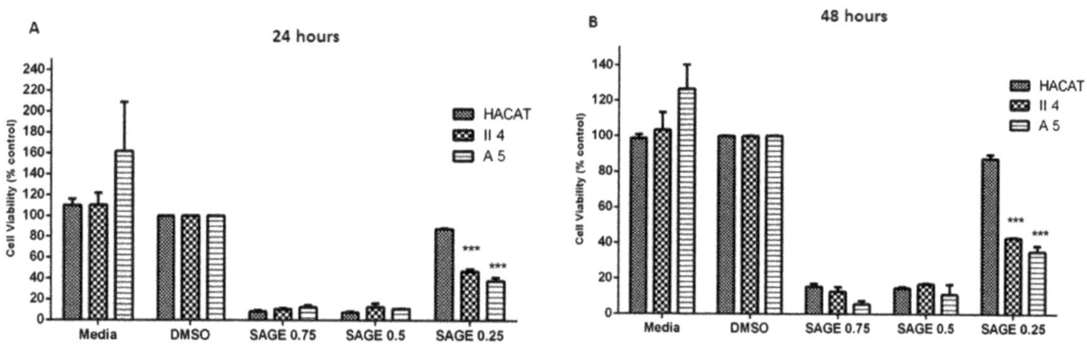

Figure 6. Comparison of *Salvia fruticosa* essential oil effect on metabolic activity (MTT assay), used as an indicator of cell viability among keratinocytes representative of different skin cancer stages following 24 and 48 h treatments as shown in (**A**,**B**) respectively. Data are presented as mean ± SEM of three experiments, performed in triplicate. Essential oil treatments of tumorigenic A5 and II4 keratinocytes were compared in reference to essential oil treatments of control non-tumorigenic HaCaT keratinocytes. *** $p < 0.001$ compared to control. Significance was detected at ($p < 0.5$).

Although established as dose dependent, the *S. fruticosa* essential oil effect on HaCaT, A5, and II4 keratinocyte cell proliferation exhibited time-independent features whereby prolonging the treatment exposure period from 24 h to 48 h failed to induce any significant alterations in the growth suppression effects against any examined cell line (Figure 6).

Of particular importance, at low essential oil concentrations, *S. fruticosa* oil's effect on cell proliferation strikingly expressed itself throughout our investigation as highly discriminant between the probed cell lines with a preferential growth inhibitory trend towards cancerous keratinocytes, wherein low concentrations, which were shown to be ineffective in normal cells, exhibited potent growth inhibitory effects on the human skin cancer cells examined. Indeed, 0.25 mg/mL *S. fruticosa* essential oil induced a relatively slight proliferation decline in the control non-tumorigenic HaCaT keratinocytes whose percentage cell viability was 87.90%, in contrast to a comparatively preeminent cell proliferation reduction in benign A5 keratinocytes whose percentage cell viability was 37.86% and in low-grade malignant II4 keratinocytes whose percentage cell viability was 46.91% (Figure 6). Accordingly, at this essential oil concentration, *S. fruticosa*'s growth inhibitory effects on A5 and

II4 keratinocytes were revealed to encompass a prominent grade of statistical significance upon comparison with the % cell viability attributed to the HaCaT control cell line. This significance in the discriminant and preferential cell death effect was preserved without any further noteworthy adjustments upon treatment exposure prolongation to 48 h (Figure 6).

3. Materials and Methods

3.1. Chemicals and Reagents

Penicillin Streptomycin, L-Glutamine, HEPES(4-(2-hydroxyethyl)-1-piperazineethan-esulfonic acid), heat-inactivated fetal bovine serum (FBS), Phosphate Buffer Saline (PBS), Dimethyl Sulfoxide (DMSO), Trypsin EDTA, and Trypan Blue Exclusion Dye were purchased from Sigma-Aldrich Co. (Burlington, MA, USA); Dulbecco Modified Eagle Media (DMEM) was obtained from Gibco Life Technologies (Carlsbad, CA, USA); the Vybrant MTT 3-(4,5-dimethylthiazol-2,5-diphenyltetrazoliumbromide) Cell Proliferation Assay Kit V-13154 was obtained from Molecular Probes subdivision of Thermo Fisher Scientific. 168 Third Avenue. Waltham, MA, USA.

3.2. Plant Collection and Essential Oil Isolation and Analysis

Essential oils from *Juniperus excelsa* M. Bieb. (Cupressaceae) berries, *Lavandula vera* DC. (Lamiaceae) flowers, and *Salvia fruticose* Mill. (Lamiaceae) leaves collected from Kartaba and Nahr Ibrahim during the months of October (*J. excelsa*) and May (*L. vera* and *S. fruticosa*) 2011 were kindly supplied by Dr. Marc Beyrouthy (USEK University). Aerial plant fragments were immediately pulverized and subjected to steam distillation for 3 h using a Clevenger type apparatus for the isolation of their essential oils. Then, the obtained essential oils were collected, dried over anhydrous sodium sulfate, filtrated, and stored at 4 °C in tightly closed, dark glass vials wrapped in aluminum foil to prevent oxygen and light exposure for a maximum period of one week until submission to Mass Spectrometry (MS), Co-injection Gas Chromatography (CoGC), and Linear Retention Index (Ri) analyses for the detection of their major constituents.

The botanical identity of the plant materials was confirmed by Dr. Marc Beyrouthy (USEK University), using regional floristic references. However, no voucher specimens were deposited in a public herbarium. The plant materials were collected and authenticated based on morphological features, and representative samples are archived in the private botanical collection of Dr. Beyrouthy for future reference.

3.3. Essential Oil Dissolution

In order to facilitate the dispersion of the oils in the aqueous culture medium prior to each employment, essential oils, sterilized with 0.45 µm Millipore filters, were dissolved in 100% DMSO to prepare 10% (*v*/*v*) stock solutions. Working concentrations of 0.75, 0.5, and 0.25 mg/mL were freshly prepared by direct dilution into complete culture medium, maintaining a final DMSO concentration of ≤0.1% (*v*/*v*) in all treated wells. These concentrations were further screened with the purpose of eluting three optimal concentrations (0.75 mg/mL, 0.5 mg/mL, and 0.25 mg/mL) to be employed in the targeted investigations.

3.3.1. GC Analyses

Analytical gas chromatography was performed on a Thermo Electron Corporation (subdivision of Thermo Fisher Scientific. 168 Third Avenue. Waltham, MA, USA) gas chromatograph fitted with a flame ionization detector (FID), a DB-5 MS capillary column (30 m × 0.25 mm) with 0.1 µm film thickness, or a fused silica HP Innowax polyethylene glycol capillary column (50 m × 0.20 mm, film thickness 0.20 µm). Helium was the carrier gas (0.7 mL/min). The column temperature was initially set to 35 °C before being gradually increased to 85 °C at 5 °C/min, held for 20 min at 85 °C, raised to 300 °C at 10 °C/min, and finally held for 5 min at 300 °C. Diluted 1 μL samples (1/100, *v/v*) were injected at 250 °C manually and in the splitless mode. Flame ionization detection (FID) was performed at 310 °C.

3.3.2. GC–MS Analyses

The GC/MS analyses were performed using an Agilent 6890 gas chromatograph coupled with a 5975 Mass Detector (by Agilent 5301 Stevens Creek Blvd Santa Clara, CA, USA). The 7683 B auto sampler injected 1 μL of each diluted oil sample (1/100, *v/v*). A fused silica capillary column DB-5 MS (30 m × 0.25 mm internal diameter, film thickener 0.1 μm) or a fused silica HP Innowax polyethylene glycol capillary column (50 m × 0.20 mm, film thickness 0.20 μm) was used. Helium was the carrier gas (0.7 mL/min). The oven temperature program was identical to that described above (cf. GC Analysis). The mass spectra were recorded at 70 eV with an ion source temperature of 310 °C and a transfer line heated to 320 °C. The acquisition was recorded in full scan mode (50–400 m/z).

3.3.3. Qualitative and Quantitative Analysis

Most constituents were identified by gas chromatography by comparing their retention indices (RIs) with those from the literature [12,13] or with those of authentic compounds obtained from Sigma-Aldrich (Lebanon and France). The retention indices were determined relative to a homologous series of n-alkanes (C_8 to C_{24}) analyzed under the same operating conditions. Concomitantly, their mass spectra on both columns were compared with those provided in the NIST and Wiley 275 libraries, our home-made library constructed with pure compounds and EOs of known composition, or with mass spectra from the literature [14,15]. The relative concentration of each component was calculated based on the GC peak areas without using correction factors.

3.4. Cell Culture

The formerly described spontaneously immortalized non-tumorigenic Human Keratinocyte (HaCaT) cell line and its benign (A5) and malignant (II4) clones were kindly provided by Dr. Marwan Sabban (American University of Beirut). The provided cells were cultured at normal 2 mM Calcium concentrations in Dulbecco's modified Eagle's medium (DMEM) supplemented with 10% heat-inactivated fetal bovine serum (FBS), 100 μg/mL penicillin–streptomycin, L-Glutamine (final concentration: 2 mM), and 1% HEPES and maintained in a humidified incubator containing 95% air and 5% CO_2 at 370 °C. Throughout the investigation course, the cell lines had their medium renewed and replenished every 2 days and were passaged upon attaining 70% confluence (every 7 to 10 days) via Trypsin–EDTA treatment. The passage methodology consisted of washing the attached cells with 5 mL (0.05%) of EDTA in order to remove the serum, Ca^{2+}, Mg^{2+}, and all factors that inhibit trypsin, followed by incubation with 5 mL (0.05%) of EDTA for a maximum of 6 min in order to accomplish desmosome destruction displayed as microscopically visible wide intercellular spaces. Once this was achieved, the cells were then incubated with 1:1 Trypsin (0.025%)–EDTA (0.05%) for up to 3 min until the cells came off the plastic upon shaking. Subsequent to cell detachment, an equal volume of fresh medium was added to the culture flask to inactivate trypsin, and the cells were then suspended, transferred into falcon tubes, and centrifuged at 900× g for 5 min at 40 °C. Finally, the resulting cell pellet was re-suspended in fresh medium and split into a maximal ratio of 1:5 into cell culture flasks for future seeding, treatment, and analysis.

3.5. Cell Viability Analysis: Trypan Blue Exclusion Test

Prior to the viability assays, HaCaT, A5, and II4 cells were plated in 24-well plates at a density of 1×10^6 cells/well in 1 mL of medium to ensure consistent conditions. After seeding, the cells were incubated for 24 h at 37 °C in 5% CO_2 to allow attachment. The medium was then replaced with fresh medium containing essential oils of *J. excelsa*, *L. vera*, or *S. fruticose Mil* at final concentrations of 0.75, 0.5, or 0.25 mg/mL. Treatments were performed in triplicate alongside a negative control (0.01% DMSO) and a positive control (culture only). Cells were incubated for an additional 24 or 48 h under the same conditions. Cytotoxicity was assessed using the Trypan Blue Exclusion Test: cells were harvested, mixed with an equal volume of Trypan Blue, and counted using a hemocytometer. Percent viability (unstained cells/total cells × 100) and percent cell death (stained cells/total cells × 100) were calculated accordingly.

3.6. Cell Proliferation Examination: MTT Assay

The MTT assay measures mitochondrial metabolic activity, which is proportional to the number of metabolically active cells and is commonly used as an indicator of cell proliferation. For proliferation analysis, HaCaT, A5, and II4 cells were seeded in 96-well plates at 1000 cells/well in 200 µL of medium and then incubated for 24 h at 37 °C with 5% CO_2 to allow cell attachment. The medium was then replaced with fresh medium containing *J. excelsa*, *L. vera*, or *S. fruticosa* essential oils at final concentrations of 0.75, 0.5, or 0.25 mg/mL. All treatments were performed in triplicate alongside a negative control (0.01% DMSO) and a positive control (media only). After 24 and 48 h of incubation, cell proliferation was assessed using the MTT assay [16], which measures metabolic activity via conversion of yellow tetrazolium to blue formazan. The assay was conducted with the Vybrant MTT kit. Briefly, the medium was replaced with 100 µL of fresh medium, and 10 µL of 12 mM MTT stock solution was added, followed by a 4 h incubation at 37 °C. A test blank (MTT without cells) was included. After incubation, 85 µL of medium was removed and replaced with 50 µL of DMSO to solubilize the formazan crystals, followed by a 10 min incubation. The absorbance was read at 540 nm, and the % viability was calculated as: (mean absorbance of treated wells/mean absorbance of untreated control) × 100.

3.7. Statistical Data Analysis

All triplicate experimental values obtained are presented as the mean ± standard error mean (SEM). The experimental data were analyzed using Graph Pad Prism 5 statistical software (GraphPad Software Inc., San Diego, CA, USA), and the statistical significance was assessed via one-way Anova Tukey test for single step multiple comparisons, one-way Anova Dunnet test for comparison with control, and two-way Anova Bonferroni test for comparisons among cell lines, with * p values < 0.05 considered as statistically significant.

4. Discussion

Skin cancer, similar to many malignancies, often results from a combination of chronic inflammation, oxidative DNA damage, and impaired regulation of cell proliferation. Exposure to ultraviolet (UV) radiation not only directly damages DNA in epidermal cells, particularly keratinocytes and melanocytes, but also generates reactive oxygen species (ROS) that trigger chronic oxidative stress and inflammation—two hallmarks of tumor initiation and progression. Furthermore, UV radiation can induce local immunosuppression, allowing mutated cells to evade immune surveillance and continue proliferating [17–20].

Essential oils offer a promising countermeasure to these pathological processes. Many have demonstrated potent antioxidant capabilities, such as neutralizing ROS and protecting skin cells from UV-induced damage. Others exhibit anti-inflammatory effects by modulating signaling pathways that are known to be upregulated in both melanoma and non-melanoma skin cancers. Several essential oils also promote apoptosis (programmed cell death) in malignant cells by activating intrinsic mitochondrial pathways or extrinsic death receptor pathways while leaving normal skin cells relatively unharmed—a feature that makes them attractive candidates for targeted cancer therapy [17,21].

Notably, plant-derived essential oils such as those from *Juniperus excelsa* M. Bieb. (Cupressaceae), *Lavandula vera* DC. (Lamiaceae), and *Salvia fruticosa* (Mill.) (Lamiaceae) have been historically used in Middle Eastern folk medicine for treating skin conditions, wounds, and infections. Recent in vitro studies suggest that extracts from these plants may also exert cytotoxic effects on human carcinoma cell lines, including keratinocytes transformed by UV exposure. For instance, components such as linalool, carvacrol, and α-pinene—found in these oils—have been shown to inhibit cell proliferation, induce apoptosis, and reduce tumor angiogenesis. Their dual role in enhancing skin barrier function and exerting anticancer activity makes them particularly suitable for therapeutic exploration in cutaneous carcinomas [9,22].

The comparative analysis of the three essential oils highlights a critical distinction in their therapeutic potential. While *J. excelsa* demonstrates potent anticancer activity, its non-selective cytotoxicity diminishes its suitability for clinical use (Figures 1 and 6). In contrast, *L. vera* and *S. fruticosa* show promising selective toxicity at lower doses, preferentially targeting malignant cells while sparing normal keratinocytes (Figures 2, 4 and 5). This cancer-selective behavior is a key criterion for developing effective and safe alternative therapies, particularly in dermatologic oncology where preserving healthy tissue integrity is paramount.

Furthermore, the observed dose-dependent but time-independent nature of *L. vera* and *S. fruticosa* suggests that therapeutic efficacy may be optimized through concentration modulation rather than prolonged application—an advantageous property when considering topical formulations. These findings underscore the importance of precise dosing in the therapeutic use of plant-derived compounds and provide a foundation for further investigation into the molecular mechanisms driving the selectivity of these essential oils.

Future studies should aim to identify the specific bioactive compounds responsible for these effects, elucidate their mechanisms of action, and validate these findings in in vivo models. If further confirmed, *L. vera* and *S. fruticosa* could emerge as promising candidates in integrative or adjunctive treatments for skin cancer, potentially enhancing efficacy while reducing the side effects associated with conventional therapies.

The anticancer properties of *J. excelsa*, *L. vera*, and *S. fruticosa* essential oils are not merely due to their cytotoxicity but also involve complex molecular mechanisms that modulate key pathways involved in cancer progression. By targeting these molecular pathways, these oils may help inhibit tumor growth, promote apoptosis, and reduce metastasis, offering a multifaceted approach to cancer treatment [23].

As mentioned earlier, skin cancer, particularly melanoma and non-melanoma skin cancers (NMSCs), is driven by UV-induced oxidative stress, which results in DNA damage and mutagenesis [17,18]. A major molecular mechanism by which the essential oils exert their anticancer effects is through the reduction of oxidative stress. Both Lavandula vera and *S. fruticosa* have demonstrated antioxidant properties by neutralizing free radicals and reducing the formation of reactive oxygen species (ROS). This helps maintain cellular integrity and prevents DNA mutations, which are crucial in the initiation and progression of skin cancer [24–27].

As another example, α-pinene, found in *J. excelsa*, has been shown to enhance the activity of antioxidant enzymes such as superoxide dismutase (SOD) and catalase, which play pivotal roles in scavenging ROS. By mitigating oxidative damage, these oils not only protect healthy cells but also inhibit the genetic alterations that drive carcinogenesis [28,29].

Another critical mechanism by which essential oils exert their anticancer effects is by modulating the cell cycle and inducing apoptosis in transformed cells. Cancer cells often acquire mutations that allow them to bypass cell cycle checkpoints and resist apoptosis, leading to uncontrolled proliferation.

Both *L. vera* and *S. fruticosa* essential oils influence key regulators of the cell cycle, including cyclins, cyclin-dependent kinases (CDKs), and tumor suppressor proteins such as p53. Studies have shown that linalool (in Lavandula vera) and carnosic acid (in *S. fruticosa*) can induce G0/G1 arrest, preventing malignant cells from progressing through the cell cycle and dividing. This results in reduced cell proliferation in cancerous tissues [24,30,31].

Furthermore, these oils also promote apoptosis through mitochondrial and caspase-dependent pathways. For instance, linalool has been demonstrated to activate caspases-3 and -9, key enzymes that execute the apoptotic process by cleaving vital cellular components. This activation of programmed cell death helps eliminate malignant cells without affecting healthy keratinocytes [30,32].

One of the hallmarks of cancer is its ability to stimulate angiogenesis, the formation of new blood vessels, which supplies nutrients to the growing tumor and facilitates metastasis. Essential oils from *L. vera* and *S. fruticosa* have been found to inhibit angiogenesis by downregulating key pro-angiogenic factors, such as vascular endothelial growth factor (VEGF). By inhibiting VEGF signaling, these oils disrupt the blood supply to tumors, thereby impeding their growth and spread [32].

Moreover, both oils have demonstrated the ability to suppress matrix metalloproteinases (MMPs), enzymes involved in the degradation of the extracellular matrix. This action may reduce the capacity of cancer cells to invade surrounding tissues and metastasize to other organs. These effects are crucial in preventing tumor progression and the spread of skin cancer to distant sites [31,32].

5. Conclusions

This study assessed the anticancer effects of essential oils from *Juniperus excelsa* M. Bieb. (Cupressaceae), *Lavandula vera* DC. (Lamiaceae), and *Salvia fruticose* (Mill.) (Lamiaceae) on non-tumorigenic HaCaT, benign A5, and low-grade malignant II4 keratinocytes. All three oils showed significant cytotoxic and antiproliferative effects. While *J. excelsa* was active against both normal and cancerous cells, its lack of selectivity limits its clinical relevance. In contrast, *L. vera* and *S. fruticosa* displayed selective toxicity toward malignant cells, particularly at low concentrations, sparing normal keratinocytes, which suggests their potential for targeted skin cancer therapies. Overall, essential oils from these Mediterranean plants emerge as promising antitumor agents. Future work should identify active compounds and investigate mechanisms such as mitochondrial membrane potential disruption, cytochrome c release, and caspase activation. In vivo studies and clinical trials are needed to evaluate safety, dosing, and possible synergy with standard treatments.

Author Contributions: Conceptualization, M.T.Z. and M.K.; methodology, R.O. and V.M.; validation, D.A., R.R., M.B. and Z.F.; formal analysis, M.T.Z. and M.K.; investigation, M.B. and J.-M.S.; writing—original draft preparation, R.O., V.M. and M.T.Z.; writing—review and editing, J.-M.S. and Z.F.; visualization, M.B. and R.R.; project administration, M.K. All authors have read and agreed to the published version of the manuscript.

Funding: This research received no external funding.

Institutional Review Board Statement: Not applicable.

Informed Consent Statement: Not applicable.

Data Availability Statement: The original contributions presented in this study are included in the article. Further inquiries can be directed to the corresponding author(s).

Conflicts of Interest: The authors declare no conflicts of interest.

References

1. Gordon, R. Skin Cancer: An Overview of Epidemiology and Risk Factors. *Semin. Oncol. Nurs.* **2013**, *29*, 160–169. [CrossRef] [PubMed]
2. Hyeraci, M.; Papanikolau, E.S.; Grimaldi, M.; Ricci, F.; Pallotta, S.; Monetta, R.; Minafò, Y.A.; Di Lella, G.; Galdo, G.; Abeni, D.; et al. Systemic Photoprotection in Melanoma and Non-Melanoma Skin Cancer. *Biomolecules* **2023**, *13*, 1067. [CrossRef] [PubMed]
3. Arda, O.; Göksügür, N.; Tüzün, Y. Basic histological structure and functions of facial skin. *Clin. Dermatol.* **2014**, *32*, 3–13. [CrossRef]
4. Abbas, O.; Miller, D.D.; Bhawan, J. Cutaneous Malignant Melanoma. *Am. J. Dermatopathol.* **2014**, *36*, 363–379. [CrossRef] [PubMed]
5. Ahmed, B.; Qadir, M.I.; Ghafoor, S. Malignant Melanoma: Skin Cancer-Diagnosis, Prevention, and Treatment. *Crit. Rev. Eukaryot. Gene Expr.* **2020**, *30*, 291–297. [CrossRef]
6. Pavithra, P.S.; Mehta, A.; Verma, R.S. Essential oils: From prevention to treatment of skin cancer. *Drug Discov. Today* **2019**, *24*, 644–655. [CrossRef]
7. Aziz, Z.A.A.; Ahmad, A.; Setapar, S.H.M.; Karakucuk, A.; Azim, M.M.; Lokhat, D.; Rafatullah, M.; Ganash, M.; Kamal, M.A.; Ashraf, G.M. Essential Oils: Extraction Techniques, Pharmaceutical And Therapeutic Potential—A Review. *Curr. Drug Metab.* **2018**, *19*, 1100–1110. [CrossRef]
8. Emami, S.A.; Abedindo, B.F.; Hassanzadeh-Khayyat, M. Antioxidant Activity of the Essential Oils of Different Parts of *Juniperus excelsa* M. Bieb. subsp. *excelsa* and J. *excelsa* M. Bieb. subsp. *polycarpos* (K. Koch) Takhtajan (Cupressaceae). *Iran. J. Pharm. Res.* **2011**, *10*, 799–810.
9. Bayala, B.; Bassole, I.H.; Scifo, R.; Gnoula, C.; Morel, L.; Lobaccaro, J.M.; Simpore, J. Anticancer activity of essential oils and their chemical components—A review. *Am. J. Cancer Res.* **2014**, *4*, 591–607.
10. Prashar, A.; Locke, I.C.; Evans, C.S. Cytotoxicity of lavender oil and its major components to human skin cells. *Cell Prolif.* **2004**, *37*, 221–229. [CrossRef]
11. Miastkowska, M.; Kantyka, T.; Bielecka, E.; Kałucka, U.; Kamińska, M.; Kucharska, M.; Kilanowicz, A.; Cudzik, D.; Cudzik, K. Enhanced Biological Activity of a Novel Preparation of *Lavandula angustifolia* Essential Oil. *Molecules* **2021**, *26*, 2458. [CrossRef] [PubMed]
12. Jennings, W.; Shibamoto, T. *Qualitative Analysis of Flavour and Fragrance Volatiles by Glass Capillary Gas Chromatography*; Academic Press: New York, NY, USA, 1980.
13. Adams, R.P. *Identification of Essential Oil Components by Gas Chromatography/Mass Spectroscopy*; Allured Publishing Corporation: Carol Stream, IL, USA, 2007.
14. Davies, N.W. Gas chromatographic retention indices of monoterpenes and sesquiterpenes on methyl silicone and Carbowax 20M phases. *J. Chromatogr. A* **1990**, *503*, 1–24. [CrossRef]
15. Mosmann, T. Rapid colorimetric assay for cellular growth and survival: Application to proliferation and cytotoxicity assays. *J. Immunol. Methods* **1983**, *16*, 55–63. [CrossRef]
16. Ciążyńska, M.; Olejniczak-Staruch, I.; Sobolewska-Sztychny, D.; Narbutt, J.; Skibińska, M.; Lesiak, A. Ultraviolet Radiation and Chronic Inflammation—Molecules and Mechanisms Involved in Skin Carcinogenesis: A Narrative Review. *Life* **2021**, *11*, 326. [CrossRef]
17. Halliday, G.M. Inflammation, gene mutation and photoimmunosuppression in response to UVR-induced oxidative damage contributes to photocarcinogenesis. *Mutat. Res.* **2005**, *571*, 107–120. [CrossRef] [PubMed]
18. Gieniusz, E.; Skrzydlewska, E.; Łuczaj, W. Current Insights into the Role of UV Radiation-Induced Oxidative Stress in Melanoma Pathogenesis. *Int. J. Mol. Sci.* **2024**, *25*, 11651. [CrossRef]
19. Ahn, C.; Lee, J.; Park, M.; Kim, J.; Yang, J.; Yoo, Y.; Jeung, E. Cytostatic effects of plant essential oils on human skin and lung cells. *Exp. Ther. Med.* **2020**, *19*, 2008–2018. [CrossRef]
20. Ghorbanzadeh, A.; Ghasemnezhad, A.; Sarmast, M.K.; Ebrahimi, S.N. An analysis of variations in morphological characteristics, essential oil content, and genetic sequencing among and within major Iranian Juniper (*Juniperus* spp.) populations. *Phytochemistry* **2021**, *186*, 112737. [CrossRef] [PubMed]
21. An, Q.; Ren, J.-N.; Li, X.; Fan, G.; Qu, S.-S.; Song, Y.; Li, Y.; Pan, S.-Y. Recent updates on bioactive properties of linalool. *Food Funct.* **2021**, *12*, 10370–10389. [CrossRef]

22. Tayarani-Najaran, Z.; Amiri, A.; Karimi, G.; Emami, S.A.; Asili, J.; Mousavi, S.H. Comparative studies of cytotoxic and apoptotic properties of different extracts and the essential oil of Lavandula angustifolia on malignant and normal cells. *Nutr. Cancer* **2014**, *66*, 424–434. [CrossRef]
23. Bethesda (MD): National Institute of Child Health and Human Development. Available online: https://www.ncbi.nlm.nih.gov/books/NBK501865/ (accessed on 25 April 2025).
24. Li, K.; Yang, W.; Li, Z.; Jia, W.; Li, J.; Zhang, P.; Xiao, T. Bitter apricot essential oil induces apoptosis of human HaCaT keratinocytes. *Int. Immunopharmacol.* **2016**, *34*, 189–198. [CrossRef] [PubMed]
25. Bahri, S.; Jameleddine, S.; Shlyonsky, V. Relevance of carnosic acid to the treatment of several health disorders: Molecular targets and mechanisms. *Biomed. Pharmacother.* **2016**, *84*, 569–582. [CrossRef] [PubMed]
26. de Oliveira Lima, M.; de Medeiros, A.A.; Silva, K.S.; Cardoso, G.; de Oliveira Lima, E.; de Oliveira Pereira, F. Investigation of the antifungal potential of linalool against clinical isolates of fluconazole resistant Trichophyton rubrum. *J. Mycol. Med.* **2017**, *27*, 195–202. [CrossRef] [PubMed]
27. Iobbi, V.; Parisi, V.; Bernabè, G.; De Tommasi, N.; Bisio, A.; Brun, P. Anti-Biofilm Activity of Carnosic Acid from *Salvia rosmarinus* against Methicillin-Resistant *Staphylococcus aureus*. *Plants* **2023**, *12*, 3679. [CrossRef]
28. Wińska, K.; Mączka, W.; Łyczko, J.; Grabarczyk, M.; Czubaszek, A.; Szumny, A. Essential Oils as Antimicrobial Agents—Myth or Real Alternative? *Molecules* **2019**, *24*, 2130. [CrossRef] [PubMed]
29. Cavanagh, H.M.A.; Wilkinson, J.M. Biological activities of Lavender essential oil. *Phytother. Res.* **2002**, *16*, 301–308. [CrossRef]
30. Dhar, P.; Chan, P.; Cohen, D.T.; Khawam, F.; Gibbons, S.; Snyder-Leiby, T.; Dickstein, E.; Rai, P.K.; Watal, G. Synthesis, Antimicrobial Evaluation, and Structure-Activity Relationship of α-Pinene Derivatives. *J. Agric. Food Chem.* **2014**, *62*, 3548–3552. [CrossRef]
31. Singh, J.; Luqman, S.; Meena, A. Carvacrol as a Prospective Regulator of Cancer Targets/Signalling Pathways. *Curr. Mol. Pharmacol.* **2023**, *16*, 542–558. [CrossRef]
32. Birtić, S.; Dussort, P.; Pierre, F.-X.; Bily, A.C.; Roller, M. Carnosic acid. *Phytochemistry* **2015**, *115*, 9–19. [CrossRef]

molecules

MDPI

Article

Evaluation of Antiproliferative Potentials Associated with the Volatile Compounds of *Lantana camara* Flowers: Selective In Vitro Activity

Jennifer El Hajj [1,†], Louna Karam [2,†], Ali Jaber [3,*], Edmond Cheble [1], Elias Akoury [3], Philippe Hussein Kobeissy [2], José-Noel Ibrahim [2] and Ali Yassin [3,*]

1 RDMPN Laboratory, Faculty of Pharmacy, Lebanese University, Beirut BP 14/6573, Lebanon; jenniferhajj0@gmail.com (J.E.H.); edmond.cheble@ul.edu.lb (E.C.)

2 Department of Biological Sciences, School of Arts and Sciences, Lebanese American University, Beirut 1102-2801, Lebanon; louna.karam@lau.edu.lb (L.K.); philippehussein.kobeissy@lau.edu.lb (P.H.K.); josenoel.ibrahim@lau.edu.lb (J.-N.I.)

3 Department of Physical Sciences, School of Arts and Sciences, Lebanese American University, Beirut 1102-2801, Lebanon; elias.akoury@lau.edu.lb

* Correspondence: ali.jaber14@lau.edu.lb (A.J.); ali.yassin02@lau.edu.lb (A.Y.)

† These authors contributed equally to this work.

Abstract: Probing the chemical profiles and biological activities of medicinal plants is important for the discovery of new potent therapeutic products. Our study deciphers the chemical composition of the essential oils (EOs) obtained from three different flowers of *Lantana camara* and evaluates their antioxidant and anticancer activities. This work represents the first study of EOs obtained from this plant and is based particularly on the difference in flower color. In addition, no other reports dealing specifically with the antitumor effects of such flower-derived EOs have been described in the literature. The collected flowers, white, pink, and orange, were extracted by hydrodistillation to yield EO1, EO2, and EO3 respectively. Gas chromatography–mass spectroscopy was primarily employed to identify the existing volatile compounds in the samples. Their antioxidant activities were screened through both DPPH (2,2-diphenyl-1-picrylhydrazyl) scavenging assays and FRAP (ferric-reducing antioxidant power) assays. The antiproliferative effects were evaluated on two distinct breast cancer cell lines, MCF-7 and MDA-MB-231, and compared to a normal human breast cell line, MCF-10A, using an MTT (3-(4,5-dimethylthiazol-2-yl)-2,5-diphenyltetrazolium) assay. All EOs showed notable antioxidant potential attributed to the active phytochemical compounds, with results being supported by a positive correlation between such activity and the total phenolic and flavonoid content. The most eminent, EO1, revealed a selective dose-dependent antiproliferative effect in both breast cancer cell lines, thus reflecting its potent role as an anticancer agent. We suggest that this highly selective activity is associated with the presence of bicyclogermacrene and *epi*-bicyclosesquiphellandrene in its chemical composition.

Keywords: bicyclogermacrene; *epi*-bicyclosesquiphellandrene; anticancer activity; *Lantana camara*; essential oil; chemical composition

1. Introduction

Known for its toxicity primarily due to the presence of lantadenes and other secondary metabolites [1], *Lantana camara* (*L. camara*) has a wide range of medicinal benefits [2] and biological activities [3,4]. In fact, this plant is one of the most noxious weeds in the world, while being toxic to animals and exerting allelopathic action on adjacent vegetation [5]. This dual-character perennial invasive shrub, originally native to the tropical regions of the Americas, is both an ornamental plant and a source of different bioactive compounds. Because of this, the careful analysis of its chemical composition within the scope of its biological effects has become of prime importance. On another note, the species is characterized by its vibrant and diverse flower colors, including mostly pink, white, and yellowish orange, as well as its small blackish fruits. Studies on the phytochemical profile of *L. camara* have shown a diverse array of secondary metabolites such as essential oils (EOs), phenols, flavonoids, alkaloids, glycosides, saponins, and tannins [6]. Historically, the plant has been used in folk medicine since ancient times, and correspondingly, antioxidant, antimicrobial [7], antipyretic, and insecticidal [8] effects have been reported by several groups recently in plant samples sourced from different geographical locations. Studies have shown significant discrepancies in the chemical compositions of the different extracts used to investigate *L. camara*'s biological activity [9]. While EOs are known for their complex mixtures of volatile compounds that contribute to their distinctive aroma and biological activities [6], their extraction and analysis have become crucial in identifying new bioactive compounds with potential therapeutic applications [10]. Comparably, some synthetic lantadenes which are pentacyclic triterpenoid derivatives obtained from the leaves of *L. camara* have been previously studied for their in vivo tumor inhibitory potential by Sharma et al. [11,12], while other groups presented some preliminary investigations on the use of *L. camara* stem extracts as inhibitors for leukemia cancer cells [13] or tried to investigate the underlying mechanism of its anticancer effects [14]. This is particularly important as cancer is becoming a major global health issue, ranking as the second leading cause of death after cardiovascular diseases. It is marked by the uncontrolled growth of cells, leading to malignant tumors with the potential to spread throughout the body [15–17]. Typical treatments have involved chemotherapy, radiotherapy, cytotoxic drugs, and surgery [18,19]. While these methods have shown success in treating various cancer types such as colon, pancreatic, testicular, breast, ovarian, and certain lung cancers, their overall effectiveness is often compromised by drug resistance and harmful side effects [20,21], which occur when healthy cells are unintentionally damaged during treatment. This underscores the need for ongoing research into safer, more effective treatments [8]. Interestingly, herbal medicines have been used for centuries in many developing countries and continue to serve as a primary form of healthcare [22]. Research has demonstrated that plants offer a promising source for the development of new anticancer drugs that are both effective and safe [23].

As a side note, gas chromatography–mass spectroscopy (GC-MS) is a powerful analytical technique used to identify and quantify the components of EOs, providing insights into their chemical profiles. In this study, we explored and analyzed the chemical compositions of EOs extracted from Lebanese *L. camara* flowers, more specifically from white, pink, and orange varieties. We then assessed their antioxidant activities as well as their antiproliferative effects on two different breast cancer cell lines. To the best of our knowledge, this work represents the first study of EOs obtained from *L. camara* based on different flower colors. Moreover, we are not aware of any reports dealing specifically with the antitumor effects of such flower-derived EOs. The aim of our study was to elucidate a potential correlation between chemical composition and the antitumor activity of EO, with the goal of advancing the development of new natural products with targeted therapeutic potential and possible clinical applications.

2. Results and Discussion

2.1. GC-MS Analysis Results and Yields of Extractions

Three different types of petals, namely, white, pink, and orange flowers from *L. camara* (Figure 1), were collected, dried, ground, and further extracted by hydrodistillation to yield the corresponding essential oils termed EO1, EO2, and EO3, respectively.

Figure 1. The three flowers of *L. camara* (white, pink, and orange) as collected.

GC-MS analysis revealed that a major component of the three oils is caryophyllene, more precisely (−)-β-caryophyllene (BCP), a natural sesquiterpene which is also a cannabis-derived compound known to bind directly to endocannabinoid receptors in the body (Figure 2). Many studies have already reported caryophyllene as being a major constituent [9,24,25], while a more recent study pointed out that this chemical makes up ~70% of the EO of *L. camara* collected from India [8]. EO1 (white flowers) also contained bicyclogermacrene and (+)-*epi*-bicyclosesquiphellandrene at 21.34% and 9.04%, respectively. For EO2 (pink flowers), γ-elemene and α-muurolene were the other main components, while EO3 (orange flowers) was dominated by α-humulene, α-trans-bergamotene, and α-phellandrene, along with the presence of BCP. The results of the GC-MS presented in Figure 2 (additional details are given in Supplementary Materials) show that monoterpenes and sesquiterpenes are the sole components of the EOs of the collected flowers of *L. camara*. Such chemical compositions with profiles differing between each sample can explain the varying antioxidant and antiproliferative activities that will be presented herein. It is worth noting that β-caryophyllene is reported in the literature to have several biological activities such as antimicrobial and anti-inflammatory, but most notably, it functions as an antioxidant [26]. Other compounds like bicyclogermacrene, (+)-*epi*-bicyclosesquiphellandrene, γ-elemene, α-muurolene, α-humulene, and α-trans-bergamotene are sesquiterpenes that have been linked to cytotoxic, antimicrobial, antitumor, and many other potential activities [27,28].

Compared to other studies on EOs of *L. camara*, our findings highlight the critical difference in chemical composition based on geographic origin/location. For example, in Egypt, α-curcumene (10.26%), β-copene (12.29%), davanone (23.27%), caryophyllene (22.96%), and humulene (14.32%) are the major components [29]. A non-exhaustive table that attempts to compile the major compounds in the EOs from different studies can be found in the Supplementary Information. Similarly, considering the yields of extraction, a study in Brazil showed that there are significant differences between extraction yields obtained even at different collection times [30]. The lowest yield (0.01%) was recorded at 7:00 a.m., and the highest one (0.09%) was obtained at 7:00 p.m. In another study, the yield of an oil obtained in India was found to be 0.032% (*w*/*w*) [31]. In Africa, one group reported that the yield of EO of *L. camara* ranged from 0.25 to 0.37% *w*/*w* [32]. Our obtained yields are slightly above 0.04%, which is a similar value to those obtained in Brazil and in India. In general, yields of EOs are very low in most studies and vary according to the extraction method, time of collection, climate, seasonal time, and geographic origins.

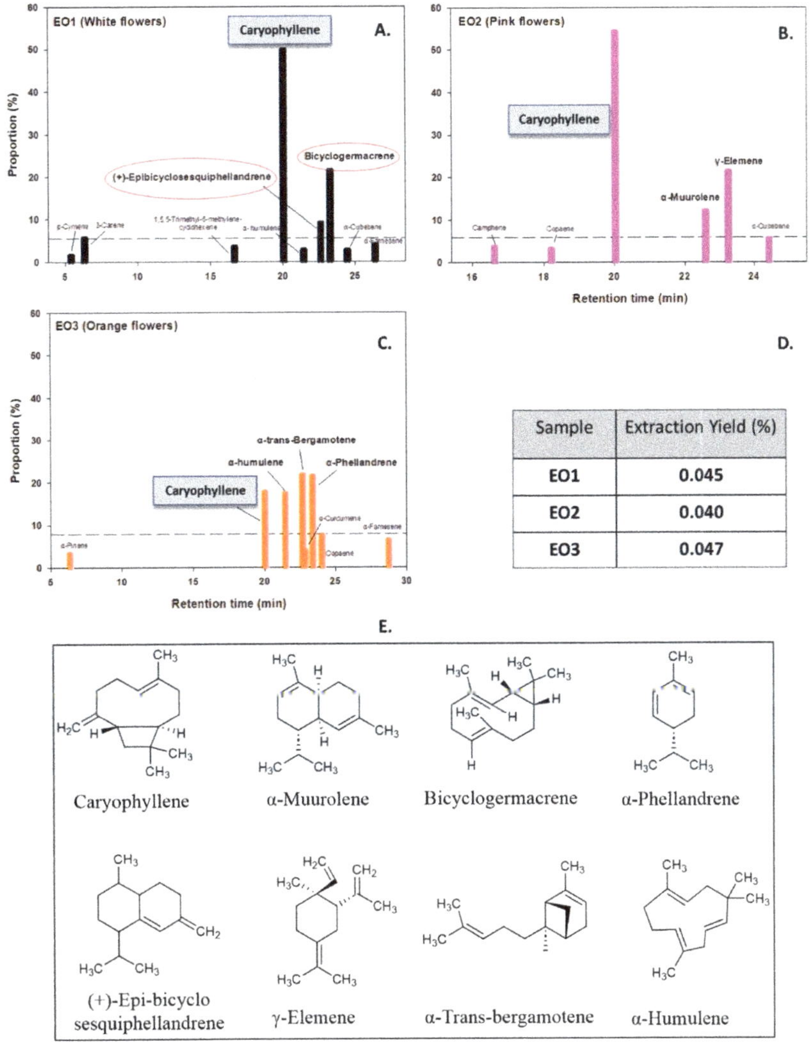

Sample	Extraction Yield (%)
EO1	0.045
EO2	0.040
EO3	0.047

Figure 2. (**A**–**C**): Main constituents of the EOs of *L. camara* obtained from GC-MS analysis. (**D**): Yields of extraction of the three samples. (**E**): Chemical structures of the different compounds.

2.2. Total Flavonoid Content (TFC) and Total Phenol Content (TPC)

It is known that EOs do not often contain flavonoids, as here, indeed, an extremely low content between 2 and 5 mg per 1 g of EO is observed. However, several reports in the literature have already indicated that some phenolic contents can be effectively extracted with essential oil [33,34]. In fact, the presence of phenolic terpenes or hydroxyl groups leads to the overestimation of the spectrophotometric test for *TFC* [35]. Meanwhile, the reason why flavonoids are not detected by GC-MS is due to the fact that they need derivatization in the sample preparation and the pre-injection step [36]. The analysis performed for *TFC* and TPC (as described in Sections 3.4 and 3.5, and further in the Supplementary Materials) showed that EO3 from orange flowers had the greatest amount of flavonoids at 5.12 mg of rutin equivalent RE per 1 g of extract, while EO1 and EO2, from white and pink flowers, respectively, had contents of less than 3 mg of RE.g^{-1} (Table 1).

Table 1. Results of the *TFC*, TPC, and IC50 of the three sample with a dilution factor *DF* = 1.

Sample	*TFC* (mg of RE/g of Extract)	TPC (mg of GAE/g of Extract)	IC_{50} (mg/mL) [1]
EO1	2.41	16.80	4.64
EO2	2.93	24.63	2.79
EO3	5.12	26.71	1.21

[1] Ascorbic acid value = 0.004.

EO3 had also the highest amount of phenol content at 26.71 mg of GAE/g extract. EO2 showed a slightly lower TPC (24.63 mg of GAE/g extract), while that of EO1 was considerably low at 16.80 mg of GAE/g of extract (Table 1). A study reported in Madagascar revealed that the TPC of the flowers' EO was 43.50 mg GAE/g of dry weight [37], while another study conducted in Nepal showed the highest phenol content to be 10.20 $\pm$ 0.34 mg GAE/g extract [38]. These findings further support our conclusion on the vast variability in phytochemical contents with regard to the geographical location of this flowering plant species, in addition to other factors including the extraction method and the storage procedure.

2.3. Antioxidant Activities

Given the fact that TPC and *TFC* are reported to be highly coupled to antioxidant capacity and are considered sources of natural antioxidants [39], we evaluated the in vitro antioxidant activities of the three EOs by both the 2,2-diphenyl-1-picrylhydrazyl (DPPH) assay and the ferric-reducing antioxidant power (FRAP) assay. Results obtained from the two tests were found to be dose-dependent. EO3 had the most potent antioxidant activity marked by its lowest IC_{50} = 1.21 mg.mL^{-1}, followed by EO2 and EO1, which presented IC_{50} values of 2.79 mg.mL^{-1} and 4.64 mg.mL^{-1}, respectively (Table 1). Nevertheless, all three EOs had much lower antioxidant activity than that of ascorbic acid with IC_{50} = 0.004 mg.mL^{-1} (Figure 3). According to one study conducted in Egypt, IC_{50} values were found to be 55.43 and 48.36 μg.mL^{-1} for the EOs of leaves and flowers, respectively [40]. In another study carried out in Pakistan, the DPPH (IC_{50}) value was found to be 5.45 μg.mL^{-1} [41], while in Malaysia, the fruit extract showed an IC_{50} of 90.11 μg.mL^{-1} [42].

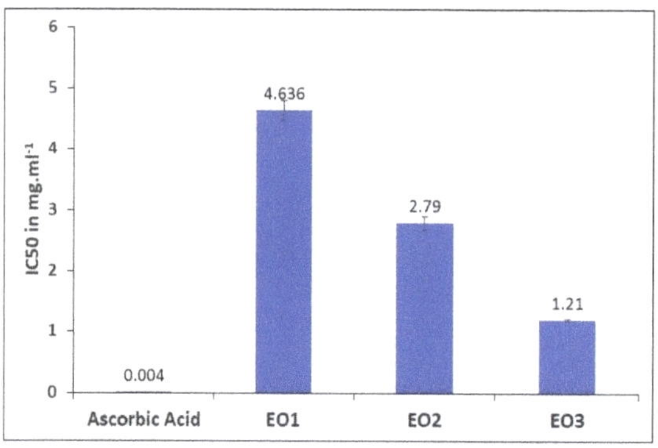

Figure 3. Results of IC50 of the three EOs as compared to ascorbic acid.

DPPH results were further validated by the FRAP assay, whereby EO3 indeed showed the highest antioxidant potential, and EO1 showed the lowest (Figure S10 in Supplementary Materials). This supports the reduction potential of the EOs transforming Fe^{3+} to Fe^{2+}, and thus, their electron-donating ability. The differences in activity reported in the current work, and in the literature, can be ascribed to various factors. The difference in the chemical compositions of each sample is the most important among such factors. The higher antioxidant activity in EO3 is mostly due to the presence of trans-α-bergamotene, which is the predominant compound in this sample. The latter is a derivative of bergamotene, which has been shown to possess diverse biological activities such as antioxidant, anti-inflammatory, immunosuppressive, cytotoxic, antimicrobial, antidiabetic, and insecticidal effects [43]. The other major derivative, α-phellandrene, is known for its antioxidant and antibacterial activities, as well as being effective against liver cancer. Both molecules are not available in the other EOs, which may explain the potency of EO3 as an antioxidant [44]. In general, sesquiterpenes are reported to be responsible for several biological activities [45], a fact that can elucidate their high antioxidant effect. As a side note, differences in IC_{50} values can also be due to ecological factors such as the age of the plant, humidity, height, temperature, and water, all which influence the quality and the quantity of the secondary metabolites, which are responsible for different biological activities, such as the phenols, flavonoids [46], monoterpenes, and sesquiterpenes that are present in the EO of *L. camara*.

2.4. Reduction in MCF-7 and MDA-MB-231 Cell Proliferation in a Dose-Dependent Manner

Antitumor activities were explored on two different breast cancer (BC) cell lines: MCF-7, a non-invasive ER+ PR+ cell line, and MDA-MB-231, an invasive triple negative BC cell line. At the same time, MCF-10A, an immortalized epithelial breast cell line, was used to assess toxicity in normal cells. As shown in Figure 4, both cancer cell lines displayed dose-dependent sensitivities to increasing concentrations of the three tested EOs. Interestingly, EO1 from white flowers exhibited the strongest antiproliferative effect on both cell lines 24 h post-treatment. The calculation of the IC_{50} revealed an approximate value of 0.3 g.mL^{-1} in both cell lines and at all time-points. EO2 was significantly less potent in inhibiting proliferation, particularly in MCF7, where cell proliferation remained greater than 50% even 72 h post-treatment, while EO3 demonstrated intermediate inhibitory potential compared to the other samples. Though EO3 showed greater potency than EO2, both were still significantly less effective than EO1, in particular at higher concentrations. Cytotoxicities are detailed in Figure 4, with dose-dependent inhibitory activities presented as a bar plot for each BC cell line. The results were further supported by the higher IC_{50} values for EO2 and EO3 at different time-points, as shown in Table 2.

More importantly, all three EOs had a less significant cytotoxic effect on the normal human breast cell line, MCF-10A, proliferation, where the inhibitory effect was only noted when cells were treated with very high concentrations, up to 0.5 mg.mL^{-1}. The calculation of the IC_{50} revealed estimated values higher than 1 mg.mL^{-1}, which are at least 2–3 times greater than those observed in MCF-7 and MDA-MB-231 cells (Table 2), hence indicating the greater selectivity of these natural compounds against cancer cells.

Our findings conclude that the EO isolated from the white flowers of *L. camara* demonstrates selective antitumor activity against breast cancer, while showing no significant antiproliferative effect on normal cells. To the best of our knowledge, no study to date has explored the antitumor potential of an EO obtained from *L. camara* flowers. EOs from the leaves, however, have been shown to account for strong anticancer activity against U-266, A-549, HCT-116, SCC-4, MiaPaCa 2, and KBM-5 cancer cell lines [39].

Table 2. Mean IC_{50} values of EOs on MCF-7, MDA-MB-231, and MCF-10A cell lines at 24-, 48-, and 72 h post-treatment.

Cell Line	Time (h)	IC_{50} (g.mL^{-1})		
		EO1	**EO2**	**EO3**
MCF-7 (BC)	24	0.3061 ± 0.091	0.3630 ± 0.057	0.5927 ± 0.03
	48	0.3021 ± 0.043	0.4101 ± 0.109	0.3937 ± 0.09
	72	0.3179 ± 0.002	0.4144 ± 0.09	0.3397 ± 0.027

Table 2. *Cont.*

Cell Line	Time (h)	IC_{50} (g.mL^{-1})		
		EO1	EO2	EO3
MDA-MB-231 (BC)	24	0.3099 ± 0.001	0.7398 ± 0.012	0.6473 ± 0.005
	48	0.2820 ± 0.022	0.4015 ± 0.01	0.4150 ± 0.06
	72	0.2800 ± 0.023	0.3931 ± 0.058	0.3951 ± 0.08
MCF-10A (normal)	24	>1	>1	>1
	48	>1	>1	>1
	72	>1	>1	>1

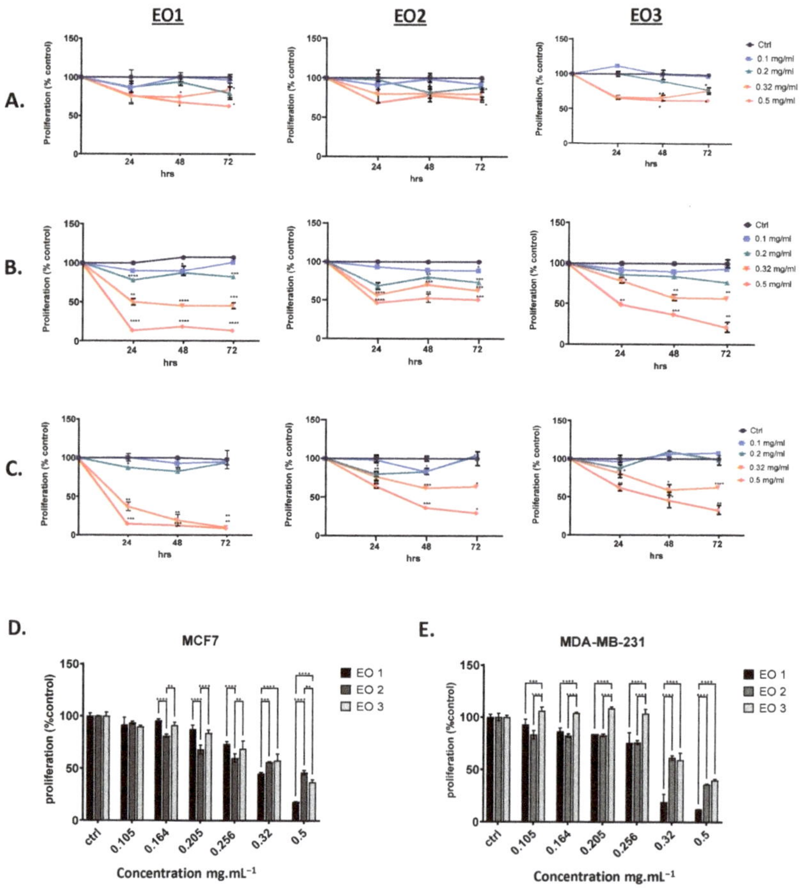

Figure 4. In vitro cytotoxic effect of EO1, EO2, and EO3 on the MCF-10A, MCF-7, and MDA-MB231 cell lines. Representative graphs of the time-dependent cytotoxic effect of different concentrations of the oils on MCF-10A (**A**), MCF-7 (**B**), and MDA-MB-231 (**C**) cells. Dose-dependent inhibitory activity for EO1, EO2, and EO3 48 h post-treatment on MCF-7 (**D**) and MDA-MB-231 (**E**) is presented as a bar plot. Data are presented as mean ± SD (n = 3) with all treatments compared to each other by a two-way ANOVA test followed by the post hoc Tukey test for multiple comparison analysis; * $p < 0.05$, ** $p < 0.01$, *** $p < 0.001$, **** $p < 0.0001$.

The differences in growth inhibitory effect could be explained by the results of GC-MS analysis which revealed distinct chemical profiles for each sample. We attribute the antiproliferative activity of EO1 against the two BC cell lines reported herein to the presence of *epi*-bicyclosesquiphellandrene (Figure 5) found exclusively, and in significant proportions, in the EO of white flowers. A study on the oily fractions of Teucrium alopecurus showed that the *epi*-bicyclosesquiphellandrene molecule had antitumor activity against colon cancer cells [47]. Another major component present exclusively in EO1 is bicyclogermacrene, which can also contribute to the significant antitumoral activity. Indeed, one research paper conducted in Brazil has shown that bicyclogermacrene plays a major role in the anticancer activity of the Myrcia genus plant against lung cancer cells [48].

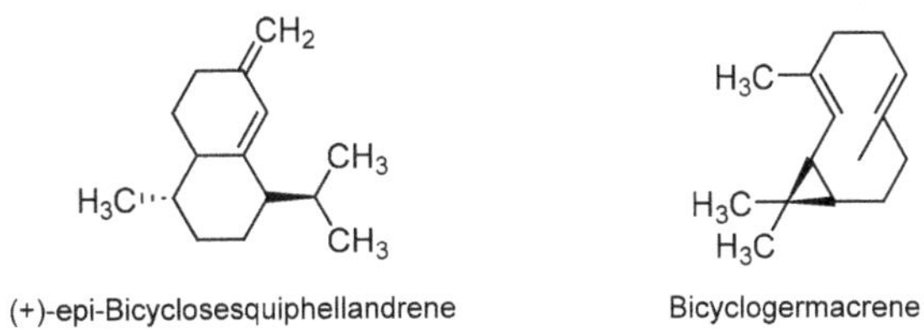

Figure 5. Chemical structures of the possible antitumor-active phytochemicals in EO1.

2.5. Pearson's Correlation Coefficient Analysis

The relationship between the TPC and *TFC* present in the EOs, and their in vitro antioxidant and antitumor activities, was expressed by adopting Pearson's correlation coefficients (PCCs), also referred to as Pearson's r, which are represented in Table 3.

Table 3. Pearson's correlation coefficients for *TFC*, TPC, DPPH, MCF-7, and MDAMB-231 in the three EOs under consideration.

	TFC	TPC	DPPH	MCF-7	MDA-MB-231 *
TFC	1				
TPC	0.7869	1			
DPPH	−0.9260	−0.9617	1		
MCF-7	0.99997	0.7913	−0.9286	1	
MDA-MB-231	0.4790	0.9186	−0.7750	0.4853	1

* Relevant IC_{50} values at 24 h post-treatment.

The PCCs of −0.9260 and −0.9617 imply a strong negative correlation between *TFC*/DPPH and TPC/DPPH, respectively. In general, a sample with high levels of *TFC* and TPC gives a high level of DPPH where phenolic compounds are classified according to the hydroxyl groups attached to the benzene ring. Such hydroxyl groups are usually good hydrogen donors and can react and neutralize reactive oxygen such as DPPH [49,50]. Consequently, a high concentration of phenolic compounds in an extract is often followed by high antioxidant activity. On the other hand, the IC_{50} values of MCF-7, at 24 h post-treatment, showed an extremely high positive correlation to the total flavonoid content (correlation coefficient: 0.99997), with a strong negative correlation with the IC_{50} of DPPH (correlation coefficient: −0.9286). Based on such findings, two conclusions can be suggested: First, flavonoids seem to be an important contributor to the antitumor effect against MCF-7, and, second, the radical scavenging mechanism follows a highly similar mechanism through which the antitumor effect against MCF-7 cells takes place. We note similarly that the content of phenols shows a strong positive correlation (0.9186) with the antitumor effect over MDA-MB-231 and its IC_{50} value at 24 h post-treatment.

3. Materials and Methods

3.1. Collection of the Petals of L. camara

Three different flower colors of L. camara were freshly collected from the Lebanese University at Hadath, Beirut (33°49′39″ N, 35°31′17″ E) during June and July 2023. Every time, 100 to 150 g of fresh white, pink, or orange petals was collected and divided into three flasks of 1 L capacity in order to perform the extraction process.

3.2. Hydrodistillation with Clevenger

Ground flowers of *L. camara* were subjected to hydrodistillation with the Clevenger apparatus under optimal operational conditions. First, 40 g of the flowers was mixed with 400 mL of distilled water. The distillation process was performed for 3 h, and the obtained essential oil was collected and dehydrated using anhydrous Na_2SO_4.

3.3. GC-MS Analysis

GC-MS analysis was carried out using the electron ionization method. The GC capillary column used was an Agilent 19091S-433 (Agilent Technologies, Santa Clara, CA, USA), HP-5MS with 5% phenyl methyl siloxane, film thickness of 0.25 μm, a length of 30 m, and an internal diameter of 250 μm. Helium was used as the carrier gas with a column head pressure of 1.09 bar, flow rate of 1 $mL.min^{-1}$, and 1 μL injections in split mode (1:50). The initial column temperature applied was 65 °C, which then increased to reach 450 °C. The GC oven temperature program ranged between an initial temperature of 65 °C and final temperature of 200 °C, with a run time of 45 min. The mass spectrometer (Agilent Technologies, Santa Clara, CA, USA) was operated in the EI mode at 70 eV with a mass scanning range of 50–500 and a source temperature of 230 °C. The identity of each compound was determined based on their retention indices and by comparison of their mass spectral fragmentation patterns with those reported in the NIST library database (NIST). The quantitative analysis expressing the percentage of the identified components in each volatile oil was obtained by the integration of the peak areas. Only fully identified compounds are reported in this study.

3.4. TFC Analysis

TFC was determined by the aluminum chloride colorimetric method [51]. In a test tube, 1 mL of the diluted plant extract solution and 1 mL of $AlCl_3.6H_2O$ (2% solution) were mixed together and left in the dark at room temperature for 30 min to react. After incubation, the absorbance of the developed yellow color was measured at λ_{max} = 410 nm using a double-beam UV-VIS spectrophotometer against a blank solution. The same procedure was carried out using rutin as a reference standard (0–0.09 $mg.mL^{-1}$). A standard curve of absorbance versus different concentrations of rutin was plotted. Results were reproduced in triplicate for each analytical trial, from which the mean and standard deviation values were calculated [52]. *TFC* was determined from the linear equation of a standard curve prepared with rutin and expressed in mg of rutin equivalent (RE) per g of plant extract using the following equation:

$$TFC = \frac{C \times V \times DF}{m}$$

where C is the concentration of rutin calculated by the calibration curve regression equation in $mg.mL^{-1}$, *V* is the volume of plant extract solution in mL, *DF* is the dilution factor, and *m* is the mass of extract in g used to prepare the plant extract solution.

3.5. TPC Analysis

TPC was determined by the Folin-Ciocalteu reagent (FCR) method. In this procedure, 100 μL of the diluted plant extract was added to 500 μL of FCR and incubated for 5 min in the dark. Then, 2 mL of Na_2CO_3 was added and samples were shaken and left in the dark at room temperature for 30 min. The absorbance of the developed blue color was measured at 760 nm using a UV-6300PC double-beam Spectrophotometer (VWR, Leuven,

Belgium) against a blank solution. The same procedure was carried out using gallic acid as a reference standard (0–0.27 mg.mL^{-1}); then, a standard curve of absorbance versus different concentrations of gallic acid was plotted [53]. The TPC was determined from the linear equation of a standard curve prepared with gallic acid and expressed in mg gallic acid equivalent (GAE) per g of plant extract using the following equation:

$$x = \frac{C \times V \times DF}{m}$$

where C is the concentration of gallic acid calculated by the calibration curve regression equation in mg.mL^{-1}, V is the volume of plant extract solution in mL, DF is the dilution factor, and m is the mass of extract in g used to prepare the plant extract solution.

3.6. DPPH Assay

The free radical scavenging activity of the three samples and of the ascorbic acid solution was investigated using the DPPH method [54]. In this procedure, 1 mL of the PE solution of different concentrations was added to 1 mL of the DPPH methanolic solution. A control consisting of 1 mL of DPPH solution with 1 mL methanol was also prepared. The mixtures were shaken vigorously and then incubated in the dark at room temperature for 30 min to reach a steady state. The color change, from violet at a low concentration to yellow at a high concentration, was determined by measuring the absorbance at 520 nm against a blank solution using a double-beam UV-VIS spectrophotometer. The same procedure was carried out for ascorbic acid, a pure antioxidant compound, which was used as a standard reference [55]. The percentage of the scavenging activity of each extract on the DPPH radical was calculated as % inhibition of DPPH using the following equation:

$$\%\ scavenging = \frac{A\ (control) - A\ (sample)}{A\ control} \times 100$$

where A (*control*) is the absorbance of DPPH alone, and A (*sample*) is the absorbance of DPPH with different concentrations of extracts. The concentration of the extract required to scavenge 50% of the DPPH free radical (IC_{50}) was determined from the percentage curve of DPPH inhibitions versus extract concentration. The antioxidant activity of all samples and their IC_{50} was compared to that of ascorbic acid.

3.7. FRAP Assay

A serial dilution of extract solutions (0.35–2.1 mg.mL^{-1}) and ascorbic acid (0.015–0.15 mg.mL^{-1}) was prepared in ethanol. First, 200 µL of each extract solution was mixed with 200 µL of 0.2 M phosphate buffer (pH 6.6) and 200 µL of potassium ferricyanide (1%). The reaction mixtures were incubated at 50 °C for 20 min. After cooling, 200 µL of trichloroacetic acid (10%) was added and the mixtures were centrifuged at 1000 rpm for 8 min. The upper layer (800 µL) was mixed with 800 µL of distilled water and 160 µL of ferric chloride (0.1%). After a 10 min reaction time, the spectrometric absorbance was recorded at 700 nm and compared with ascorbic acid as the positive control. The absorbance values were plotted against the concentration, and a linear regression analysis was carried out. Higher absorbance readings indicate higher reducing power [56].

3.8. Cell Culture

The human breast cancer MCF-7 and MDA-MB-231 cell lines were purchased from the American Type Culture Collection (ATCC, Manassas, VA, USA). MCF-7 and MDA-MB-231 were cultured as previously described [57,58]. MCF-10A cells were cultured in DMEM-F12 supplemented with horse serum (5%), hydrocortisone (0.5 µg.mL^{-1}), EGF (20 ng.mL^{-1}), insulin (10 µg.mL^{-1}), 100 U.mL^{-1} of penicillin, and 100 µg.mL^{-1} of streptomycin. Cells were either left untreated (control) or treated with different concentrations of EOs for 24, 48, and 72 h.

3.9. Cell Proliferation Assay

The antiproliferative effect of the EOs was assessed using the MTT (3-(4,5-dimethylthiazol-2-yl)-2,5-diphenyltetrazolium bromide) assay (Sigma-Aldrich, Darmstadt, Germany). This assay determines the cell viability by measuring the amount of formazan dye that is produced upon cellular reduction in MTT by metabolically active cells. MCF-7, MDA-MB-231, and MCF-10A cells were seeded in triplicate for each condition in 96-well plates at a density of 7×10^3 cells/0.1 mL, 5×10^3 cells/0.1 mL, and 10×10^3 cells/0.1 mL respectively. After adherence, cells were left untreated or treated with a concentration range of 0–500 mg.mL^{-1} of EO1, EO2, or EO3 for 24, 48, and 72 h. At each time-point, 10 µL/well of MTT reagent (5 mg.mL^{-1}) was added, and cells were incubated at 37 °C for 3h. The reaction was stopped and the formazan crystals were dissolved by adding 100 µL of 10% SDS in 0.01 N HCl. Absorbance was measured at 570 nm using a Varioskan Flash plate reader (Thermo Fisher Scientific, Waltham, MA, USA). Cell proliferation was represented as the mean percent control plotted against the log of concentration and incubation time of three independent experiments repeated for each sample. IC_{50} values were generated from the non-linear regression with variable-slope dose-response curves.

3.10. Statistical Analyses

Statistical analyses were performed using the GraphPad Prism 5.04 software (GraphPad, San Diego, CA, USA). Values are expressed as mean ± SD (standard deviation). Differences between control and treated groups were assessed for statistical significance by the two-way ANOVA test followed by the post hoc Tukey test for multiple comparison analysis. Each experiment was repeated at least three times, and each condition was carried out in triplicate.

4. Conclusions and Outlook

In a nutshell, we are reporting on the first study of EOs obtained from *L. camara* flowers based on their distinct blossom color. The work deals specifically with the antitumor effect of such flower-derived samples. GC-MS analysis revealed different chemical profiles for the EOs of white, pink, and orange flowers. EO1 (white flowers) was shown to be distinctively rich in (+)-*epi*-bicyclosesquiphellandrene and bicyclogermacrene, both of which are known for their antitumor activities. Interestingly, EO3 (orange flowers) showed the highest antioxidant activity in comparison to a very low performance for EO1. On the other hand, the highly potent anticancer activity of the latter was perceived on MCF-7 and MDA-MB-231 breast cancer cell lines. EO1 exhibited such strong activity by significantly inhibiting the cell proliferation at lower concentrations, while selectively targeting cancer cells over normal breast cells. This study underscores the importance of exploring the detailed chemical diversity of plant-derived essential oils in relation to their biological activities, paving the way for the discovery of new bioactive compounds with potential clinical applications. However, further research is needed to isolate and characterize the specific active compounds and to elucidate their mechanisms of action in cancer therapy. Additionally, we are interested in the evaluation of the antioxidant and antitumor effects of some commercially available compounds that are components of the three EOs and that could be tested in their pure form.

Supplementary Materials: The following supporting information can be downloaded at https://www.mdpi.com/article/10.3390/molecules29225431/s1, Figure S1: Petals of the *Lantana Camara* flowers: White, Pink, and Orange, respectively; Table S1: Main constituents of *L. Camara* white flowers essential oil (EO1); Figure S2: GC-MS chromatogram of EO1; Table S2: Main constituents of *L. Camara* pink flowers essential oil (EO2); Figure S3: GC-MS chromatogram of EO2; Table S3: Main constituents of *L. Camara* orange flowers essential oil (EO3); Figure S4: GC-MS chromatogram of EO3; Table S4: Main constituents of EOs of *Lantana Camara* from different geographical origins previously reported in literature; Table S5: Results of the phytochemical screening of *L. Camara* flowers; Figure S5: Variation of the Absorbance in function of the concentration of the Rutin in µg/mL; Table S6: Results of the *TFC* of the different samples; Figure S6: Comparison of the *TFC* in different extracts of *Lantana Camara*;

Figure S7: Variation of the Absorbance in function of the concentration of Gallic Acid in mg/mL; Table S7: Results of TPC of the different samples; Figure S8: Comparison of the TPC in different essential oil extracts of *Lantana Camara*; Figure S9: Results of the DPPH scavenging assay in the three EOs and the ascorbic acid measuring the % Scavenging in function of its different concentrations in mg/mL; Table S8: Results of DPPH Assay of the different samples; Figure S10: Results of the FRAP assay of the three EO samples, measuring absorbance as function of the different concentrations in mg.mL^{-1} (see [9,29,40,59–61]).

Author Contributions: J.E.H.: Investigation, Writing—Original Draft; L.K.: Investigation, Formal analysis; A.J.: Conceptualization, Methodology, Project Administration, Writing—Review and Editing; E.C.: Resources, Supervision; E.A.: Resources, Writing—Review and Editing; P.H.K.: Data Curation, Supervision, Writing—Review and Editing; J.-N.I.: Data Curation, Supervision, Writing—Review and Editing; A.Y.: Conceptualization, Project Administration, Validation, Writing—Original Draft, Funding Acquisition. All authors have read and agreed to the published version of the manuscript.

Funding: This research was in part funded by the Lebanese American University, President's Intramural Research Fund PIRF (grant number I0042).

Data Availability Statement: The original contributions presented in the study are included in the article/Supplementary Materials; further inquiries can be directed to the corresponding authors.

Acknowledgments: This paper is dedicated to the loving memory of *Mariam Hussein Kobeissy* (2021–2024).

Conflicts of Interest: The authors declare no conflicts of interest.

Abbreviations

The following is a list of abbreviations in the sequence of their appearance. EO: essential oils; DPPH: 2,2-diphenyl-1-picrylhydrazyl; FRAP: ferric-reducing antioxidant power; MTT: 3-(4,5-dimethylthiazol-2-yl)-2,5-diphenyltetrazolium; *L. camara*: *Lantana camara*; GC-MS: gas chromatography–mass spectroscopy; BCP: (−)-β-caryophyllene; *TFC*: total flavonoid content; TPC: Total Phenol Content; BC: breast cancer; PCC: Pearson's correlation coefficients; RE: rutin equivalent; FCR: Folin–Ciocalteu reagent; GAE: gallic acid equivalent; ATCC: American Type Culture Collection; SD: standard deviation.

References

1. Parimoo, H.A.; Sharma, R.; Patil, R.D.; Patial, V. Sub-acute toxicity of lantadenes isolated from *Lantana camara* leaves in guinea pig animal model. *Comp. Clin. Pathol.* **2015**, *24*, 1541–1552. [CrossRef]
2. Kalita, S.; Kumar, G.; Karthik, L.; Rao, K.V.B. A Review on Medicinal Properties of *Lantana camara* Linn. *Res. J. Pharm. Technol.* **2012**, *5*, 711–715.
3. Nayak, B.S.; Raju, S.S.; Eversley, M.; Ramsubhag, A. Evaluation of wound healing activity of *Lantana camara* L.—A Preclinical study. *Phytother. Res.* **2009**, *23*, 241–245. [CrossRef] [PubMed]
4. Ganesh, T.; Saikat, S.; Thilagam, E.; Thamotharan, G.; Loganathan, T.; Chakraborty, R. Pharmacognostic and anti-hyperglycemic evaluation of *Lantana camara* (L.) var. aculeate leaves in alloxan-induced hyperglycemic rats. *Int. J. Res. Pharm. Sci.* **2010**, *1*, 247–252.
5. Sharma, O.P.; Makkar, H.P.S.; Dawra, R.K. A review of the noxious plant *Lantana camara*. R.K. *Toxicon* **1988**, *26*, 975–987. [CrossRef]
6. Shah, M.; Alharby, H.F.; Hakeem, K.R. *Lantana camara*: A Comprehensive Review on Phytochemistry, Ethnopharmacology and Essential Oil Composition. *Lett. Appl. NanoBioSci.* **2020**, *9*, 1199–1207. [CrossRef]
7. Mansoori, A.; Singh, N.; Dubey, S.K.; Thakur, T.K.; Alkan, N.; Das, S.N.; Kumar, A. Phytochemical Characterization and Assessment of Crude Extracts From *Lantana camara* L. for Antioxidant and Antimicrobial Activity. *Front. Agron.* **2020**, *2*, 582268. [CrossRef]
8. Aisha, K.; Visakh, N.U.; Pathrose, B.; Mori, N.; Baeshen, R.S.; Shawer, R. Extraction, Chemical Composition and Insecticidal Activities of *Lantana camara* Linn. Leaf Essential Oils against *Tribolium castaneum*, *Lasioderma serricorne* and *Callosobruchus chinensis*. *Molecules* **2024**, *29*, 344. [CrossRef]
9. Nea, F.; Kambiré, D.; Genva, M.; Tanoh, E.; Wognin, E.; Martin, H.; Brostaux, Y.; Tomi, F.; Lognay, G.C.; Tonzibo, Z.; et al. Composition, Seasonal Variation, and Biological Activities of *Lantana camara* Essential Oils from Côte d'Ivoire. *Molecules* **2020**, *25*, 2400. [CrossRef]

10. El Kantar, S.; Yassin, A.; Nehmeh, B.; Labaki, L.; Mitri, S.; Naser Aldine, F.; Hirko, A.; Caballero, S.; Monck, E.; Garcia-Maruniak, A.; et al. Deciphering the therapeutical potentials of rosmarinic acid. *Sci. Rep.* **2022**, *12*, 15489. [CrossRef]
11. Sharma, M.; Sharma, P.D.; Bansal, M.P.; Singh, J. Synthesis, Cytotoxicity, and Antitumor Activity of Lantadene-A Congeners. *Chem. Biodivers.* **2007**, *4*, 932–939. [CrossRef] [PubMed]
12. Sharma, M.; Sharma, P.D.; Bansal, M.P. Lantadenes and Their Esters as Potential Antitumor Agents. *J. Nat. Prod.* **2008**, *71*, 1222–1227. [CrossRef]
13. Babar, V.B.; Khapale, P.R.; Nagarale, S. Preliminary investigation and in-vitro anticancer activity of *Lantana camara* L. (Verbenaceae). *J. Pharmacogn. Phytochem.* **2019**, *8*, 2524–2527.
14. Han, E.B.; Chang, B.Y.; Jung, Y.S.; Kim, S.Y. *Lantana camara* Induces Apoptosis by Bcl-2 Family and Caspases Activation. *Pathol. Oncol. Res.* **2015**, *21*, 325–331. [CrossRef]
15. Wang, H.; Khor, T.O.; Shu, L.; Su, Z.-Y.; Fuentes, F.; Lee, J.-H.; Kong, A.-N.T. Plants vs. Cancer: A Review on Natural Phytochemicals in Preventing and Treating Cancers and Their Druggability. *Anti-Cancer Agents Med. Chem.* **2012**, *12*, 1281–1305. [CrossRef] [PubMed]
16. Ochwang'i, D.; Kimwele, C.; Oduma, J.; Gathumbi, P.; Mbaria, J.; Kiama, S.J. Medicinal plants used in treatment and management of cancer in Kakamega County, Kenya. *J. Ethnopharmacol.* **2014**, *151*, 1040–1055. [CrossRef] [PubMed]
17. Greenwell, M.; Rahman, P.K.S.M. Medicinal Plants: Their Use in Anticancer Treatment. *Int. J. Pharm. Sci. Res.* **2015**, *6*, 4103–4112. [CrossRef]
18. Aiello, P.; Sharghi, M.; Mansourkhani, S.M.; Ardekan, A.P.; Jouybari, L.; Daraei, N.; Peiro, K.; Mohamadian, S.; Rezaei, M.; Heidari, M.; et al. Medicinal Plants in the Prevention and Treatment of Colon Cancer. *Oxidative Med. Cell. Longev.* **2019**, *2019*, 2075614. [CrossRef]
19. O'Reilly, M.S.; Boehm, T.; Shing, Y.; Fukai, N.; Vasios, G.; Lane, W.S.; Flynn, E.; Birkhead, J.R.; Olsen, B.R.; Folkman, J. Endostatin: An Endogenous Inhibitor of Angiogenesis and Tumor Growth. *Cell* **1997**, *88*, 277–285. [CrossRef]
20. The American Cancer Society Medical and Editorial Content Team: Chemotherapy Side Effects. Available online: https://www.cancer.org/cancer/managing-cancer/treatment-types/chemotherapy/chemotherapy-side-effects.html (accessed on 7 October 2024).
21. Cragg, G.M.; Newman, D.J. Plants as a source of anti-cancer agents. *J. Ethnopharmacol.* **2005**, *100*, 72–79. [CrossRef]
22. Sivaraj, R.; Rahman, P.; Rajiv, P.; Narendhran, S.; Venckatesh, R. Biosynthesis and characterization of *Acalypha indica* mediated copper oxide nanoparticles and evaluation of its antimicrobial and anticancer activity. *Spectrochim. Acta A Mol. Biomol. Spectrosc.* **2014**, *129*, 255–258. [CrossRef] [PubMed]
23. Hassan, B. Plants and Cancer Treatment. In *Medicinal Plants—Use in Prevention and Treatment of Diseases*; IntechOpen: London, UK, 2020. [CrossRef]
24. Zoubiri, S.; Aoumeur, B. Chemical Composition and Insecticidal Properties of *Lantana camara* L. Leaf Essential Oils from Algeria. *J. Essent. Oil Res.* **2012**, *24*, 377–383. [CrossRef]
25. Chowdhury, J.U.; Nandi, N.C.; Nazrul, I.; Bhuiyan, N.I. Chemical composition of leaf essential oil of *Lantana camara* L. from Bangladesh. *Bangladesh J. Bot.* **2007**, *36*, 193–194. [CrossRef]
26. Gushiken, L.; Beserra, F.; Hussni, M.; Gonzaga, M.; Ribeiro, V.; Fernanda de Souza, P.; Campos, J.; Massaro, T.; Hussni, C.; Takahira, R.; et al. Beta-caryophyllene as an antioxidant, anti-inflammatory and re-epithelialization activities in a rat skin wound excision model. *Oxidative Med. Cell. Longev.* **2022**, *2022*, 9004014. [CrossRef]
27. Wang, J.-H.; Luan, F.; He, X.-D.; Wang, Y.; Li, M.-X. Traditional uses and pharmacological properties of Clerodendrum phytochemicals. *J. Tradit. Complement. Med.* **2018**, *8*, 24–38. [CrossRef]
28. Do, A.R.; Emami, S.A.; Akaberi, M. Phytochemical diversity and pharmacological effects of sesquiterpenes from *Artemisia* spp. *Phytochem. Rev.* **2024**, *23*, 1–38. [CrossRef]
29. Sundufu, A.J.; Shoushan, H. Chemical composition of the essential oils of *Lantana camara* L. occurring in south China. *Flavour Fragr. J.* **2004**, *19*, 229–232. [CrossRef]
30. Sousa, E.O.; Colares, A.V.; Rodrigues, F.F.G.; Campos, A.R.; Lima, S.G.; Costa, J.G.M. Effect of Collection Time on Essential Oil Composition of *Lantana camara* Linn (Verbenaceae) Growing in Brazil Northeastern. *Rec. Nat. Prod.* **2010**, *4*, 31–37.
31. Singh, R.K.; Tiwari, B.; Sharma, U.; Singh, S.P. Chemical Composition of *Lantana camara* Fruit Essential Oil. *Asian J. Chem.* **2012**, *24*, 5955–5956.
32. Liambila, R.; Wesonga, J.; Ngamau, C.; Wallyambillah, W. Chemical composition and bioactivity of *Lantana camara* L. essential oils from diverse climatic zones of Kenya against leaf miner (*Tuta absoluta* Meyrick). *Afr. J. Agric. Res.* **2021**, *17*, 1198–1208. [CrossRef]
33. Mamadaliev, A.N.; Kushiev, K.K.; Abdullaeva, Z.R. Determination of Flavonoids in Different Oils of *Cannabis sativa* L (Cannabaceae) and Evaluation of Physico-chemical Indicators. *Austrian J. Tech. Nat. Sci.* **2022**, *5–6*, 3. [CrossRef]
34. Semeniuc, C.A.; Socaciu, M.-I.; Socaci, S.A.; Mureșan, V.; Fogarasi, M.; Rotar, A.M. Chemometric Comparison and Classification of Some Essential Oils Extracted from Plants Belonging to Apiaceae and Lamiaceae Families Based on Their Chemical Composition and Biological Activities. *Molecules* **2018**, *23*, 2261. [CrossRef] [PubMed]
35. Shraim, A.M.; Ahmed, T.A.; Rahman, M.; Hijji, Y.M. Determination of total flavonoid content by aluminum chloride assay: A critical evaluation. *LWT* **2021**, *150*, 111932. [CrossRef]

36. Zhang, K.; Zuo, Y.J. GC-MS Determination of Flavonoids and Phenolic and Benzoic Acids in Human Plasma after Consumption of Cranberry Juice. *J. Agric. Food Chem.* **2004**, *52*, 222–227. [CrossRef]
37. Al Snafi, A.E. Chemical Constituents and Pharmacological Activities of *Lantana Camara*—A Review. *Asian J. Pharm. Clin. Res.* **2019**, *12*, 10–20. [CrossRef]
38. Kapali, J.; Sharma, K.R. Estimation of phytochemicals, antioxidant, antidiabetic and brine shrimp lethality activities of some medicinal plants growing in Nepal. *J. Med. Plants* **2021**, *20*, 102–116. [CrossRef]
39. Li, M.; Paré, P.W.; Zhang, J.; Kang, T.; Zhang, Z.; Yang, D.; Wang, K.; Xing, H. Antioxidant Capacity Connection with Phenolic and Flavonoid Content in Chinese Medicinal Herbs. *Rec. Nat. Prod.* **2018**, *12*, 239–250. [CrossRef]
40. El Baroty, G.S.; Goda, H.M.; Khalifa, E.A.; Abd El Baky, H.H. Antimicrobial and antioxidant activities of leaves and flowers essential oils of Egyptian *Lantana camara* L. *Der Pharma Chem.* **2014**, *6*, 246–255.
41. Sajid, A.; Manzoor, Q.; Imran, M.; Aslam, F.; Gondal, T.A.; Ahmad, R.S.; Hussain, G.; Imran, M.; Aslam, F.; Arshad, M.; et al. Essential Oil and Leaves from *Lantana camara* Significantly Ameliorate Different Cancer Cell Lines by Suppressing the NF-κB Pathway. *Sains Malays.* **2021**, *50*, 2923–2936. [CrossRef]
42. Mahdi-Pour, B.; Jothy, S.L.; Latha, L.Y.; Chen, Y.; Sasidharan, S. Antioxidant activity of methanol extracts of different parts of *Lantana camara*. *Asian Pac. J. Trop. Biomed.* **2012**, *2*, 960–965. [CrossRef]
43. Annaz, H.; El Fakhouri, K.; Ben Bakrim, W.; Mahdi, I.; El Bouhssini, M.; Sobeh, M. Bergamotenes: A comprehensive compile of their natural occurrence, biosynthesis, toxicity, therapeutic merits and agricultural applications. *Crit. Rev. Food Sci. Nutr.* **2024**, *64*, 7343–7362. [CrossRef] [PubMed]
44. Radice, M.; Durofil, A.; Buzzi, R.; Baldini, E.; Martínez, A.P.; Scalvenzi, L.; Manfredini, S. Alpha-Phellandrene and Alpha-Phellandrene-Rich Essential Oils: A Systematic Review of Biological Activities, Pharmaceutical and Food Applications. *Life* **2022**, *12*, 1602. [CrossRef] [PubMed]
45. Ivanescu, B.; Miron, A.; Corciova, A.J. Sesquiterpene Lactones from *Artemisia Genus*: Biological Activities and Methods of Analysis. *J. Anal. Methods Chem.* **2015**, *2015*, 247685. [CrossRef] [PubMed]
46. Karak, P. Biological Activities of Flavonoids: An Overview. *Int. J. Pharm. Sci. Res.* **2019**, *10*, 1567–1574. [CrossRef]
47. Guesmi, F.; Tyagi, A.K.; Prasad, S.; Landoulsi, A. Terpenes from essential oils and hydrolate of *Teucrium alopecurus* triggered apoptotic events dependent on caspases activation and PARP cleavage in human colon cancer cells through decreased protein expressions. *Oncotarget* **2018**, *9*, 32305–32320. [CrossRef]
48. Montalvão, M.M.; Felix, F.B.; Propheta dos Santos, E.W.; Santos, J.F.; de Lucca Júnior, W.; Farias, A.S.; de Souza Ribeiro, A.; Cavaleiro, C.; Machado, S.M.F.; Scher, R.; et al. Cytotoxic activity of essential oil from Leaves of *Myrcia splendens* against A549 Lung Cancer cells. *BMC Complement. Med. Ther.* **2023**, *23*, 139. [CrossRef]
49. Pereira, R.P.; Fachinetto, R.; de Souza Prestes, A.; Puntel, R.L.; Santos da Silva, G.N.; Heinzmann, B.M.; Boschetti, T.K.; Athayde, M.L.; Bürger, M.E.; Morel, A.F.; et al. Antioxidant Effects of Different Extracts from *Melissa officinalis*, *Matricaria recutita* and *Cymbopogon citratus*. *Neurochem. Res.* **2009**, *34*, 973–983. [CrossRef]
50. San Miguel-Chávez, R. Phenolic Antioxidant Capacity: A Review of the State of the Art. In *Phenolic Compounds*; Soto-Hernandez, M., Palma-Tenango, M., Garcia-Mateos, M., Eds.; IntechOpen: London, UK, 2017.
51. Jaber, A.; Soukariyeh, R.; Khalil, A.; Abdel-Sater, F.; Cheble, E. Biological Activities of Total Oligomeric Flavonoids Enriched Extracts of *Nicotiana tabacum* from Eight Lebanese Regions. *Int. J. Pharm. Sci. Rev. Res.* **2020**, *61*, 70–77.
52. Barreca, D.; Laganà, G.; Leuzzi, U.; Smeriglio, A.; Trombetta, D.; Bellocco, E. Evaluation of the nutraceutical, antioxidant and cytoprotective properties of ripe pistachio (*Pistacia vera* L., variety Bronte) hulls. *Food Chem.* **2016**, *196*, 493–502. [CrossRef]
53. Singleton, V.L.; Orthofer, R.; Lamuela-Raventós, R.M. [14] Analysis of total phenols and other oxidation substrates and antioxidants by means of folin-ciocalteu reagent. *Methods Enzymol.* **1999**, *299*, 152–178. [CrossRef]
54. Awada, N.; Ayoub, A.; Jaber, A.; Ibrahim, F.; El Ghotmi, N.; Cheble, E. Evaluation of the Anticancer, Anti-Inflammatory, and Antioxidant Properties of Various Extracts of *Annona squamosa* L. *Pharm. Sci.* **2023**, *29*, 384–394. [CrossRef]
55. Oyaizu, M. Studies on Products of Browning Reactions: Antioxidative Activities of Product of Browning Reaction Prepared from Glucosamine. *Jpn. J. Nutr. Diet.* **1986**, *44*, 307–315. [CrossRef]
56. Balouiri, M.; Sadiki, M.; Ibnsouda, S.K. Methods for in vitro evaluating antimicrobial activity: A review. *J. Pharm. Anal.* **2016**, *6*, 71–79. [CrossRef] [PubMed]
57. Semaan, J.; El-Hakim, S.; Ibrahim, J.-N.; Safi, R.; Elnar, A.; El Boustany, C. Comparative effect of sodium butyrate and sodium propionate on proliferation, cell cycle and apoptosis in human breast cancer cells MCF-7. *Breast Cancer* **2020**, *27*, 696–705. [CrossRef]
58. Ibrahim, J.-N.; El-Hakim, S.; Semaan, J.; Ghosn, S.; El Ayoubi, H.; Elnar, A.A.; Tohme, N.; El Boustany, C. Sodium Butyrate (NaB) and Sodium Propionate (NaP) Reduce Cyclin A2 Expression, Inducing Cell Cycle Arrest and Proliferation Inhibition of Different Breast Cancer Subtypes, Leading to Apoptosis. *Biomedicines* **2024**, *12*, 1779. [CrossRef]
59. Sousa, E.O.; Barreto, F.S.; Rodrigues, F.F.; Campos, A.R.; Costa, J.G. Chemical composition of the essential oils of *Lantana camara* L. and *Lantana montevidensis* Briq. and their synergistic antibiotic effects on aminoglycosides. *J. Essent. Oil Res.* **2012**, *24*, 447–452. [CrossRef]

60. Walden, A.B.; Haber, W.A.; Setzer, W.N. Essential Oil Compositions of Three *Lantana* Species from Monteverde, Costa Rica. *Nat. Prod. Commun.* **2008**, *4*, 1934578X0900400123. [CrossRef]
61. Khan, M.; Mahmood, A.; Alkhathlan, H.Z. Characterization of leaves and flowers volatile constituents of *Lantana camara* growing in central region of Saudi Arabia. *Arab. J. Chem.* **2016**, *9*, 764–774. [CrossRef]

molecules

Article

Antioxidant and Cytotoxic Properties of *Berberis vulgaris* (L.) Stem Bark Dry Extract

Ionuț Mădălin Ivan [1], Octavian Tudorel Olaru [1,*], Violeta Popovici [2,*], Carmen Lidia Chițescu [3], Liliana Popescu [1], Emanuela Alice Luță [1], Elena Iuliana Ilie [1], Lorelei Irina Brașoveanu [4], Camelia Mia Hotnog [4], George Mihai Nițulescu [1], Rica Boscencu [1] and Cerasela Elena Gîrd [1]

1 Faculty of Pharmacy, University of Medicine and Pharmacy "Carol Davila", Traian Vuia 6, 020956 Bucharest, Romania; ionut.ivan@drd.umfcd.ro (I.M.I.); liliana.popescu22@umfcd.ro (L.P.); emanuela.luta@umfcd.ro (E.A.L.); elena.ionita@drd.umfcd.ro (E.I.I.); george.nitulescu@umfcd.ro (G.M.N.); rica.boscencu@umfcd.ro (R.B.); cerasela.gird@umfcd.ro (C.E.G.)

2 Center for Mountain Economics, "Costin C. Kirițescu" National Institute of Economic Research (INCE-CEMONT), Romanian Academy, 725700 Vatra-Dornei, Romania

3 Faculty of Medicine and Pharmacy, "Dunărea de Jos" University of Galați, A.I. Cuza 35, 800010 Galați, Romania; carmen.chitescu@ugal.ro

4 Center of Immunology, "Stefan S. Nicolau" Institute of Virology, Romanian Academy, 285 Mihai Bravu Ave., 030304 Bucharest, Romania; lorelei.brasoveanu@virology.ro (L.I.B.); camelia.hotnog@virology.ro (C.M.H.)

* Correspondence: octavian.olaru@umfcd.ro (O.T.O.); violeta.popovici@ce-mont.ro (V.P.)

Abstract: *Berberis vulgaris* (L.) has remarkable ethnopharmacological properties and is widely used in traditional medicine. The present study investigated *B. vulgaris* stem bark (Berberidis cortex) by extraction with 50% ethanol. The main secondary metabolites were quantified, resulting in a polyphenols content of 17.6780 ± 3.9320 mg Eq tannic acid/100 g extract, phenolic acids amount of 3.3886 ± 0.3481 mg Eq chlorogenic acid/100 g extract and 78.95 μg/g berberine. The dried hydro-ethanolic extract (BVE) was thoroughly analyzed using Ultra-High-Performance Liquid Chromatography coupled with High-Resolution Mass Spectrometry (UHPLC–HRMS/MS) and HPLC, and 40 bioactive phenolic constituents were identified. Then, the antioxidant potential of BVE was evaluated using three methods. Our results could explain the protective effects of Berberidis cortex EC_{50}FRAP = 0.1398 mg/mL, IC_{50}ABTS = 0.0442 mg/mL, IC_{50}DPPH = 0.2610 mg/mL compared to ascorbic acid (IC_{50} = 0.0165 mg/mL). Next, the acute toxicity and teratogenicity of BVE and berberine—berberine sulfate hydrate (BS)—investigated on *Daphnia* sp. revealed significant BS toxicity after 24 h, while BVE revealed considerable toxicity after 48 h and induced embryonic developmental delays. Finally, the anticancer effects of BVE and BS were evaluated in different tumor cell lines after 24 and 48 h of treatments. The MTS assay evidenced dose- and time-dependent antiproliferative activity, which was higher for BS than BVE. The strongest diminution of tumor cell viability was recorded in the breast (MDA-MB-231), colon (LoVo) cancer, and OSCC (PE/CA-PJ49) cell lines after 48 h of exposure (IC_{50} < 100 μg/mL). However, no cytotoxicity was reported in the normal epithelial cells (HUVEC) and hepatocellular carcinoma (HT-29) cell lines. Extensive data analysis supports our results, showing a significant correlation between the BVE concentration, phenolic compounds content, antioxidant activity, exposure time, and the viability rate of various normal cells and cancer cell lines.

Keywords: *Berberis vulgaris* (L.) stem bark; dry hydro-ethanolic extract; phenolic secondary metabolites; berberine; antioxidant activity; cytotoxicity

1. Introduction

Numerous pharmaceutical companies are focused on researching and developing new formulations based on herbal sources, which can help manage chronic diseases. The World Health Organization also supports conventional plant-based treatments due to their accessibility, safety for long-term uses, and relatively low production costs. This shift toward natural remedies occurred mainly because some synthetic pharmaceutical drugs may have harmful side effects when used for the long-term treatment of chronic diseases [1]. Therefore, based on traditional medical systems (Ayurvedic and Chinese), phytotherapy in chronic disorders is currently used as an alternative treatment worldwide. Berberis is a significant plant genus with approximately 500 species worldwide. It belongs to the *Berberidaceae* family and has considerable potential applications in the food and pharmaceutical industries [2]. *Berberis* species are native to central and southern Europe, Asia (including the northern zones of Pakistan and Iran), and the north-eastern area of the United States. *Berberis vulgaris* (L.), known as European barberry, common barberry, or Épine-Vinette, has an essential role in herbal therapy; its different parts (fruits, leaves, roots, stem, branches, stem/root bark) have been used in traditional medicine for more than 2500 years. This species can be helpful in various inflammations, high blood pressure, gastrointestinal diseases, hepatic disorders, and diabetes. Numerous studies show that *B. vulgaris* has valuable pharmacological properties, such as antioxidant, antihyperglycemic, anticholinergic, hypolipidemic, anti-inflammatory, anticancer, and antimicrobial properties. In homeopathy, *B. vulgaris* is mainly used in urinary lithiasis, dermatology, rheumatism, and liver diseases [3]. Berberine, the specific isoquinoline alkaloid mainly extracted from common barberry root and stem barks, is formulated for oral administration alone or in various combinations. The administration of berberine-based phytotherapeutics could have a beneficial impact on lipid and carbohydrate metabolism, particularly on glucose homeostasis, being helpful in weight loss, diabetes mellitus [4] and endocrine disorders, liver diseases [5], cardiovascular diseases, atherosclerosis, neurodegenerative diseases, rheumatic diseases, and infectious diseases [1]. Several studies reported berberine-induced toxicity in humans and mice [1]. However, toxic phenomena could be diminished through berberine combination with other phytochemicals or plant extracts. Synergistic effects would also be expected in adequate combinations [1]. Moreover, *B. vulgaris* and berberine display anticancer effects through various cell signaling pathways' modulation [6], diminishing tumor cell viability and reducing their multiplication in various neoplasia (lung, breast, ovary, gastric cancer) [1].

In the present study, we aimed to investigate the hydro-ethanolic dry extract of *B. vulgaris* stem bark (BVE), obtained using a reflux extraction process in 50% ethanol, rotary evaporation, and freeze-drying. Here, 50% ethanol was used as an extraction solvent for obtaining the dry plant extract because of its effectiveness in extracting a broad range of phytochemicals (polar, moderately polar, and some nonpolar compounds); in addition, it has a low toxicity profile compared to other solvents (methanol, acetone, hexane, ethyl acetate or chloroform). Reflux extraction is a low-cost and efficient tool that ensures a high content of bioactive constituents through the consequent rotary evaporation and freeze-drying. The lyophilization (freeze-drying) process provides substantial stability to plant extracts by preserving the secondary metabolites with antioxidant activity.

A complex analysis of BVE's phenolic compounds was performed using ultra-high-performance liquid chromatography coupled with high-resolution mass spectrometry (UHPLC–HRMS/MS). Berberine was quantified through HPLC-DAD. The BVE's antioxidant potential was in vitro evaluated through the free radical scavenging (DPPH and ABTS) and reducing power (FRAP). The acute toxicity of BVE and berberine—berberine sulfate hydrate (BS)—was assessed in vivo on two *Daphnia* species. In contrast, their teratogenic potential was evaluated by applying the embryo test to *Daphnia magna* embryos. Moreover, the antitumor potential of BVE and BS was investigated in several human cancer cell lines: hepatocellular (HEP G2), colon (LoVo and HT-29), breast (MDA-MB-231), ovary (SK-OV-3), and tongue (PE/CA-PJ49), using classical oncolytic drugs as positive controls. Extensive data analyses support our results, showing significant correlations between the BVE concentration, exposure time, phenolic constituent content, antioxidant activity, and cytotoxicity.

2. Results and Discussion

2.1. Phenolic Compounds (Polyphenols and Phenolic Acids) Quantification

Berberidis cortex dry hydro-ethanolic extract was obtained with a yield of 16.35%. Other studies reported similar yields: 18.7% for roots and 14.7% for leaf extracts in ethanol [6]. Ethanol's availability and regulatory approval make it an obvious choice due to its balance of effectiveness, safety, and applicability. Moreover, ethanol's moderate boiling point makes it easy to remove by evaporation, simplifying the process of concentrating the extract and falling within the trend of implementing green technology and using green solvents, which are much safer for the environment. The rotary evaporator could be connected with a vacuum pump, which decreases the boiling point of ethanol (78.2 °C) and facilitates its evaporation. After ethanol collection, it could be subjected to fractional distillation to remove the moisture content and increase its purity.

The standard calibration curves are displayed in Figure S1 in the Supplementary Materials, while the TPC and TPA values are presented in Table 1. BVE is rich in total polyphenols (TPC = 17.6780 ± 3.9320 mg Eq tannic acid/100 g extract); however, it has shown a phenolic acid content (TPA) of only 3.3886 ± 0.3481 mg Eq chlorogenic acid/100 g extract.

Table 1. Total polyphenols content, total phenolic acids, and antioxidant activity of BVE.

Phenolic Compounds		
Total Polyphenols (mg Eq Tannic Acid/100 g Extract)		Total Phenolic Acids (mg Eq Chlorogenic Acid/100 g Extract)
17.6780 ± 3.9320		3.3886 ± 0.3481
Antioxidant Activity		
IC_{50}DPPH (mg/mL)	IC_{50}ABTS (mg/mL)	EC_{50}FRAP (mg/mL)
0.2610	0.0442	0.1398

DPPH—2,2-diphenyl-1-picryl-hydrazine; ABTS—2,20-azinobis-3-ethylbenzotiazoline-6-sulfonic acid; FRAP—ferric reducing antioxidant power.

The literature data show that the TPC in various *B. vulgaris* extracts is very different. Our hydro-ethanolic extract of Berberidis cortex has a TPC of 1767.80 mg/g, while El-Zahar et al. [7] reported much lower TPC levels in ethanol extracts of roots (147.2 mg/g) and leaves (120.7 mg/g). Och et al. [8] indicated similar TPC values quantified in 80% methanol extracts of various *B. vulgaris* parts: 58.5 mg/g for the leaf extract, 57.7 mg/g for the stem one, and 52.8 mg/g for the fruit extract.

2.2. Identification and Quantification of BVE Phytoconstituents by UHPLC–HRMS/MS and HPLC-DAD

Table 2 registers all the phytochemicals identified in BVE.

Some constituents were quantified, with gallic acid having the highest amount (540.00 µg/g). It is followed, in decreasing order, by naringenin (90.41 µg/g), berberine (78.95 µg/g), rutin (72.41 µg/g), kaempferol (68.24 µg/g), and galangin (67.21 µg/g).

Figure 1A displays the chromatogram of the primary phytochemicals identified in BVE by UHPLC-MS, and Figure 1B shows the HPLC-DAD chromatogram of BVE, where berberine has an RT = 32.513.

Table 2. The phytochemicals identified in *B. vulgaris* stem bark dry extract (BVE) by UHPLC–HRMS/MS and HPLC-DAD.

Nr. Crt.	Identified Compound	Phytochemical Classification	Chemical Formula	Adduct Ion/ Monitored Negative Ion (*m/z*)	Retention Time (min)	Content (µg/g)
1	Quercetin	Flavonoid	$C_{15}H_{10}O_7$	301.0354	15.01	28.42
2	Rutin (quercetin 3-O-rutinoside)	Flavonoid	$C_{27}H_{30}O_{16}$	609.14613	12.39	72.41
3	Apigenin	Flavonoid	$C_{15}H_{10}O_5$	269.04502	16.71	10.45
4	Kaempferol	Flavanol	$C_{15}H_{10}O_6$	285.04049	16.51	68.74

Table 2. *Cont.*

Nr. Crt.	Identified Compound	Phytochemical Classification	Chemical Formula	Adduct Ion/ Monitored Negative Ion (*m*/*z*)	Retention Time (min)	Content (μg/g)
5	6-Methoxyluteolin (Nepetin)	Flavonoid	$C_{16}H_{12}O_7$	315.05105	16.75	-
6	Naringenin	Flavanone	$C_{15}H_{12}O_5$	271.06122	15.46	90.41
7	Hesperitin	Flavonoid	$C_{16}H_{14}O_6$	301.07179	13.71	44.00
8	Galangin	Flavonoid	$C_{15}H_{10}O_5$	269.04557	16.71	67.21
9	Genistein	Isoflavone	$C_{15}H_{10}O_5$	269.04502	16.73	-
10	Glycitein	Isoflavone	$C_{16}H_{12}O_5$	283.06122	11.15	19.21
11	Gallic acid	Hydroxybenzoic acid	$C_7H_6O_5$	169.01427	1.70	540.00
12	Chlorogenic acid/Neochlorogenic	Cinnamate ester	$C_{16}H_{18}O_9$	353.08783	6.08	10.54
13	Ferulic acid	Hydroxycinnamic acid	$C_{10}H_{10}O_4$	193.05066	9.94	39.36
14	AbsCisPtic acid	Terpenoid	$C_{15}H_{20}O_4$	263.12891	14.76	8.61
15	*p*-Coumaric acid	Hydroxycinnamic acid	$C_9H_8O_3$	163.03954	8.80	30.33
16	Syringic acid	Hydroxybenzoic acid	$C_9H_{10}O_5$	197.04555	8.73	3.35
17	Afrormosin	Isoflavone	$C_{17}H_{14}O_5$	297.07687	17.17	-
18	Kaempferol-3-O-rutinoside	Flavonol glycoside	$C_{27}H_{30}O_{15}$	593.15122	9.35	-
19	Kaempferol (luteolin)-O-glucoside/ isomers	Flavonoid	$C_{21}H_{20}O_{11}$	447.09331	13.56	-
20	Vitexin (apigenin 8-C-glucoside)/isovitexin	Flavonol glycoside	$C_{21}H_{20}O_{10}$	431.09839	11.98	-
21	Azelaic acid	Dicarboxylic acid	$C_9H_{16}O_4$	187.09761	13.99	-
22	Apigenin 7-O-glucosylglucoside	Flavonoid	$C_{27}H_{30}O_{15}$	593.15122	9.45	-
23	Rosmarinic acid	Ester of caffeic acid	$C_{18}H_{16}O_8$	359.07727	13.42	-
24	Carnasol	Diterpene	$C_{20}H_{26}O_4$	329.17586	18.83	-
25	Rosmadial/Isomeri	Diterpene lactone	$C_{20}H_{24}O_5$	343.15510	20.38	-
26	Rosmanol methyl ether	Diterpene	$C_{21}H_{28}O_5$	359.18640	22.19	-
27	Quercetin 3-O-glucuronide	Flavonol glucuronide	$C_{21}H_{18}O_{13}$	477.06749	12.16	23.04
28	Narirutin (naringenin-7-O-rutinoside)	Flavonol glycoside	$C_{27}H_{32}O_{14}$	579.17195	12.13	-
29	Apigenin-7-O-glucuronide	Flavonoid-7-O-glucuronides	$C_{21}H_{18}O_{11}$	445.07763	13.29	-
30	Procyanidine B1/B2	Flavonoid	$C_{30}H_{26}O_{12}$	577.13515	16.24	-
31	Sinapic acid	Hydroxycinnamic acid	$C_{11}H_{12}O_5$	223.06122	10.33	-
32	Hidroxyferulic acid/Isomers	Hydroxycinnamic acid	$C_{16}H_{20}O_{10}$	371.09839	4.81	-
33	Valerenic acid	Sesquiterpenoid	$C_{15}H_{22}O_2$	233.15473	21.33	-
34	Lehmannin	Flavanone	$C_{25}H_{28}O_5$	407.18642	26.40	-
35	Ginkgetin	Flavone	$C_{32}H_{22}O_{10}$	565.11404	7.25	-
36	Taxifolin 3-O-rhamnoside	Flavonoid	$C_{21}H_{22}O_{11}$	449.10896	12.24	-
37	Piceatannol	Stilbenoid	$C_{14}H_{12}O_4$	243.06631	13.08	-
38	Lignan	Polyphenolic compound	$C_{25}H_{30}O_8$	457.18682	27.33	-
39	Cyanidin 3-O-arabinoside	Anthocyanidin-3-O-glycoside	$C_{20}H_{19}ClO_{10}$	453.05942	7.17	-
40	Berberine	Isoquinoline alkaloid	$C_{20}H_{18}NO_4$	-	32.51	78.95

In Argentinian/Patagonian barberry fruit (*Berberis microphylla*) ethanol extract, Boeri et al. [9] reported the highest amounts of quercetin (1134.54 μg/g), caffeic acid (1092.75 μg/g), and syringic acid (368.55 μg/g). In contrast, gallic acid was 48.17 μg/g. Berberidis cortex is a significant and frequently used crude drug registered in the "Drug Standards of Tibetan Medicines" since 1995. The specific bioactive compounds are alkaloids (berberine, magnoflorine, jatrorrhizine, palmatine), and the most known is berberine, quantified in our BVE (78.95 μg/g). HPLC analysis of methanolic extract of Berberidis cortex from China, harvested from different geographical zones, recorded a range of 21.12–37.5 μg/g berberine [10]. Our results indicated a value twice as high as the first one.

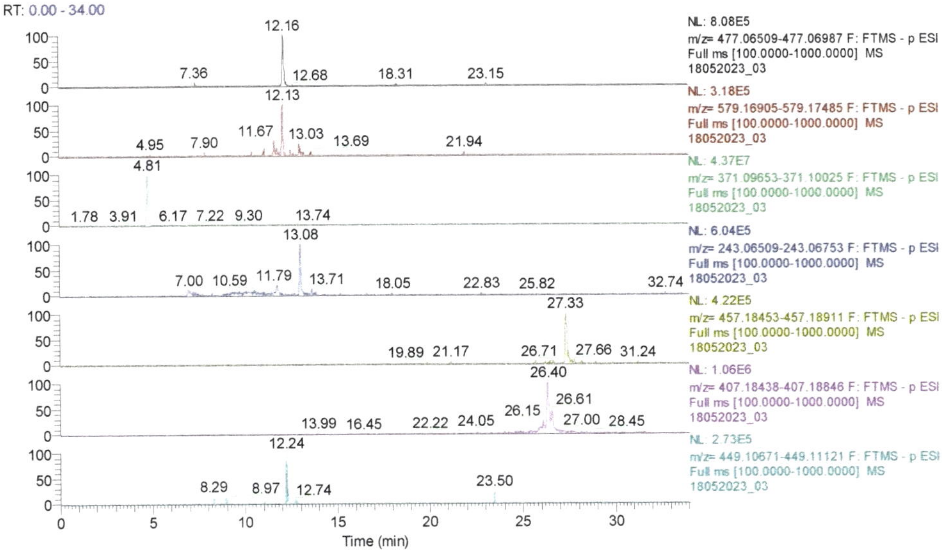

(A)

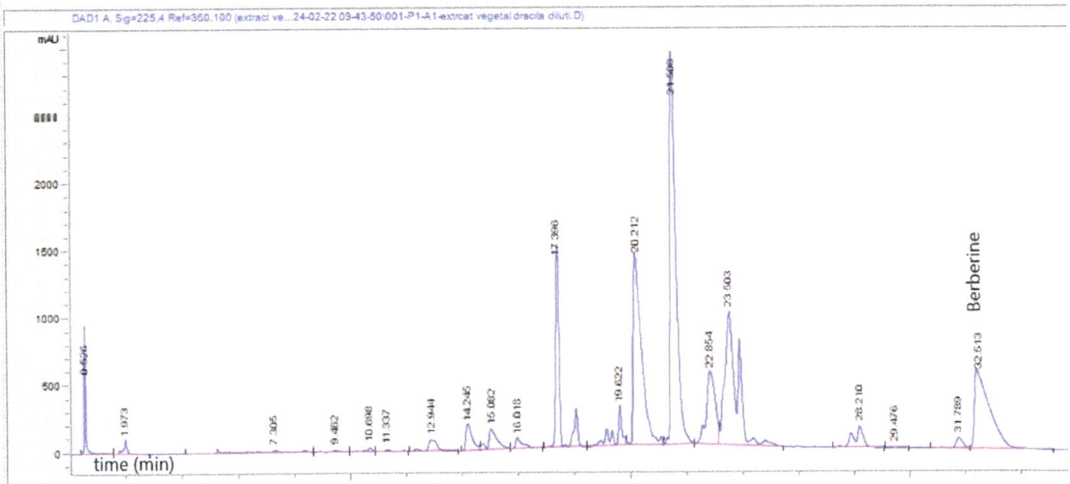

(B)

Figure 1. (**A**). UHPLC–HRMS/MS chromatogram of phytochemicals identified in BVE; from top to bottom: quercetin 3-O-glucuronide (*m*/*z* = 477.06749, RT = 12.16), narirutin (naringenin-7-O-rutinoside) (*m/z* = 579.17195, RT = 12.13), hydroxyferulic acid (*m/z* = 371.09839, RT = 4.81), piceatannol (*m/z* = 243.06631, Rt = 13.08), lignan (*m/z* = 457.18682, RT = 27.33), lehmannin (*m/z* = 407.18642, RT = 26.40), taxifolin 3-O-rhamnoside (*m/z* = 449.10896, RT = 12.24). (**B**). HPLC-DAD chromatogram of BVE; berberine has an RT = 32.513.

2.3. Antioxidant Activity

Table 2 shows significant differences between the IC_{50}/EC_{50} values determined by all three methods, IC_{50}DPPH = 0.2610 mg/mL, IC_{50}ABTS = 0.0442 mg/mL, and EC_{50}FRAP = 0.1398 mg/mL, compared to ascorbic acid (IC_{50} = 0.0165 mg/mL). Similar values were reported for *B. microphylla* ethanol extract [9]: ABTS IC_{50} = 0.26 mg/mL and DPPH IC_{50} = 0.38 mg/mL. The substantial antioxidant potential is underlined by the phenolic constituent content and berberine and other alkaloids, which are known for their protective activity [11]

2.4. 48-h Acute Toxicity Test Using Daphnia Magna and Daphnia Pulex

After 24 h, *D. magna*'s total lethality was recorded at concentrations ≥ 25 μg/mL BVE and ≥10 μg/mL BS. Similarly, *D. pulex*'s total lethality occurred at concentrations ≥ 50 μg/mL BVE and ≥25 μg/mL BS. Our results revealed that the BS toxicity is higher than BVE for both *Daphnia* species, with *D. magna* being more vulnerable than *D. pulex*. The time-dependent toxicity is more evident in *D. magna* than in *D. pulex*. (Figure 2). The lethality curves' analysis revealed that, in the *D. magna* bioassay (Figure 2A), BS exhibited a lower LC_{50} value at 24 h compared to BVE, suggesting the higher toxicity of the pure alkaloid, which was expected. However, at 48 h, the BVE toxicity significantly increases, almost to the same potency as BS. After 48 h of exposure, the concentration–response curves for both tested solutions in *D. pulex* displayed similar trends to those recorded in *D. magna* but with differences in the magnitude of lethality (Figure 2B).

Time	**24 h**		**48 h**	
Parameter	**LC_{50} (μg/mL)**	**95%CI**	**LC_{50} (μg/mL)**	**95%CI**
		Daphnia magna		
BS	8.7	6.5–11.8	~5.3	NA *
BVE	30.5	22.6–41.1	6.7	3.7–11.9

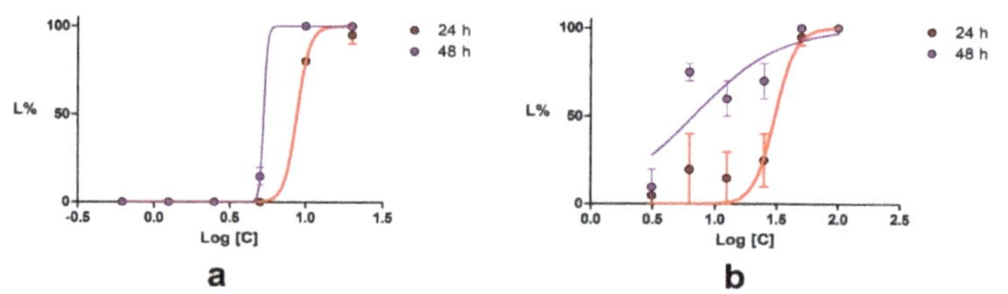

(A)

Time	**24 h**		**48 h**	
		Daphnia pulex		
BS	NA **	NA **	6.6	4.6–9.5
BVE	~143.0	NA *	37.4	28.8–48.7

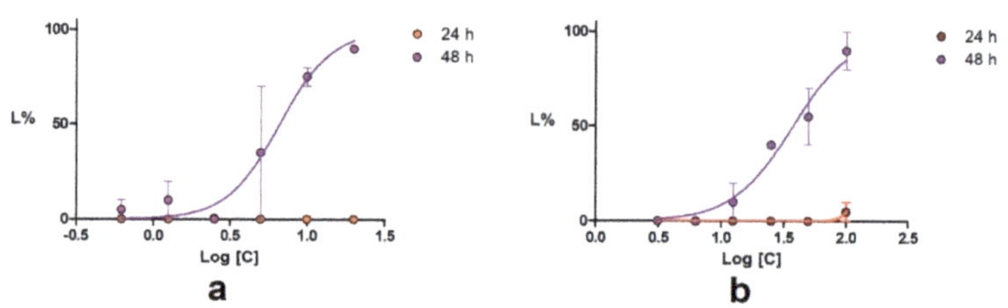

(B)

Figure 2. The results of the 48 h acute toxicity test using *Daphnia* sp. Lethality curves obtained after 48 h exposure of *Daphnia* sp. to BS (a) and BVE (b): *D. magna* (**A**) and *D. pulex* (**B**). BS—berberine sulfate hydrate; BVE—*B. vulgaris* stem bark dry hydro-ethanolic extract; NA—the values could not be calculated: * The interval is vast. ** The values could not be calculated as the maximum L% was 10%.

2.5. Daphnia Magna Embryonic Development Assay

Following the acute toxicity test results, the embryo assay was performed at non-lethal concentrations (2.5 μg/mL BS and 3.125 μg/mL BVE). Minor differences were observed after 24 h (Figure 3a,b). After 48 h, BS stimulated the development of all the embryos

(Figure 3c), while BVE exhibited a significant inhibitory effect (Figure 3d), which could be due to the extract's complex composition. Only 20% of the embryos treated with BVE were fully developed, compared to 90% recorded for those exposed to BS. The mobility and viability of neonates developed in BS solution were similar to those of the control. However, they all failed to form the compound eye, even after 48 h of exposure, suggesting a potential developmental risk.

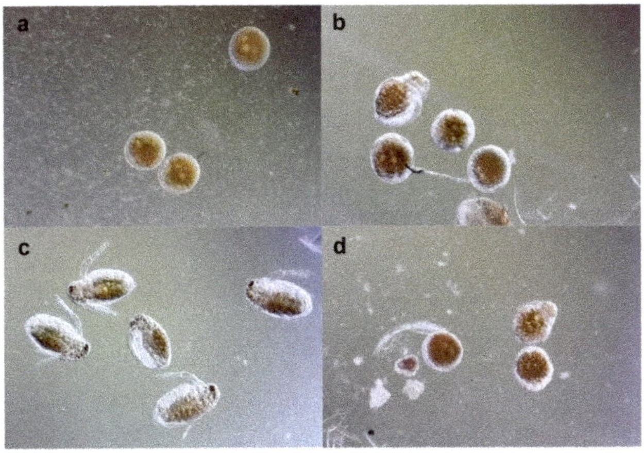

Figure 3. *Daphnia magna* embryonic development assay: (**a**) embryos before testing; (**b**) embryo development after 24 h in BS 2.5 µg/mL; (**c**) embryo development after 48 h in BS 2.5 µg/mL; (**d**) embryo development after 48 h in BVE 3.125 µg/mL.

In *Daphnia magna*, Vesela et al. [12] reported that the berberine chloride toxicity recorded an LC_{50} of 0.903 µg/mL after 24 h and 0.822 µg/mL at 48 h exposure. Our berberine sulfate hydrate recorded 9.7 µg/mL and 5.3 µg/mL, respectively. In another study on another crustacean [13], 7 µg/mL berberine chloride induced 100% lethality in *Artemia salina* larvae. These differences could be explained by berberine salt, the animal model species, and the provenance. The *D. magna* embryos failed to form compound eyes after BS and BVS treatment. Natural berberine also affects cardiovascular system morphogenesis and functionality in Zebrafish embryos [14]. Based on these findings, in the *Medicinal Plants Monograph Volume 4* [15], the WHO mentions the potential side effects of berberine on humans after consuming more than 500 mg.

2.6. In Vitro Anticancer Activity

The antiproliferative activity induced by BVE was evaluated in vitro through several cytotoxic assays by applying different BVE and BS concentrations (6.25–400 µg/mL) to cells derived from six tumor cell lines of different histological origin: HEP G2, LoVo, HT-29, MDA-MB-231, SK-OV-3, and PE/CA-PJ49. Human umbilical vein endothelial cells (HUVECs) were selected as the reference normal cells.

The BVE and BS antiproliferative capacities, as tested on normal human cells and tumor cell lines, are shown in Table 3.

Table 3. The cytotoxicity of BVE and BS in normal cell and tumor cell lines (expressed as cell viability %) after 24 and 48 h of exposure.

Concentration (µg/mL)	24 h						48 h					
	BVE			BS			BVE			BS		
	V (%)	SD	IC_{50} (µg/mL)	V (%)	SD	IC_{50} (µg/mL)	V (%)	SD	IC_{50} (µg/mL)	V (%)	SD	IC_{50} (µg/mL)
HUVEC												
6.25	109.38	5.13		100.60	4.15		104.59	5.04		107.52	5.67	
12.5	107.10	5.38		97.10	5.91		100.54	4.35		103.36	6.91	
25	104.60	1.06		96.28	4.58		103.16	3.56		105.17	3.38	
50	101.16	0.09	>>400	99.53	4.33	>>400	98.71	5.03	>>400	96.18	2.98	>400
100	102.78	3.54		95.35	2.21		96.38	3.59		82.99	6.04	
200	99.10	7.94		96.72	1.86		89.95	4.76		77.15	4.99	
400	86.53	4.69		84.60	2.30		75.48	0.09		56.10	0.09	
HEP G2												
6.25	99.57	6.47		97.46	6.76		98.31	0.33		96.91	5.81	
12.5	97.94	4.25		91.71	8.04		97.60	2.06		84.48	0.54	
25	96.43	0.12		90.43	4.08		96.03	6.67		65.45	2.39	
50	94.49	1.98	>400	79.68	5.30	>200	91.33	1.90	>100	50.48	2.28	>50
100	91.83	2.97		71.19	7.51		81.87	3.20		38.32	0.22	
200	73.62	0.52		53.77	5.48		46.76	1.95		25.69	2.44	
400	52.86	0.93		42.16	2.80		22.62	3.36		21.82	6.02	
HT-29												
6.25	100.78	2.80		100.08	1.27		99.90	4.08		98.23	5.95	
12.5	98.86	7.60		98.63	2.90		97.39	0.00		95.90	2.72	
25	98.62	4.80		94.72	3.54		94.81	1.53		92.29	1.93	
50	97.18	3.59	>>400	83.49	4.75	>400	92.31	3.34	>>400	86.48	5.16	>400
100	95.08	1.16		70.77	0.60		90.62	1.25		72.18	2.44	
200	91.06	0.69		66.59	4.75		86.62	0.28		61.43	2.21	
400	85.04	3.69		60.80	6.76		77.31	5.45		54.02	6.56	
LoVo												
6.25	98.74	4.77		97.08	5.37		92.46	3.67		95.64	4.80	
12.5	93.91	0.74		91.98	3.52		89.06	6.32		91.86	2.73	
25	90.85	7.54		88.50	4.96		84.68	8.47		82.13	6.31	
50	85.13	4.94	>200	79.91	0.87	>50	72.68	4.55	>50	65.02	5.69	>50
100	75.97	2.35		38.60	1.85		48.39	0.63		28.19	1.39	
200	53.72	4.45		26.22	2.10		14.72	0.00		15.77	1.14	
400	25.05	4.57		15.57	8.53		3.09	2.66		8.08	2.91	
MDA-MB-231												
6.25	96.83	4.41		87.73	6.43		75.86	4.48		71.40	3.52	
12.5	90.33	2.57		76.14	0.55		73.74	3.67		62.02	0.95	
25	87.71	0.37		66.43	5.51		70.81	0.15		45.93	2.57	
50	82.02	2.20	>100	49.76	1.10	>25	60.88	1.84	>50	37.83	6.68	>12.5
100	69.16	2.94		23.19	2.02		45.27	3.97		20.29	2.13	
200	53.71	3.67		20.67	0.37		25.65	1.47		14.39	2.06	
400	17.21	3.49		10.43	4.04		2.99	0.44		8.54	0.81	
PE/CA-PJ49												
6.25	99.85	6.60		97.37	4.32		96.00	2.18		91.19	5.00	
12.5	93.17	6.89		91.01	8.44		87.26	0.51		77.49	6.21	
25	86.25	7.68		78.14	5.90		78.68	7.12		62.58	0.51	
50	80.89	2.54	>100	63.65	2.81	>50	69.09	5.95	>50	50.24	4.70	>50
100	61.92	1.17		42.84	0.34		49.39	5.35		35.37	1.91	
200	43.51	2.40		23.86	0.34		23.20	2.06		12.96	0.88	
400	12.25	2.54		9.75	1.71		4.70	0.51		3.95	0.66	
SK-OV-3												
6.25	99.72	4.94		99.03	4.61		93.80	6.76		91.34	5.80	
12.5	92.28	5.89		90.99	4.63		90.86	7.09		82.43	5.30	
25	89.38	6.28		83.36	5.25		83.56	0.55		78.78	6.68	
50	83.35	5.84	>400	79.62	6.04	>200	73.17	2.56	>100	67.19	7.40	>50
100	78.30	2.95		66.62	1.87		58.24	3.52		33.37	6.76	
200	72.10	7.07		57.21	4.81		30.87	3.38		26.46	2.47	
400	55.83	3.83		39.90	1.25		16.29	0.37		10.05	0.80	

BVE = *B. vulgaris* stem bark dry hydro-ethanolic extract, 24 and 48 h = cell line exposure time (hours) to the different BVE concentrations (µg/mL). HUVEC—human umbilical endothelial cell; HEP G2 —human hepatocellular carcinoma; HT-29 and LoVo—human colon adenocarcinomas; MDA-MB-231—human breast adenocarcinoma; PE/CA-PJ49—human squamous tongue carcinoma; SK-OV-3 —human ovary adenocarcinoma; SD—standard deviation. Interpretation of the IC_{50} values is based on that of the National Cancer Institute [16]: $IC_{50} \leq 20$ µg/mL—strong cytotoxic properties, IC_{50} = 21–200 µg/mL—moderate cytotoxicity, IC_{50} = 201–500 µg/mL—low cytotoxicity and $IC_{50} \geq 500$ µg/mL—no cytotoxic activity. Data are expressed as the mean values ± standard deviations (SD) of three experiments (n = 3).

The IC_{50} values displayed in Table 3 could be interpreted according to Hidayat et al. [16], resulting in an overview of the BVE and BS cytotoxicity in various cell lines. In normal endothelial cells (HUVEC), they have no cytotoxicity after 24 and 48 h (IC_{50} >> 400 µg/mL). The same interpretation is also available for HT-29 tumor cells, which showed no significant decrease in viability after the BVE/BS treatments.

Generally, BVE exhibited moderate cytotoxicity in the other tumor cells. The most substantial effect, with the lowest IC_{50} values after 24 and 48 h (IC_{50} > 100 µg/mL, respectively, IC_{50} > 50 µg/mL) was seen in breast cancer cells (MDA-MB-231) and OSCC ones (PE/CA-PJ49). In LoVo cells (colon cancer), the cytotoxicity at 24 h was appreciably lower (IC_{50} > 200 µg/mL) but after 48 h of exposure was moderate, similar to the previous ones (IC_{50} > 50 µg/mL). BVE exhibited the lowest effect on human ovary cancer (SK-OV-3 cells) and hepatocellular carcinoma (HEP G2 cells) after 24 and 48 h (IC_{50} > 400 µg/mL and, respectively, >100µg/mL).

Globally, the antiproliferative effects of BS are more potent than those of BVE. In MDA-MB-231 cells, moderate to high cytotoxicity was registered (IC_{50} > 25 and, respectively, >12.5 µg/mL after 24 and 48 h, $p < 0.05$). BS showed similar activity on PE/CA-PJ49 and LoVo (IC_{50} > 50 µg/mL for both exposure times). Moreover, it recorded low toxicity after 24 h contact with SK-OV-3 and HEP G2 cells and a moderate one after 48 h (IC_{50} > 200 µg/mL, respectively, >50 µg/mL).

The results of in vitro studies are detailed and presented in Figure S2 and File S1 in the Supplementary Materials.

After 24 h, BVE cytotoxicity at the selected concentration range (6.25–400 µg/mL) in normal and tumor cell lines showed significant differences (at $\alpha < 0.05$, p-value was established at 0.0024) between PE/CA-PJ49 and HUVEC, and MDA-MB-231 and HUVEC ($p = 0.001$, Figure S1a). Substantial differences ($p < 0.05$) were also observed in the case of HUVEC and all the other tumor cells, except the HT-29 ones, and MDA-MB-231 and HUVEC compared to HT-29 cells (Figure S2a in the Supplementary Materials).

After 48 h of exposure to BVE, the percentual values of cell vitality significantly differed between the HUVEC and MDA-MB-231 cells ($p < 0.001$). Appreciable differences ($p < 0.05$, Figure S2b) were reported between HUVEC and LoVo, PE/CA-PJ49, and SK-OV-3. Moreover, there were notable differences between HT-29 and LoVo, MDA-MB-231 and PE/CA-PJ49, and HEP G2 and MDA-MB-231 ($p < 0.05$).

In the case of BS, significant differences ($p = 0.0001$ and $p = 0.000$) were recorded exclusively between the HUVEC and MDA-MB-231 cell viability in both periods of exposure (Figure S2c,d). After 24 h, remarkable differences were reported between HUVEC and LoVo, PE/CA-PJ49, and SK-OV-3, and HT-29 and MDA-MB-231 and PE/CA-PJ49 ($p < 0.05$, Figure S1c). Moreover, after 48 h, there were notable differences between HUVEC and HEP G2, PE/CA-PJ49, LoVo, and SK-OV-3 ($p < 0.05$); the same was true for HT-29 and PE/CA-PJ49 and MDA-MB-231 (Figure S2d from Supplementary Materials).

No statistically significant differences were reported between the BVE and BS cytotoxicity for the same exposure period in the same cell line (Figure S2 in the Supplementary Materials).

The cytotoxic activity of BVE was compared to that induced by several drugs (5-Fluorouracil, Cisplatin, and Doxorubicin) [17] that are commonly used in oncological treatments and were applied throughout all the experiments as positive controls. The concentration range used for Cisplatin (CisPt) and 5-Fluorouracil (5-FU) was 3.125–200 µM, while for Doxorubicin (DOX) it was between 0.625 and 40 µM [17], as illustrated in Figures 4 and 5.

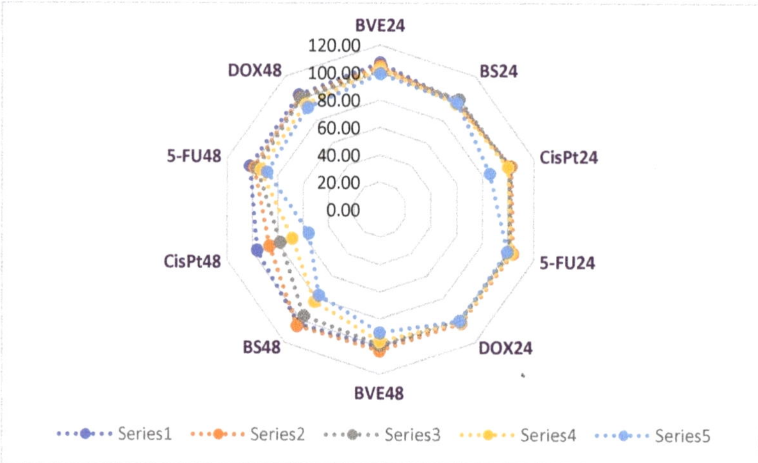

Figure 4. The effects of BVE and BS on normal endothelial cell (HUVEC) viability (%) compared to oncolytic drugs after 24 and 48 h. Series 1–5 = concentration range: 12.5–200 µg/mL for BVE and BS, 12.5–200 µM for 5-FU and CisPt, and 1.25–20 µM for DOX. BVE—dry hydro-ethanolic extract of *Berberis vulgaris* stem bark; BS—berberine sulfate hydrate; CisPt—Cisplatin; DOX—Doxorubicin, 5-FU—5-Fluorouracil; 24 and 48—treatment time (24 and 48 h).

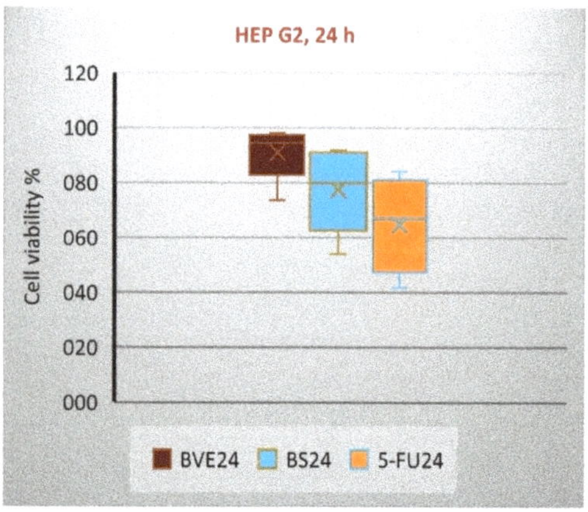

(A)

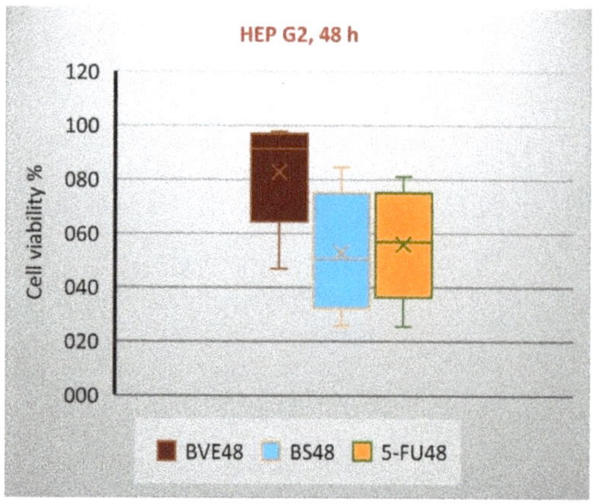

(B)

Figure 5. *Cont.*

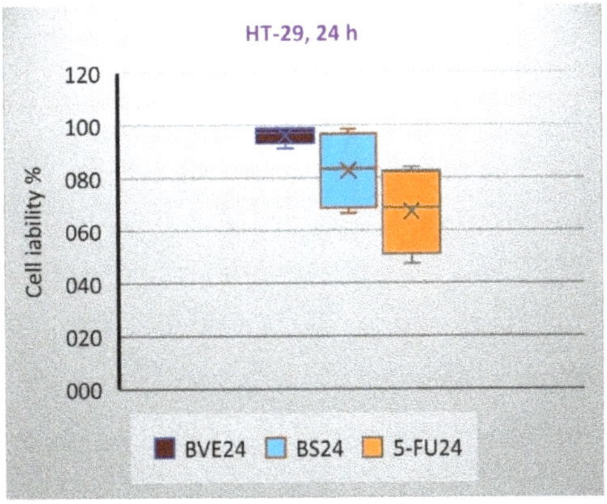

(C)

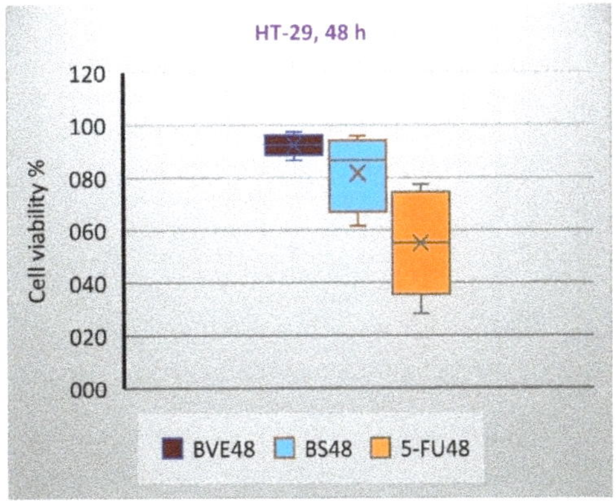

(D)

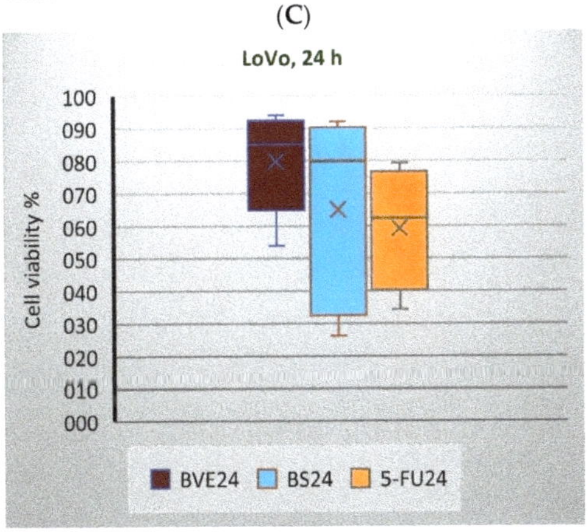

(E)

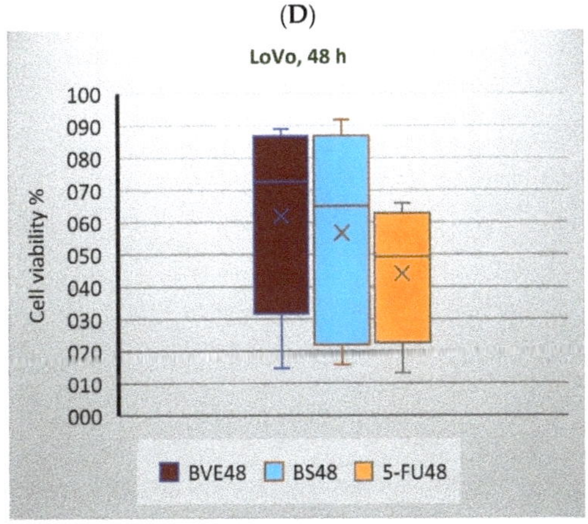

(F)

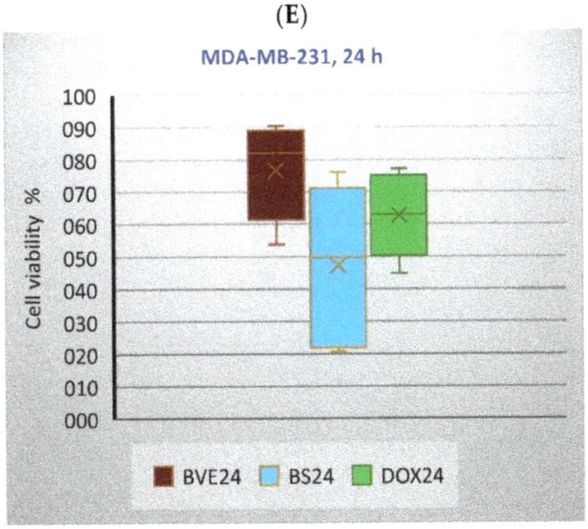

(G)

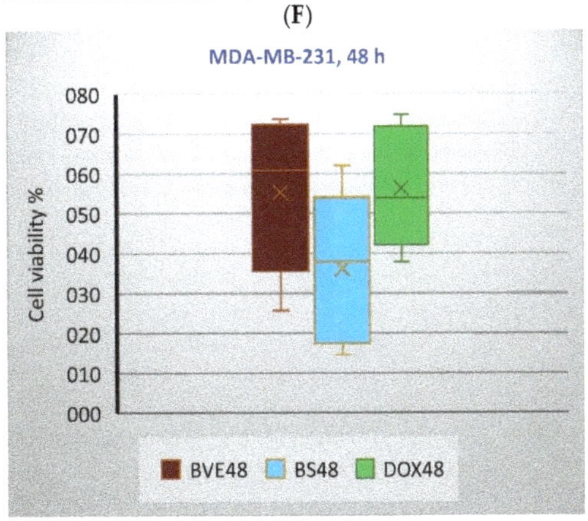

(H)

Figure 5. *Cont.*

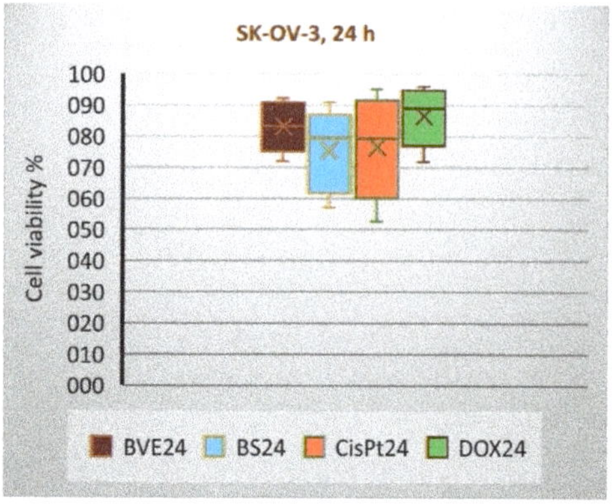

(I)

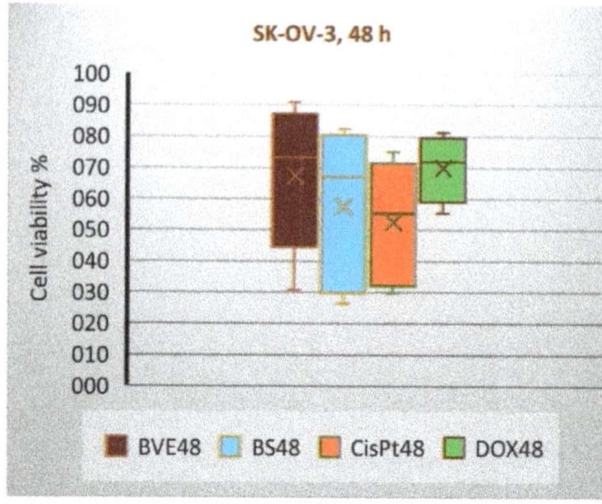

(J)

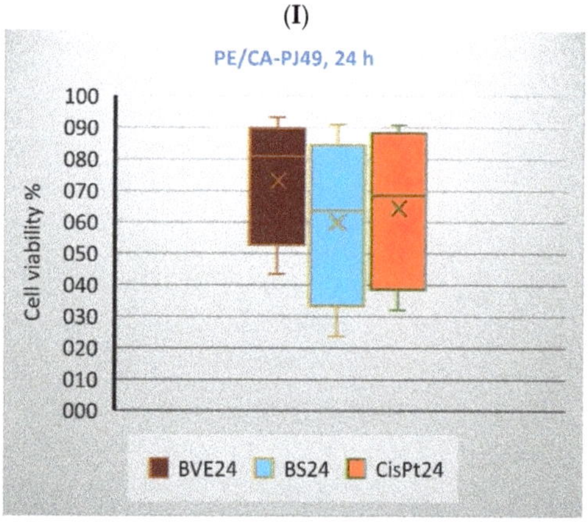

(K)

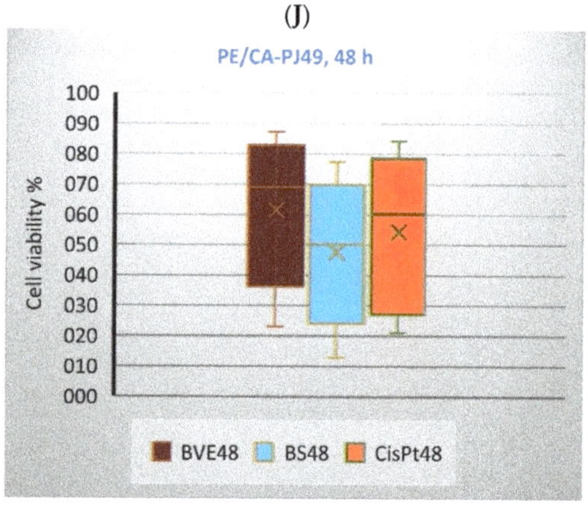

(L)

Figure 5. Box and Whisker plots displaying tumor cell viability % (F1 axis) after 24 h (**A**,**C**,**E**,**G**,**I**,**K**) and 48 h (**B**,**D**,**F**,**H**,**J**,**L**) following treatments with BVE, BS, and standard anticancer drugs: (**A**,**B**): HEP G2; (**C**,**D**): HT-29; (**E**,**F**): LoVo; (**G**,**H**): MDA-MB-231; (**I**,**J**): SK-OV-3; (**K**,**L**): PE/CA-PJ49. HEP G2—human hepatocellular carcinoma; HT-29 and LoVo—human colon adenocarcinomas; MDA-MB-231—human breast adenocarcinoma; PE/CA-PJ49—human squamous tongue carcinoma; SK-OV-3—human ovary adenocarcinoma; BVE—dry hydro-ethanolic extract of *Berberis vulgaris* stem bark; BS—berberine sulfate hydrate; CisPt—Cisplatin; DOX—Doxorubicin, 5-FU—5-Fluorouracil; 24 and 48—treatment time (24 and 48 h).

The effects of BVE and BS compared to anticancer drugs on normal endothelial cells' (HUVEC) viability after 24 and 48 h of exposure are displayed in Figure 4.

The highest HUVEC cell viability diminution was recorded after 24 h at 200 μM CisPt (85.58%), while, at the corresponding concentrations, BVE, BS, and both other drugs did not significantly affect it. At 48 h, 200 μM CisPt reduced the normal cell viability to 55.80%, followed by BS at 200 μg/mL, with 77.15%. BVE at 200 μg/mL (89.95%) acted similarly, with 200 μM 5-FU (88.63%) and 20 μM DOX (91.97%).

Table 1 and Figure 4 show that, in normal cells (HUVEC), both berberine and BVE have no cytotoxic effects (IC50 >> 400 μg/mL). Our results are similar to those from the scientific literature [18].

Figure 5 shows the influence of BVE and BS on cancer cell viability compared to standard oncolytic drugs in the same concentration range (12.5–200 μg/mL for BVE and BS, 12.5–200 μM for 5-FU and CisPt, and 1.25–20 μM for DOX) [17].

Generally, the tumor cell viability diminution was higher after 48 h than 24 h. In all cases, BS showed a higher cytotoxicity than BVE. In HT-29 and LoVo cells, the BS activity was lower than 5-FU after both exposure times, while in HEP G2, the BS activity was higher than 5-FU after 24 h and lower after 48 h (Figure 5A–F). The PE/CA-PJ49 cell viability decreased in the following order: be > CisPt > BS after 24 and 48 h of treatment (Figure 5K,L). In MDA-MB-231, the cell viability after 24 h decreased in the order of BVE > DOX > BS; after 48 h, BVE acted to a slightly higher extent than DOX (Figure 5G,H). After 24 h, the SK-OV-3 cell viability decreased in the order of DOX > BVE > CisPt > BS, while after 48 h, the previously mentioned order changed: DOX > BVE > BS > CisPt (Figure 5I,J). These effects are due to the potential synergism between the phytochemicals in BVE [19].

2.7. Statistical Analysis

The Pearson correlation shows that BVE24 is highly correlated with BS24, BVE48 and BS48 (r = 0.906, r = 0.910, r = 0.866, $p < 0.05$). BS24 exhibits a strong correlation with 5-FU24, BVE48 and BS48 (r = 0.813, r = 0.935, r = 0.928, $p < 0.05$). BVE48 has a considerable correlation with BS48 (r = 0.955) and a moderate one with 5-FU48 (r = 0.788), $p < 0.05$. Moreover, CisPt24 significantly correlates with DOX24, CisPt48 and DOX48 (r = 0.934, r = 0.997, r = 0.837, $p < 0.05$), CisPt48 with DOX48 (r = 0.830, $p < 0.05$), DOX24 with CisPt48 and DOX48 (r = 0.918, r = 0.920, $p < 0.05$), and 5-FU24 with BVE48 and BS48 (r = 0.866, r = 0.853, $p < 0.05$); 5FU24 also shows a moderate correlation with 5-FU48 (r = 0.788, $p < 0.05$). The place of each cytotoxic agent linked to the cell type is shown in Figure 6A, and the similarities between them are displayed in Figure 6B.

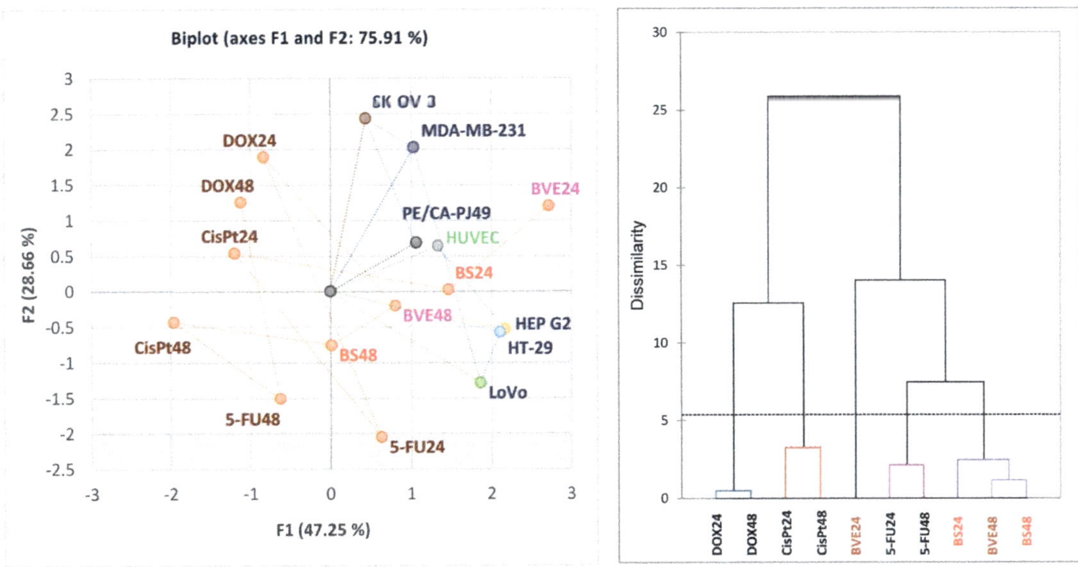

(**A**) (**B**)

Figure 6. (**A**). Symmetric biplot displaying the correlation between the cytotoxic effects of BVE, BS, and anticancer drugs on normal and tumor cells after 24 and 48 hr of treatment. (**B**). AHC-Dendrogram. HEP G2—human hepatocellular carcinoma; HT-29 and LoVo—human colon adenocarcinomas; MDA-MB-231—human breast adenocarcinoma; PE/CA-PJ49—human squamous tongue carcinoma; SK-OV-3—human ovary adenocarcinoma; BVE—*B. vulgaris* dry extract; BS —berberine sulfate hydrate; CisPt—Cisplatin; DOX—Doxorubicin, 5-FU—5-Fluorouracil; 24 and 48—treatment time (24 and 48 h).

The correlations between the bioactive phytoconstituents—total phenolic content (TPC) and total phenolic acid (TPA)—and their pharmacological potential are detailed in Table S3 in the Supplementary Materials. Their dual redox behavior could explain the antiproliferative effect on tumor cells, leading to decreasing viability; the prooxidant effect of phytochemicals is responsible for the BVE cytotoxicity. The antioxidant effect, measured by three methods, shows a substantial positive correlation with the TPC and TPA (r = 0.972–0.994, $p < 0.05$). Moreover, the variable parameters determined by all three methods (DPPH, FRAP, ABTS) are intercorrelated (r = 0.997–0.998, $p < 0.05$) and show a substantial negative correlation with the antiproliferative activity (r = −[0.951–0.999], $p < 0.05$). Similarly, the TPC and TPA display a significant negative correlation with the cell viability diminution (r = −[0.970–0.997], $p < 0.05$ (Table S3). The outstanding capacity of *B. vulgaris* for scavenging ABTS, hydroxyl radicals, and DPPH is due to berberine and phenolic compounds with dual redox behavior that can act synergistically in the extract [2]. Recent studies showed that galangin and berberine in a synergic combination might induce esophageal carcinoma cells' apoptosis through cell cycle arrest in the G2/M phase via oxidative stress [19]. Moreover, apigenin, gallic acid, and berberine have immunomodulatory potential and could be helpful as immune checkpoint inhibitors and fight cancers via multiple targets [20].

The accessed literature data regarding the cytotoxic effects of berberine evaluated in vitro in various tumor cells, potential mechanisms, and IC_{50} values are synthesized in Table 4.

Table 4. In vitro cytotoxicity of berberine in various tumor cell lines, based on the literature data.

Cancer Cell Line	Cytotoxic Responses	Berberine Concentration	IC_{50} Value	Reference
Liver cancer HEP G2 SMMC-7721 Bel-7402	- Decreases the cell viability in a time- and dose-dependent manner.	3.125, 6.25, 12.5, 25, 50 and 100 μM	HEP G2—34.5 μM, SMMC-7721—25.2 μM Bel-7402—53.6 μM	[21]
Ehrlich ascites carcinoma EAC	- Increases apoptotic cells (at 10 μg/mL); - Inhibits DNA synthesis; - Changes the morphology of dsDNA; - Induces cell death (at 50 and 100 μg/mL).	10, 50 and 100 μg/mL	<1 μg/mL	[22]
Dalton's lymphoma ascites DLA	- Induces cytotoxicity of 44% at a concentration of 1 mg/mL; - At lower concentrations, it caused dose-dependent cytotoxicity in DLA cells.	100–1000 mg/mL	NA	[23]
Breast cancer MCF-7 MDA-MB-231	Dose- and time-dependent inhibitory effects of cancer cell proliferation: - Increases apoptotic ratio; - Stimulates caspase-3 activity and alteration in cell morphology; - Increases ROS generation; - Induces overexpression of p53.	10–100 μM 10–100 μg/mL	NA MCF7—15.93 ug/mL	[24]
Ovarian cancer CsSki, SiHa, HeLa	- Inhibits the invasion of CsSki, HeLa, and SiHa cells in a dose-dependent manner; - Inhibits the migration of CsSki, SiHa, and HeLa cells; - Decreases the SiHa cell motility.	20 μM	NA	[25]
Prostate cancer LNCaP PC-82	- Decreases the cell viability and induces programmed necrosis and apoptosis in a dose-dependent manner.	1–100 μM	NA	[26]
Rat glioma C6	Cytotoxic effects occur in a time- and dose-dependent manner, as follows: - Alters the cell morphology; - Promotes the caspase-3, -8, and -9 activity; - Increases the production of ROS; - Induces apoptotic cell death.	100 μM	NA	[27]

Table 4. *Cont.*

Cancer Cell Line	Cytotoxic Responses	Berberine Concentration	IC_{50} Value	Reference
Colorectal carcinoma HCT116, SW480 LoVo	- In a concentration- and time-dependent manner, the cancer cell growth was inhibited via programmed death.	0–100 μM for 24–72 h	NA	[28]
Human prostate cancer LNCaP, PC-3	- Blocks growth and proliferation in cancer cells in a time- and concentration-dependent manner; - Induces apoptotic cell death.	0, 5, 10, 20, 50, and 100 μM	LNCap cells: 60 μM PC-3 cells: ≥100 μM	[26,29]
Lung cancer A549	- Did not show a cytotoxic effect on the A549 cells (up to 24 h); - Slight cytotoxicity was observed after 48 h of exposure (at 20 and 40 μM).	2.5–40 μM	NA	[30]
Human esophageal cancer YES-2	- Reduces cell viability and proliferation; - Inhibits production of interleukin 6; - All effects are dose- and time-dependent.	8–32 μM	NA	[31]
Oral cancer: OC2 KB	- Inhibits activator protein 1; - Exhibits anti-inflammatory effects by reducing the production of cyclooxygenase-2 (COX-2) and prostaglandin E2 (PGE2).	1, 10, and 100 μM (2–12 h)	NA	[32]
Human OSCC: HSC-2, HSC-3, HSC-4 Human Promyelocytic Leukemia: HL 60	- Increases apoptotic cells; - Induces DNA fragmentation; - Stimulates caspase-3, -8 and -9 and proapoptotic BAD protein; - In HSC-2 cells, BAD protein increase was not available.	10, 20 and 80 μM	18–136 μM	[32]
Mouse melanoma K1735-M2	- Inhibitory effect on cell proliferation is dose- and time-dependent; - 50% of growth inhibition was observed after 72 and 96 h of exposure.	0, 10, 25, 50, 75, and 100 μM	NA	[33]

NA—Not available.

In the present study, the BS IC_{50} against HEP G2 was slightly over 50 ug/mL, being around that registered in Table 4; the same was true for the colon carcinoma (LoVo), colon cancer (SK-OV-3) and tongue squamous cell carcinoma (PE/CA-PJ49) cell lines. Moreover, Table 4 indicates that the IC_{50} of BS against OSCC was 18–136 μM, and our value belongs to this range. For the MDA-MB-231 (breast cancer) cell line, the IC_{50} was > 25 μg/mL after 24 h and 12.5 after 24 h. Similar studies investigated the anticancer effects *of B. vulgaris* extract and berberine chloride in other cancer cell lines, evaluating the cell viability after 24, 48, and even 72 h and reporting various IC_{50} values [18,34,35]. In HEP G2 (liver cancer), Caco-2 (colon cancer), and MCF-7 (breast cancer), the IC_{50} values for barberry extract were 68.02 > 49.96 > 15.61 μg/mL, and for berberine chloride, lower values were recorded: 65.86 > 17.64 > 15.93 μg/mL [34]. After 48 h, the IC_{50} values drastically decreased: 5.55, 3.84, and 4.44 μg/mL for berberis extract vs. 11.49, 5.1, and 4.43 μg/mL for berberine chloride. Moreover, in HEP G2 and CaCo2, the antitumor activity of berberis extract was stronger than that of berberine chloride [34]. In our study, both BVE and BS had moderate cytotoxicity. Another research team analyzed the cytotoxicity of *B. vulgaris* extract in 70% ethanol on breast cancer cell lines (MCF-7) after 24, 48, and 72 h and obtained significantly higher IC50 values, respectively, 4000, 2000 and 1000 μg/mL [35]. Och et al. investigated the cytotoxic and proapoptotic properties of *B. thunbergii* extract and berberine on various hematopoietic cancer cell lines: acute promyelocytic leukemia (HL-60, HL-60/MX1, HL-60/MX2), myeloma (U266B1), acute lymphoblastic leukemia (CCRF/CEM and CEM/C1) and acute T cell leukemia (J45.01) [18]. After 24 h, the extract did not show cytotoxic effects in the tested cells, and the IC_{50} value of berberine was 80–250 μM [18]. However, tumor cells' exposure to a high concentration of *B. thunbergii* extract influenced the activity of proapoptotic genes (upregulation of B2M, downregulation of BAD and BNIP2, and increased expression of BAX, BAK1, BIK, and CASP9i) in all the leukemia cell lines [18]. These phenomena suggest the potential detection of cellular apoptosis after an exposure longer than 24 h, and further experiments in the 72 and 96 h models are requested [18].

3. Materials and Methods

3.1. Materials

3.1.1. Chemicals

All the chemicals were of analytical grade. Analytical standards of 31 compounds were purchased from Sigma-Aldrich, Schnelldorf, Germany. Methanol and ethyl alcohol, HPLC grade, were purchased from Merck, Bucharest, Romania; formic acid (98%) and ultrapure water (LC-MS grade) were also purchased from Merck (Merck Romania, Romania). The Pierce LTQ Velos ESI positive and negative ion calibration solutions (Thermo Fisher Scientific, Dreieich, Germany) calibrated the Orbitrap Mass Spectrometer.

The standard phenolic compounds (8 phenolic acids, 7 isoflavones, and 15 flavonoids), berberine sulfate hydrate, ethanol, sodium acetate, $AlCl_3$, DPPH, ABTS ammonium salt, trichloroacetic acid, phosphate buffer (pH = 6.6), ascorbic acid, $K_3(FeCN)_6$ and $FeCl_3$ were purchased from Sigma-Aldrich, Germany. Methanol and ethanol, potassium persulfate, formic acid (98%), and ultrapure water (LC-MS grade) were provided by Merck (Merck Romania SRL, Bucharest, Romania). The Pierce LTQ Velos ESI positive and negative ion calibration solutions (Thermo Fisher Scientific, Germany) calibrated the Orbitrap Mass Spectrometer.

In the in vitro studies in cell lines, various materials were used: Dulbecco's Modified Eagle's Medium (DMEM, PAN Biotech, Aidenbach, Germany), cell-washing medium HBSS (Hanks' Balanced Buffer Solution), 200 mM L-glutamine, fetal bovine serum (FBS), 100 mM ethylenediaminetetraacetic acid (EDTA), phosphate-buffered saline (TFS), dimethyl sulfoxide (DMSO, Sigma-Aldrich, St. Louis, MO, USA), antibiotic mixture (10,000 U/mL penicillin and 10,000 μg/mL streptomycin) (Biochrom GmbH, Berlin, Germany), and Trypan Blue and CellTiter 96® AQueous One Solution Cell Proliferation Assay (MTS) kit (Promega, Madison, WI, USA). The anticancer drugs (5-Fluorouracyl, Cisplatin, and Doxorubicin) and berberine sulfate hydrate were purchased from Sigma-Aldrich Chemie GmbH, Schnelldorf, Germany. Human adherent cancer cell lines of various histological origins were obtained from the international cell banks "European Collection of Authenticated Cell Cultures" (ECACC, Porton Down, Wiltshire, UK) or "American Type Culture Collection" (ATCC, Manassas, VA, USA) as follows: (a) MDA-MB-231 breast adenocarcinoma (ECACC, cat. no 92020424), (b) SK-OV-3 ovarian adenocarcinoma (ECACC, cat. no 91091004), (c) HEP G2 hepatocyte carcinoma (ECACC, cat. no 85011430), (d) PE/CA-PJ49 oral (tongue) squamous cell carcinoma (ECACC, cat. no 00060606), (e) HT-29 colon adenocarcinoma (ECACC, cat. no 91072201), and (f) LoVo colon adenocarcinoma (ATCC, cat. no CCL-229) [36]. As normal controls during the in vitro investigations, immortalized cells from human umbilical vein endothelial cells were used (adherent HUVEC cell line, kindly provided by Dr. Viviana Roman, Center of Immunology, "Stefan S. Nicolau" Institute of Virology, Bucharest, Romania).

The *Daphnia magna* Straus for the in vivo studies originated from a culture maintained parthenogenetically at the Department of Pharmaceutical Botany and Cell Biology, Faculty of Pharmacy, "Carol Davila" University of Medicine and Pharmacy Bucharest, since 2012.

3.1.2. *B. vulgaris* Extract Preparation

B. vulgaris (L.) cortex was harvested in March 2023 from a local ecological crop in Oratia—Lat/Long (in decimal degrees): 45.445199, −27.013190—Buzau County, Romania. It was identified by Prof. Octavian Tudorel Olaru, Department of Pharmaceutical Botany and Cell Biology, and Prof. Cerasela Elena Gîrd, Department of Pharmacognosy, Phytochemistry and Phytotherapy, Faculty of Pharmacy, "Carol Davila" University of Medicine and Pharmacy, Bucharest. The voucher specimen is also preserved in the Department of Pharmacognosy, Phytochemistry, and Phytotherapy collection. Morphological peculiarities: the vegetable product is presented as flat or slightly recurved fragments; the inner face shows a bright yellow–green fluorescence in UV light (due to berberine). Organoleptic characteristics include a brown–gray color on the outside and a golden-yellow on the inside (due to berberine), which becomes brown through preservation (Figure S4 in the Supplementary Materials), bitter taste, and no smell. As previously described, 50 g of powdered stem bark was subjected to reflux extraction with 50% ethanol (Sigma-Aldrich, Darmstadt, Germany) [37]. After filtration, the obtained extract (BVE) was concentrated in a rotary evaporator R100 with a vacuum pump V-700 (BUCHI Corporation, New Castle, DE, USA) and lyophilized (Christ Alpha 1-2/B Braun, BiotechInt, New Delhi, India).

3.2. *Total Polyphenol Content (TPC)*

The Folin–Ciocalteu reagent was used following a spectrophotometric method described extensively in a previously published article [38]. The absorbances were measured at 725 nm (Jasco V-530 spectrophotometer, JASCO, Tokyo, Japan), and tannic acid was the standard for the calibration curve in a linear concentration range of 2–9 μg/mL. The TPC is expressed as mg Eq tannic acid/100 g BVE.

3.3. *Total Phenolic Acid (TPA)*

The quantification method was based on the phenolic acids that form nitro derivatives with nitrous acids. Our previously published article detailed it [37]. The absorbance was immediately measured at 525 nm (Jasco spectrophotometer, Japan) and compared to a sample that lacked the Arnow reagent. Chlorogenic acid (Sigma-Aldrich, Germany) was used as a standard for the calibration curve in the linear range of 11–53 μg/mL, with $R^2 = 0.9998$. The total phenolic acid (TPA) content was expressed as mg chlorogenic acid equivalents per gram of extract (mg Eq chlorogenic acid/g BVE).

3.4. *Identification and Quantification of Phenolic Constituents and Berberine*

3.4.1. Ultra-High-Performance Liquid Chromatography Coupled with High-Resolution Mass Spectrometry (UHPLC–HRMS/MS)

The phenolic profile of BVE was established based on non-targeted tandem mass spectrometry (MS-MS) using the hyphenated technique represented by Ultra-High-Performance Liquid Chromatography (UHPLC) coupled with the Q-Exactive High-Resolution Mass Spectrometer (HRMS). The same method was used to quantify selected phenolic compounds for each available analytical standard (Sigma-Aldrich, Germany). Our previously published study describes all the detailed data [17].

3.4.2. High-Performance Liquid Chromatography

The separation was achieved on a reverse-phased analytical column (octadecylsilyl silica gel—C18, [25 × 0.4] mm i.d., 5 μm particle). The mobile phase consisted of a mixture of water and phosphoric acid = 0.1% (v/v) (solvent A) and 0.1% phosphoric acid in the acetonitrile (solvent B). The gradient used was as follows: 90%A/10%B, 0 min.; 90 → 78A/10 → 22/B, 0–15 min; 78 → 60A/28 → 40/B, 15–25 min.; 60 → 30%A/40 → 70%B, 15–40 min; 30 → 20%A/70 → 80%B, 40–55 min. The flow rate was 1.0 mL /min, associated with an injection volume of 20 μL and a monitoring wavelength of 330 nm.

3.5. *Antioxidant Activity*

3.5.1. Diphenyl-1-Picrylhydrazyl Free Radical Scavenging Assay (DPPH)

Under an antioxidant, the purple free radical 2,2-diphenyl-1-picrylhydrazyl (DPPH) formed its corresponding yellow hydrazine. The absorbance value was measured at $\lambda = 515$ nm. The IC_{50} value was determined from the inhibition curves and their linear equations [39].

3.5.2. Azinobis-3-Ethylbenzotiazoline-6-Sulfonic Acid Assay (ABTS)

The turquoise-colored ABTS radical resulted from a potent oxidizing agent (potassium persulfate) reaction with the ammonium salt of 2,2′-azino-bis(3-ethylbenzothiazoline-6-sulfonic acid). Under the action of the antioxidant, the intensity of the color was reduced to colorless. The absorbance was determined at λ = 734 nm, and the IC_{50} value was calculated from the inhibition curves and their linear equations [37].

3.5.3. Ferric-Reducing Antioxidant Power Assay (FRAP)

The antioxidant analyte reacted with Fe^{3+}, reducing to Fe^{2+}, and imprinting blue. The coloration intensity was directly proportional to the antioxidant activity. The absorbance values were measured at λ = 700 nm (spectrophotometer Jasco V-530) and compared to the control (prepared under the same conditions without sample solution). It was expressed as an EC_{50} value; it represented the sample concentration at which the absorbance had a value of 0.5 or half the concentration at which the antioxidant activity was at a maximum, as determined by the trendline equation [40].

3.6. 48-h Acute Toxicity Test Using Daphnia Magna and Daphnia Pulex

The daphnids belonging to the species *Daphnia magna* and *Daphnia pulex* were chosen based on their size from parthenogenetic cultures maintained in an artificial medium for 24 h before testing [41,42]. The assay was performed in 24-well culture plates (Greiner Bio-One, Kremsmünster, Austria), with each well containing around 10 organisms. The samples were tested in six concentrations, ranging from 3.125 µg/mL to 100 µg/mL for BVE; as a positive control, BS was used from 0.625 to 20.0 µg/mL. The tests were duplicated, and lethality was assessed at 24 and 48 h. The 50% lethal concentrations (LC_{50}) and the 95% confidence interval (CI95%) of the LC_{50} values were determined using GraphPad Prism v 5.1.2008 software (GraphPad Software, Boston, MA, USA) [17].

3.7. Daphnia Magna Embryonic Development Assay

The following concentrations were chosen for testing: BVE at 3.125 µg/mL and BS at 2.5 µg/mL, based on the results obtained in the viability test. The embryos were exposed to the sample solutions in the dark, maintaining a constant temperature and humidity of 25 °C and 75% RH, respectively. The experiments were carried out on culture plates with 48 wells (Greiner Bio-One, Kremsmünster, Austria). Every 24 h, the embryos were examined at a magnification of 80× under a microscope (bScope® microscope, Euromex Microscope BV, Arnhem, The Netherlands) to assess the developmental stages and detect abnormalities compared to the untreated control [17].

3.8. In Vitro Anticancer Activity

3.8.1. Cell Cultures and Treatments

The antiproliferative effect of the BVE hydro-ethanolic extract and BS standard was evaluated in vitro in six tumor cell lines (SK-OV-3, LoVo, HEP-G2, HT-29, MDA-MB-231, PE/CA-PJ49), with normal HUVEC cell, used as the control. All the cell lines were cultured in DMEM/F12 medium enriched with 2 mM L-glutamine and 10% fetal calf serum and antibiotics mixture (100 U/mL penicillin and 100 µg/mL streptomycin). They were incubated at 37 °C in a 5% CO_2 humidified atmosphere. For the cytotoxicity assays, the cells were detached from the culture flasks and then cultivated in 96-well flat-bottom plates for 24 h until they reached around 70% confluence. Then, the cells were treated for various periods (24 h and 48 h) with different concentrations of BVE, BS, or oncolytic drugs (5-FU, CisPt, DOX) used as positive controls [43]. The BVE and BS stock solutions were prepared by dissolving them in a minimal amount of DMSO and preserved at 4 °C; all the working solutions were prepared from the stocks by serial dilutions with culture medium before each treatment assay [17].

3.8.2. MTS Assay

The cytotoxic potential of BVE and BS was evaluated by a colorimetric cell viability method, the MTS assay, and it was assessed in both tumor and normal cells and compared with the action of oncolytic drugs: DOX, CisPt, and 5-FU [44].

All the assays were performed in triplicate using the CellTiter 96® AQueous One Solution Cell Proliferation Assay (MTS) kit (Promega, USA). It contains a reagent mixture of two components: MTS [3-(4,5 dimethylthiazol 2 yl) 5 (3 carboxymethoxy phenyl) 2 (4 sulfophenyl) 2H tetrazolium] and PES (phenazine ethosulphate), a cationic dye with high chemical stability, which may be combined with MTS to form a stable solution [45]. The method's principle is based on the ability of metabolically active cells to reduce MTS (a yellow tetrazolium salt) to the colored formazan, which is soluble in the culture medium and can be spectrophotometrically quantified at a 492 nm wavelength. Briefly, 1.5×10^4 cells/well were cultured in 100 µL of the medium; after 24 h, the culture supernatants were discarded, and the cells were treated with increasing concentrations of BVE, BS, or reference drug solutions for 24 h or 48 h. At the expiration of the contact time, 20 µL of reagent mixture was added to each well, and the culture dishes were incubated for an additional 4 h at 37 °C, with gentle shaking every 20 min. Absorbance was read at λ = 492 nm with the Dynex ELISA reader (DYNEX Technologies—MRS, Chantilly, VA, USA) [46].

The cell viability was expressed as a percentage, was compared to the untreated cells (considered 100% viable), and was calculated according to the following formula:

$$\text{Cell viability } \% = 100 \times \frac{T - B}{U - B} \quad (1)$$

where T = optical density of treated cells, B = optical density of the blank (culture medium, in the absence of cells), and U = optical density of untreated cells.

The obtained results were expressed as the mean values from three different experiments (n = 3) ± standard deviation (SD) [47]. For the assessment of the DMSO cytotoxicity, the same experimental determinations were performed as in the MTS assay, and no impairment of cell viability was observed at concentrations lower than 1%.. Also, to observe the possible nonspecific reactions between BVE, BS, or drugs and MTS, their absorbance was determined without cells, and the values were extracted during the calculations.

3.9. Data Analysis

The statistically significant differences (at $\alpha < 0.05$) between the various experimental groups were established by multiple pairwise comparisons using Dunn's procedure from XLSTAT 2023.1.4. by Lumivero (Denver, CO, USA) [48].

The correlations between the bioactive constituents of the extracts and their antioxidant activity and cytotoxicity were determined using Principal Component Analysis [49] performed with XLSTAT 2023.1.4. by Lumivero (Denver, CO, USA) through Pearson correlation. A probability value $p < 0.05$ indicated a statistically significant difference [50].

4. Conclusions

This research investigated the autochthonous Berberidis cortex, obtaining a dry extract in 50% ethanol through successive reflux extraction, then solvent evaporation and freeze-drying. Through complex UHPLC–HRMS/MS and HPLC-DAD analysis of BVE, 40 phenolic constituents, including berberine, were identified. The main classes of phenolic metabolites (polyphenols and flavonoids) and bioactive representatives were also quantified. BVE's significant antioxidant potential was revealed by in vitro evaluation of the radical scavenging ability and reducing power. Then, the acute toxicity tests highlighted BVE's significant acute toxicity and teratogenicity in *Daphnia* sp. It also displayed moderate antiproliferative activity in various tumor cell lines and did not affect normal cells. Compared to BVE, berberine showed higher toxicity. It is essential to show that berberine sulfate reduced the viability in several tumor cell lines more than the standard anticancer drugs used as positive controls.

Strong and statistically significant correlations were recorded between the exposure time, concentration, phenolic metabolites content, antioxidant activity, and cytotoxicity of *B. vulgaris* stem bark dry hydro-ethanolic extract.

Our results could enrich the scientific database regarding the composition and pharmacological properties of autochthonous Berberidis cortex. Further research could explore the acute toxicity and teratogenicity of BVE and berberine using other animal models and investigate their anticancer activity mechanisms in various other tumor cell lines.

Supplementary Materials: The following supporting information can be downloaded at https://www.mdpi.com/article/10.3390/molecules29092053/s1, File S1, which includes Figure S1: Standard calibration curves for TPC (left) and TPA (right); Figure S2; The viability% (V%) of normal and tumor cells exposed to BVE (a,b) and BS (c,d) various concentrations for 24 h (a,c) and 48 h (b,d). Bonferro-ni-corrected significance level at $\alpha = 0.05$ is 0.0024; * p-value < 0.024 corresponds to $\alpha < 0.05$ and shows statistically significant differences. HUVEC—human umbilical endothelial cell; HUV—HUVEC—human umbilical endothelial cell; HEP—HEP G2—human hepatocellular carcinoma; HT—HT-29 and LO—LoVo—human colon adenocarcinomas; MDA—MDA-MB-231—human breast adenocarcinoma; PE—PE/CA-PJ49—human squamous tongue carcinoma; SK—SK-OV-3—human ovary adenocarcinoma; 24 and 48—period of exposure (hours); c—control (BS); BVE—*B. vulgaris* dry hydro-ethanol extract; BS—Berberine standard; Figure S3: Comparison between BVE and BS cytotoxicity (quantified as cell viability %) after 24 h (a) and 48 h (b) of contact; Figure S4: Berberidis cortex, Table S1: The antiproliferative effects of positive controls on normal cell and tumor cell lines after 24 and 48 h of exposure; Table S2: Pearson correlation between the cytotoxic effects of BVE, BS, and anticancer drugs (expressed as cell viability %) after 24 and 48 h of treatment on tested cell lines; and Table S3: Pearson correlation between BVE secondary metabolites content, antioxidant activity, and cytotoxicity expressed as cell viability % after 24 and 48 h of treatment.

Author Contributions: Conceptualization, I.M.I., V.P. and C.E.G.; methodology, C.L.C., L.P., E.A.L., E.I.I., L.I.B., C.M.H., O.T.O., G.M.N., R.B. and C.E.G.; software, V.P., C.L.C., L.P., E.A.L., E.I.I., L.I.B., C.M.H. and R.B.; validation, C.L.C., L.I.B., O.T.O., G.M.N. and R.B.; formal analysis, V.P., C.L.C., L.P. and R.B.; investigation, I.M.I. and E.I.I.; resources, I.M.I.; data curation, I.M.I., V.P., O.T.O. and G.M.N.; writing—original draft preparation, V.P., C.L.C., L.P., E.A.L., E.I.I., L.I.B., C.M.H., O.T.O. and G.M.N.; writing—review and editing, V.P., C.L.C., L.P., E.A.L., E.I.I., L.I.B., C.M.H., O.T.O., G.M.N., R.B. and C.E.G.; visualization, I.M.I., V.P., C.L.C., L.P., E.A.L., E.I.I., L.I.B., C.M.H., O.T.O., G.M.N., R.B. and C.E.G.; supervision, C.E.G.; project administration, C.E.G.; funding acquisition, I.M.I., V.P. and C.E.G. All authors have read and agreed to the published version of the manuscript.

Funding: Publication of this paper was supported by the University of Medicine and Pharmacy "Carol Davila" through the institutional program "Publish not Perish".

Institutional Review Board Statement: Not applicable.

Informed Consent Statement: Not applicable.

Data Availability Statement: Data are available in the manuscript and Supplementary Materials.

Conflicts of Interest: The authors declare no conflicts of interest.

References

1. Mana, T.; Devi, O.B.; Singh, Y.D. Therapeutic Application of Berberine: A Consolidated Review. *Curr. Pharmacol. Rep.* **2023**, *9*, 329–340. [CrossRef]
2. Salehi, B.; Selamoglu, Z.; Sener, B.; Kilic, M.; Kumar Jugran, A.; de Tommasi, N.; Sinisgalli, C.; Milella, L.; Rajkovic, J.; Flaviana, B.; et al. Berberis Plants—Drifting from Farm to Food Applications, Phytotherapy, and Phytopharmacology. *Foods* **2019**, *8*, 522. [CrossRef]
3. Homśopathic Materia Medica, Berberis Vulgaris (Barberry). Available online: http://homeoint.org/books/boericmm/b/berb.htm (accessed on 26 April 2024).
4. Han, Y.; Xiang, Y.; Shi, Y.; Tang, X.; Pan, L.; Gao, J.; Bi, R.; Lai, X. Pharmacokinetics and Pharmacological Activities of Berberine in Diabetes Mellitus Treatment. *Evid.-Based Complement. Altern. Med.* **2021**, *2021*, 9987097. [CrossRef]
5. Grădinariu, L.; Dediu, L.; Crețu, M.; Grecu, I.R.; Docan, A.; Istrati, D.I.; Dima, F.M.; Stroe, M.D.; Vizireanu, C. The Antioxidant and Hepatoprotective Potential of Berberine and Silymarin on Acetaminophen Induced Toxicity in *Cyprinus carpio* L. *Animals* **2024**, *14*, 373. [CrossRef]

6. Almatroodi, S.A.; Alsahli, M.A.; Rahmani, A.H. Berberine: An Important Emphasis on Its Anticancer Effects through Modulation of Various Cell Signaling Pathways. *Molecules* **2022**, *27*, 5889. [CrossRef]
7. El-Zahar, K.M.; Al-Jamaan, M.E.; Al-Mutairi, F.R.; Al-Hudiab, A.M.; Al-Einzi, M.S.; Mohamed, A.A.-Z. Antioxidant, Antibacterial, and Antifungal Activities of the Ethanolic Extract Obtained from *Berberis vulgaris* Roots and Leaves. *Molecules* **2022**, *27*, 6114. [CrossRef]
8. Och, A.; Olech, M.; Bąk, K.; Kanak, S.; Cwener, A.; Cieśla, M.; Nowak, R. Evaluation of the Antioxidant and Anti-Lipoxygenase Activity of *Berberis vulgaris* L. Leaves, Fruits, and Stem and Their LC MS/MS Polyphenolic Profile. *Antioxidants* **2023**, *12*, 1467. [CrossRef]
9. Boeri, P.; Piñuel, L.; Dalzotto, D.; Monasterio, R.; Fontana, A.; Sharry, S.; Barrio, D.A.; Carrillo, W. Argentine Patagonia Barberry Chemical Composition and Evaluation of Its Antioxidant Capacity. *J. Food Biochem.* **2020**, *44*, e13254. [CrossRef]
10. Li, Y.; Lv, X.-M.; Tang, C.; Lai, X.-R.; Zhang, Y.; Fan, G. Quality Evaluation of Cortex Berberidis from Different Geographical Origins by Simultaneous High Performance Liquid Chromatography Combined with Statistical Methods. *Trop. J. Pharm. Res.* **2016**, *15*, 1973. [CrossRef]
11. Fernández-Poyatos, M.D.P.; Ruiz-Medina, A.; Zengin, G.; Llorent-Martínez, E.J. Phenolic Characterization, Antioxidant Activity, and Enzyme Inhibitory Properties of *Berberis thunbergii* DC. Leaves: A Valuable Source of Phenolic Acids. *Molecules* **2019**, *24*, 4171. [CrossRef]
12. Veselá, Š.; Ondruška, V.; Kuča, K.; Patočka, J. Freshwater Microcrustacean Daphnia Magna Straus as an Early Screen Model to Compare Toxicity of Acetylcholinesterase Inhibitors. *J. Appl. Biomed.* **2006**, *4*, 105–110. [CrossRef]
13. Paudel, B.; Bhattarai, H.D.; Kim, I.C.; Lee, H.; Sofronov, R.; Ivanova, L.; Poryadina, L.; Yim, J.H. Estimation of Antioxidant, Antimicrobial Activity and Brine Shrimp Toxicity of Plants Collected from Oymyakon Region of the Republic of Sakha (Yakutia), Russia. *Biol. Res.* **2014**, *47*, 10. [CrossRef] [PubMed]
14. Martini, D.; Pucci, C.; Gabellini, C.; Pellegrino, M.; Andreazzoli, M. Exposure to the Natural Alkaloid Berberine Affects Cardiovascular System Morphogenesis and Functionality during Zebrafish Development. *Sci. Rep.* **2020**, *10*, 17358. [CrossRef] [PubMed]
15. The World Health Organisation. WHO Monographs on Selected Medicinal Plants Volume 4. Available online: https://iris.who.int/bitstream/handle/10665/42052/9789241547055_eng.pdf;jsessionid=A4510414DC6FA90BC17E4906B7CC489F?sequence=4 (accessed on 26 April 2024).
16. Hidayat, D.; Dwira, S. Phytochemical Analysis and In Vitro Cytotoxicity Test of Black Soybean (*Glycine soja* L.) Ethanolic Extract as a Growth Inhibitor of the HCT-116 Colon Carcinoma Cell Line. *J. Phys. Conf. Ser.* **2018**, *1073*, 032041. [CrossRef]
17. Ivan, I.M.; Popovici, V.; Chițescu, C.L.; Popescu, L.; Luță, E.A.; Ilie, E.I.; Brașoveanu, L.I.; Hotnog, C.M.; Olaru, O.T.; Nițulescu, G.M.; et al. Phytochemical Profile, Antioxidant and Cytotoxic Potential of *Capsicum annuum* (L.) Dry Hydro-Ethanolic Extract. *Pharmaceutics* **2024**, *16*, 245. [CrossRef] [PubMed]
18. Och, A.; Zalewski, D.; Komsta, Ł.; Kołodziej, P.; Kocki, J.; Bogucka-Kocka, A. Cytotoxic and Proapoptotic Activity of Sanguinarine, Berberine, and Extracts of *Chelidonium majus* L. and *Berberis thunbergii* DC. toward Hematopoietic Cancer Cell Lines. *Toxins* **2019**, *11*, 485. [CrossRef] [PubMed]
19. Ren, K.; Zhang, W.; Wu, G.; Ren, J.; Lu, H.; Li, Z.; Han, X. Synergistic Anticancer Effects of Galangin and Berberine through Apoptosis Induction and Proliferation Inhibition in Oesophageal Carcinoma Cells. *Biomed. Pharmacother.* **2016**, *84*, 1748–1759. [CrossRef] [PubMed]
20. Zhong, Z.; Vong, C.T.; Chen, F.; Tan, H.; Zhang, C.; Wang, N.; Cui, L.; Wang, Y.; Feng, Y. Immunomodulatory Potential of Natural Products from Herbal Medicines as Immune Checkpoints Inhibitors: Helping to Fight against Cancer via Multiple Targets. *Med. Res. Rev.* **2022**, *42*, 1246–1279. [CrossRef]
21. Yang, X.; Huang, N. Berberine Induces Selective Apoptosis through the AMPK-Mediated Mitochondrial/Caspase Pathway in Hepatocellular Carcinoma. *Mol. Med. Rep.* **2013**, *8*, 505–510. [CrossRef]
22. Letašiová, S.; Jantová, S.; Miko, M.; Ovádeková, R.; Horváthová, M. Effect of Berberine on Proliferation, Biosynthesis of Macromolecules, Cell Cycle and Induction of Intercalation with DNA, DsDNA Damage and Apoptosis in Ehrlich Ascites Carcinoma Cells. *J. Pharm. Pharmacol.* **2010**, *58*, 263–270. [CrossRef]
23. Anis, K.V.; Kuttan, G.; Kuttan, R. Role of Berberine as an Adjuvant Response Modifier during Tumour Therapy in Mice. *Pharm. Pharmacol. Commun.* **1999**, *5*, 697–700. [CrossRef]
24. Refaat, A.; Abdelhamed, S.; Yagita, H.; Inoue, H.; Yokoyama, S.; Hayakawa, Y.; Saiki, I. Berberine Enhances Tumor Necrosis Factor-Related Apoptosis-Inducing Ligand-Mediated Apoptosis in Breast Cancer. *Oncol. Lett.* **2013**, *6*, 840–844. [CrossRef]
25. Chu, S.-C.; Yu, C.-C.; Hsu, L.-S.; Chen, K.-S.; Su, M.-Y.; Chen, P.-N. Berberine Reverses Epithelial-to-Mesenchymal Transition and Inhibits Metastasis and Tumor-Induced Angiogenesis in Human Cervical Cancer Cells. *Mol. Pharmacol.* **2014**, *86*, 609–623. [CrossRef]
26. Choi, M.S.; Oh, J.H.; Kim, S.M.; Jung, H.Y.; Yoo, H.S.; Lee, Y.M.; Moon, D.C.; Han, S.B.; Hong, J.T. Berberine Inhibits P53-Dependent Cell Growth through Induction of Apoptosis of Prostate Cancer Cells. *Int. J. Oncol.* **2009**, *34*, 1221–1230. [CrossRef]
27. Chen, T.C.; Lai, K.C.; Yang, J.S.; Liao, C.L.; Hsia, T.C.; Chen, G.W.; Lin, J.J.; Lin, H.J.; Chiu, T.H.; Tang, Y.J.; et al. Involvement of Reactive Oxygen Species and Caspase-Dependent Pathway in Berberine-Induced Cell Cycle Arrest and Apoptosis in C6 Rat Glioma Cells. *Int. J. Oncol.* **2009**, *34*, 1681–1690. [CrossRef]

28. Zhao, Y.; Roy, S.; Wang, C.; Goel, A. A Combined Treatment with Berberine and Andrographis Exhibits Enhanced Anticancer Activity through Suppression of DNA Replication in Colorectal Cancer. *Pharmaceuticals* **2022**, *15*, 262. [CrossRef]
29. Zhang, L.; Wu, Y.; Gao, X.; Guo, F. Mitochondrial Protein Cyclophilin-D-Mediated Programmed Necrosis Attributes to Berberine-Induced Cytotoxicity in Cultured Prostate Cancer Cells. *Biochem. Biophys. Res. Commun.* **2014**, *450*, 697–703. [CrossRef]
30. Peng, P.; Kuo, W.-H.; Tseng, H.-C.; Chou, F.-P. Synergistic Tumor-Killing Effect of Radiation and Berberine Combined Treatment in Lung Cancer: The Contribution of Autophagic Cell Death. *Int. J. Radiat. Oncol. Biol. Phys.* **2008**, *70*, 529–542. [CrossRef]
31. Iizuka, N.; Miyamoto, K.; Hazama, S.; Yoshino, S.; Yoshimura, K.; Okita, K.; Fukumoto, T.; Yamamoto, S.; Tangoku, A.; Oka, M. Anticachectic Effects of Coptidis Rhizoma, an Anti-Inflammatory Herb, on Esophageal Cancer Cells That Produce Interleukin 6. *Cancer Lett.* **2000**, *158*, 35–41. [CrossRef]
32. Kuo, C.L.; Chi, C.W.; Liu, T.Y. The Anti-Inflammatory Potential of Berberine In Vitro and In Vivo. *Cancer Lett.* **2004**, *203*, 127–137. [CrossRef]
33. Pereira, G.C.; Branco, A.F.; Matos, J.A.C.; Pereira, S.L.; Parke, D.; Perkins, E.L.; Serafim, T.L.; Sardão, V.A.; Santos, M.S.; Moreno, A.J.M.; et al. Mitochondrially Targeted Effects of Berberine [Natural Yellow 18, 5,6-Dihydro-9,10-Dimethoxybenzo(*g*)-1,3-Benzodioxolo(5,6-*a*) Quinolizinium] on K1735-M2 Mouse Melanoma Cells: Comparison with Direct Effects on Isolated Mitochondrial Fractions. *J. Pharmacol. Exp. Ther.* **2007**, *323*, 636–649. [CrossRef]
34. Abd El-Wahab, A.E.; Ghareeb, D.A.; Sarhan, E.E.; Abu-Serie, M.M.; El Demellawy, M.A. In Vitro Biological Assessment of *Berberis vulgaris* and Its Active Constituent, Berberine: Antioxidants, Anti-Acetylcholinesterase, Anti-Diabetic and Anticancer Effects. *BMC Complement. Altern. Med.* **2013**, *13*, 218. [CrossRef]
35. Hoshyar, R.; Mahboob, Z.; Zarban, A. The Antioxidant and Chemical Properties of *Berberis vulgaris* and Its Cytotoxic Effect on Human Breast Carcinoma Cells. *Cytotechnology* **2016**, *68*, 1207–1213. [CrossRef]
36. Mihaila, M.; Hotnog, C.M.; Bostan, M.; Munteanu, A.C.; Vacaroiu, I.A.; Brasoveanu, L.I.; Uivarosi, V. Anticancer Activity of Some Ruthenium(III) Complexes with Quinolone Antibiotics: In Vitro Cytotoxicity, Cell Cycle Modulation, and Apoptosis-Inducing Properties in LoVo Colon Cancer Cell Line. *Appl. Sci.* **2021**, *11*, 8594. [CrossRef]
37. Costea, L.; Chitescu, C.L.; Boscencu, R.; Ghica, M.; Lupuliasa, D.; Mihai, D.P.; Deculescu-Ionita, T.; Dutu, L.E.; Popescu, M.L.; Luta, E.-A.; et al. The Polyphenolic Profile and Antioxidant Activity of Five Vegetal Extracts with Hepatoprotective Potential. *Plants* **2022**, *11*, 1680. [CrossRef]
38. Ungureanu, A.R.; Chițescu, C.L.; Luță, E.A.; Moroșan, A.; Mihaiescu, D.E.; Mihai, D.P.; Costea, L.; Ozon, E.A.; Fița, A.C.; Balaci, T.D.; et al. Outlook on Chronic Venous Disease Treatment: Phytochemical Screening, In Vitro Antioxidant Activity and In Silico Studies for Three Vegetal Extracts. *Molecules* **2023**, *28*, 3668. [CrossRef]
39. Luță, E.-A.; Biță, A.; Moroșan, A.; Mihaiescu, D.E.; Mihai, D.P.; Popescu, L.; Bejenaru, L.E.; Bejenaru, C.; Popovici, V.; Olaru, O.T.; et al. Implications of the Cultivation of Rosemary and Thyme (*Lamiaceae*) in Plant Communities for the Development of Antioxidant Therapies. *Int. J. Mol. Sci.* **2023**, *24*, 11670. [CrossRef]
40. Luță, E.A.; Biță, A.; Moroșan, A.; Mihaiescu, D.E.; Ghica, M.; Mihai, D.P.; Olaru, O.T.; Deculescu-Ioniță, T.; Duțu, L.E.; Popescu, M.L.; et al. The Influence of Phytosociological Cultivation and Fertilization on Polyphenolic Content of Menthae and Melissae Folium and Evaluation of Antioxidant Properties through In Vitro and In Silico Methods. *Plants* **2022**, *11*, 2398. [CrossRef]
41. Zanfirescu, A.; Nitulescu, G.; Stancov, G.; Radulescu, D.; Trif, C.; Nitulescu, G.M.; Negres, S.; Olaru, O.T. Evaluation of Topical Anti-Inflammatory Effects of a Gel Formulation with Plantago Lanceolata, Achillea Millefolium, Aesculus Hippocastanum and Taxodium Distichum. *Sci. Pharm.* **2020**, *88*, 26. [CrossRef]
42. Ionus, E.; Schröder, V.; Chiţescu, C.L.; Bucur, L.A.; Lupu, C.E.; Dumitrescu, D.-E.; Popescu, L.; Mihai, D.P.; Olaru, O.T.; Nițulescu, G.M.; et al. Phytochemical, In Vitro, In Vivo, and In Silico Research on the Extract of *Ajuga chamaepitys* (L.) Schreb. *Plants* **2024**, *13*, 1192. [CrossRef]
43. Ivan, B.-C.; Barbuceanu, S.-F.; Hotnog, C.M.; Olaru, O.T.; Anghel, A.I.; Ancuceanu, R.V.; Mihaila, M.A.; Brasoveanu, L.I.; Shova, S.; Draghici, C.; et al. Synthesis, Characterization and Cytotoxic Evaluation of New Pyrrolo[1,2-*b*]Pyridazines Obtained via Mesoionic Oxazolo-Pyridazinones. *Int. J. Mol. Sci.* **2023**, *24*, 11642. [CrossRef]
44. Munteanu, A.; Badea, M.; Olar, R.; Silvestro, L.; Mihaila, M.; Brasoveanu, L.I.; Musat, M.G.; Andries, A.; Uivarosi, V. Cytotoxicity Studies, DNA Interaction and Protein Binding of New Al (III), Ga (III) and In (III) Complexes with 5-hydroxyflavone. *Appl. Organomet. Chem.* **2018**, *32*, e4579. [CrossRef]
45. Munteanu, A.; Musat, M.G.; Mihaila, M.; Badea, M.; Olar, R.; Nitulescu, G.M.; Rădulescu, F.Ș.; Brasoveanu, L.I.; Uivarosi, V. New Heteroleptic Lanthanide Complexes as Multimodal Drugs: Cytotoxicity Studies, Apoptosis, Cell Cycle Analysis, DNA Interactions, and Protein Binding. *Appl. Organomet. Chem.* **2021**, *35*, e6062. [CrossRef]
46. Ivan, B.-C.; Barbuceanu, S.-F.; Hotnog, C.M.; Anghel, A.I.; Ancuceanu, R.V.; Mihaila, M.A.; Brasoveanu, L.I.; Shova, S.; Draghici, C.; Olaru, O.T.; et al. New Pyrrole Derivatives as Promising Biological Agents: Design, Synthesis, Characterization, In Silico, and Cytotoxicity Evaluation. *Int. J. Mol. Sci.* **2022**, *23*, 8854. [CrossRef]
47. Maciuca, A.-M.; Munteanu, A.-C.; Mihaila, M.; Badea, M.; Olar, R.; Nitulescu, G.M.; Munteanu, C.V.A.; Bostan, M.; Uivarosi, V. Rare-Earth Metal Complexes of the Antibacterial Drug Oxolinic Acid: Synthesis, Characterization, DNA/Protein Binding and Cytotoxicity Studies. *Molecules* **2020**, *25*, 5418. [CrossRef]
48. Ungureanu, A.R.; Popovici, V.; Oprean, C.; Danciu, C.; Schröder, V.; Olaru, O.T.; Mihai, D.P.; Popescu, L.; Luță, E.-A.; Chițescu, C.L.; et al. Cytotoxicity Analysis and In Silico Studies of Three Plant Extracts with Potential Application in Treatment of Endothelial Dysfunction. *Pharmaceutics* **2023**, *15*, 2125. [CrossRef]

49. Neagu, R.; Popovici, V.; Ionescu, L.E.; Ordeanu, V.; Popescu, D.M.; Ozon, E.A.; Gîrd, C.E. Antibacterial and Antibiofilm Effects of Different Samples of Five Commercially Available Essential Oils. *Antibiotics* **2023**, *12*, 1191. [CrossRef]
50. Popovici, V.; Matei, E.; Cozaru, G.C.; Bucur, L.; Gîrd, C.E.; Schröder, V.; Ozon, E.A.; Karampelas, O.; Musuc, A.M.; Atkinson, I.; et al. Evaluation of *Usnea barbata* (L.) Weber ex F.H. Wigg Extract in Canola Oil Loaded in Bioadhesive Oral Films for Potential Applications in Oral Cavity Infections and Malignancy. *Antioxidants* **2022**, *11*, 1601. [CrossRef]

molecules

MDPI

Article

Antitumor Mechanism and Therapeutic Potential of Cordycepin Derivatives

Linlin Cui [1,2], Li Zhao [1,2], Guanghuan Shen [1,2], Dahai Yu [1,2], Tian Yuan [1,2], Yingyu Zhang [1,2] and Bo Yang [1,2,*]

1 College of Pharmacy, Harbin University of Commerce, Harbin 150076, China; 102638@hrbcu.edu.cn (L.C.); zhaol202312@163.com (L.Z.); shengh@hrbcu.edu.cn (G.S.); hrbadr@163.com (D.Y.); 17745602736@163.com (T.Y.); zyy2433280805@163.com (Y.Z.)

2 Heilongjiang Provincial Key Laboratory of Drug Prevention and Treatment for Senile Diseases, Harbin 150076, China

* Correspondence: 102117@hrbcu.edu.cn

Abstract: Cordycepin has good antitumor activity, but its clinical application is limited due to the easy deamination of N6 in structure. In this study, a large lipolysis group was introduced at the cordycepin N6 to improve the problem, cordycepin derivatives (**3a–4c**) were synthesized, and biological evaluation of compounds was studied. In this study, the vitro antitumor activity of the compounds against MCF7 cells, HepG2 cells and SGC-7901 cells was evaluated by MTT assay. In the results, compound **4a** showed the most obvious inhibitory effect on MCF7 cells with an IC_{50} value of 27.57 ± 0.52 μM, which was much lower than cordycepin. Compound **4a** showed high selectivity between MCF7 and normal MCF-10A cells. Further biological evaluation showed that compound **4a** promoted apoptosis and blocked the cell cycle in the G0/G1 phase. Then, Western Blot was used to detect related apoptotic proteins. It was found that Compound **4a** could down-regulate the expression of Bcl-2 protein and up-regulate the expression of p53, Bax, Caspase-3 and Caspase-9 proteins. The mitochondrial membrane potential decreased continuously and the positive expression rate decreased. It was speculated that compound **4a** induced the apoptosis of MCF7 cells through the mitochondrial pathway.

Keywords: cordycepin derivatives; antitumor; structural modification

1. Introduction

According to estimates from the World Health Organization, cancer incidence and mortality are rapidly increasing worldwide. Cancer is expected to rank as the leading cause of death in the 21st century. Despite the advances in cancer clinical research, a high percentage of oncological patients will suffer from overwhelming morbi-mortality.

Cordyceps militaris has been used in traditional Chinese medicine for a long time [1]. *C. militaris* exhibits a variety of clinical health effects including immunomodulatory, anticancer, antioxidant, anti-inflammatory and anti-microbial activities [2,3], and is applied as a functional ingredient in health foods and cosmetics [4,5]. In recent years, traditional Chinese medicine has been used to treat various cancers. Cordycepin is a nucleoside antibiotic, isolated and purified from C. militaris. Cordycepin has anti-inflammatory [6,7], antitumor [8–10], antiviral and immunomodulatory effects [11–14], and plays an important role in antitumor activity. Cancer is a major public health problem worldwide. It is the second leading cause of death after cardiovascular disease and the most important obstacle to further improving life expectancy. There is clear evidence that cordycepin inhibits breast cancer (caspase, MAPK), liver cancer (caspase, PLC) and stomach cancer (PI3K/Akt) through a variety of signaling pathways [15–17]. Studies have shown the antitumor mechanism of cordycepin; that is, cordycepin is phosphorylated after entering cells, and it is a monophosphate derivative, a diphosphate derivative or a triphosphate derivative. Triphosphate derivatives can inhibit mRNA synthesis and ultimately affect the synthesis of corresponding proteins, thus inhibiting tumor growth. At the same time, studies have shown that after cordycepin enters the body, rapid deamination occurs under the catalysis of adenosine deaminase (ADA) to produce the inactive metabolite 3′-deoxyinosine. This can lead to a short half-life of cordycepin in the body, and influence clinical treatment. If structural modification can effectively improve its susceptibility to deamination and short half-life, and effectively improve its bioavailability, it will become a feasible way to develop cordycepin drugs. Many studies on the modification of cordycepin mainly focus on the modification of hydroxyl and N_6 positions. In this study, the hydroxyl group in cordycepin was modified, and heterocyclic rings were introduced in the N_6 position. A series of cordycepin derivatives were designed according to the principle of bioelectron equiplatoon, and biological activities and mechanisms were evaluated through related experiments.

2. Results and Discussion

2.1. Chemistry

The novel derivatives were synthesized according to the procedures shown in Schemes 1 and 2. As the substrate, cordycepin was connected to TBDMSCL at the 5′-OH position, which acts as an intermediate. Then, heterocyclic compounds were introduced at the N6 position, respectively. Target compounds **3a–3c** and **4a–4c** were synthesized and characterized by ^{1}H-NMR, ^{13}CNMR and HRMS.

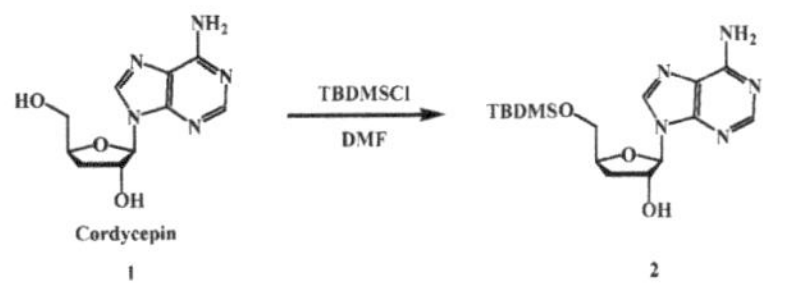

Scheme 1. Synthesis of compound **2**.

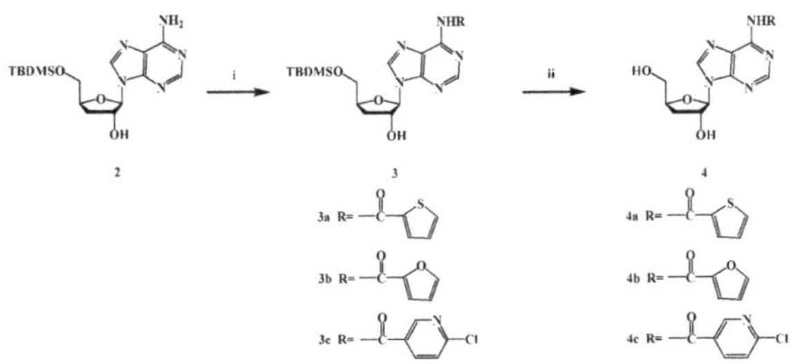

Scheme 2. Synthesis of cordycepin derivatives (**3a–4c**). (**i**) RCI, pyridine, rt; (**ii**) TBAF, THF, rt.

2.2. *Effects of Cordycepin Derivatives Inhibits the Proliferation on MCF7, HepG2 and SGC-7901*

The assay of cell growth inhibitory effects of cordycepin derivatives **3a–4c** was tested against by MTT assay. The results are summarized in Table 1. The results show that compounds **3a–c** and **4a–c** inhibit cell proliferation in a concentration-dependent manner within the range of 0 to 80 μM. Compounds **3c**, **4a**, **4b** and **4c** have better inhibitory effects on MCF7 than cordycepin. For HepG2 proliferation inhibition, the effect of **4b** and **4c** is better. Compounds **3b** and **4a** show enhanced inhibitory effects on SGC-7901 compared with cordycepin. In the results, compound **4a** shows the most obvious inhibitory effect on MCF7 with an IC_{50} value of 27.57 ± 0.52. The results show that the modified compounds in the sugar ring had no obvious effect on the improvement of the activity, the introduction of heterocyclic rings in the base could improve the antitumor activity to a certain extent, and the introduction of thiophene ring had the best effect. It can be seen from the results that the structural modification of cordycepin can improve the antitumor activity effectively. Therefore, it is a feasible way to improve its activity by modifying its structure.

Table 1. Inhibitory effect of compounds on MCF7, HepG2, SGC-7901 cells ($\bar{x} \pm s$, $n = 6$).

Compounds	R1	R2	IC_{50} [a] (μM)		
			MCF7	HepG2	SGC-7901
Cordycepin	-OH	-H	46.85 ± 1.62	51.84 ± 1.36	51.27 ± 3.77
3a	-OTBDMS	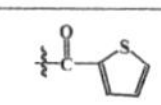	114.08 ± 2.03 **	133.75 ± 5.12 **	102.03 ± 1.08 *
3b	-OTBDMS	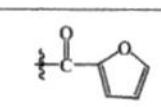	98.03 ± 0.04 **	60.43 ± 1.15 **	41.74 ± 2.36 **
3c	-OTBDMS	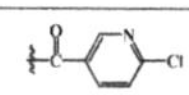	45.79 ± 1.01 **	87.01 ± 0.04 **	82.61 ± 1.48 **
4a	-OH	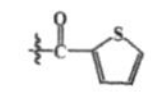	27.57 ± 0.52 ***	68.79 ± 3.34 **	38.93 ± 0.06 *
4b	-OH	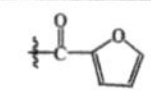	40.94 ± 1.67 **	33.37 ± 0.08 **	86.31 ± 2.35 **
4c	-OH	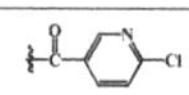	45.26 ± 0.58 **	48.39 ± 0.79 **	80.39 ± 1.05 **
10-hydroxycamptothecin (HCPT) [b]			7.56 ± 0.05 ***	6.56 ± 0.45 **	7.96 ± 0.65 ***

[a] All compounds were examined in a set of experiments repeated three times; IC_{50} values of compounds represent the concentration that caused 50% enzyme activity loss. [b] HCPT is positive control. compared with Cordycepin: * $p < 0.05$; ** $p < 0.01$; *** $p < 0.001$.

Then, compound **4a** was evaluated for its in vitro cytotoxicity against one normal breast cell line (MCF-10A), which revealed low toxicity (Figure 1). With an inhibition rate ranging from 3.278% to 28.508%, compound **4a** has higher selectivity between MCF7 and MCF-10A. Therefore, this led us to select compound **4a** for further investigations against MCF7 breast cancer cells; according to its IC_{50} value, 56 μM, 28 μM and 14 μM were taken as the high, medium and low concentrations of compound 4a, HCPT (8 μM) and cordycepin (80 μM) were used as the positive control.

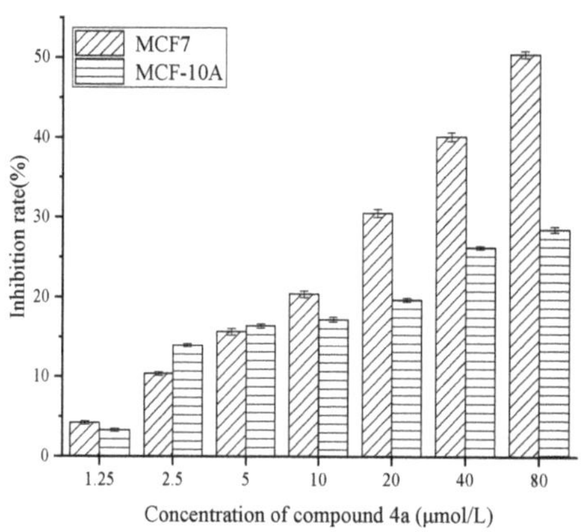

Figure 1. Inhibition rate of compound **4a** on MCF7 and MCF-10A cells.

It has been reported that cordycepin significantly reduced breast cancer cell viability in a dose- and time-dependent manner and has low toxicity against normal cells but greater toxicity against breast cancer cells [18]. This reveals IC_{50} values ranging between 135 μM for MCF7 and 70 μM for MDA-MB-453 cells. In contrast to MCF7, cordycepin appeared to be preferentially active against highly de-differentiated breast cancer cells [19].

Previous studies have indicated that we performed experiments in an optimal non-toxic concentration (50 and 100 μM) of cordycepin with no change in morphology [20].

Furthermore, inverted light microscopy showed a decrease in the number of MCF7 cells treated with compound **4a**. The fluorescence staining results are shown in Figure 2. Breast cancer MCF7 cells adhered to the wall growth, the cell outline was clear, the cell development was good, and the number was large. When treated with cordycepin (as shown in Figure 2b), compared with the blank group, the number of cells is reduced, the cell edges are not clear, and there are spherical dead cells. But when treated with the positive drug HCPT, as shown in Figure 2c, the number of cells is decreased significantly. We can find that cell morphology changed, and the dead cells are spherical.

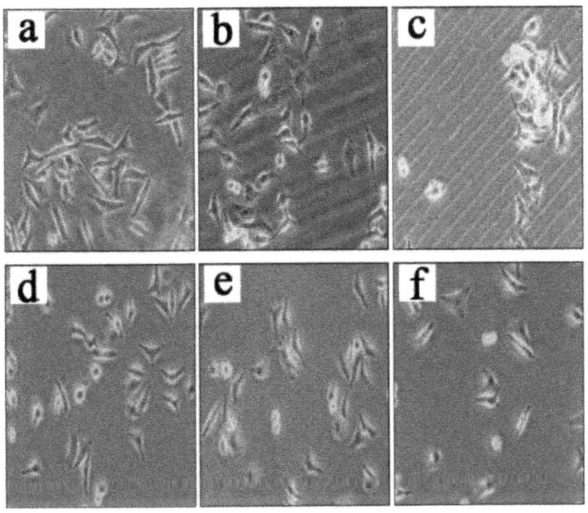

Figure 2. MCF7 cell morphology of the blank group (**a**), cordycepin group (**b**), positive control (**c**), low dose of compound **4a** (**d**), medium dose of compound **4a** (**e**), high dose of compound **4a** (**f**), (10 × 40 times).

Compared with normal breast cancer MCF7 cells, round dead cells can be seen when breast cancer MCF7 cells are treated with compound **4a** at the concentration of 14 μM in the low-dose group (as shown in Figure 2d), the number of cells in the low-dose group is reduced. The cell adherent state is damaged in the medium-dose group, and spherical dead cells could be seen. When treated with compound **4a** of 56 μM (as shown in Figure 2f), cells are not clear at the edge of the cells and had shrunken, and the number of cells is significantly less than that of the medium-dose cells, and spherical dead cells can be seen.

2.3. *Compound* **4a** *Induces Apoptosis in MCF7 Cells*

Hoechst 33258 staining solution can pass through the cell membrane and combine with DNA inside the cell to produce blue fluorescence, thus reflecting the degree of apoptosis. Hoechst 33342 staining showed a significant increase in apoptosis in the MCF7 cells following compound **4a** treatment (Figure 3). As shown in the figure, after MCF7 cells were treated with the increase in the dose, the cell density showed a concentration trend and gradually decreased, the intracellular dense staining granular fluorescence was enhanced, the cell shrank, the adhesion was damaged, and the volume decreased. Compound **4a** broke the cell membranes leading to inducing the nuclear condensation by apoptotic in comparison with the control cells.

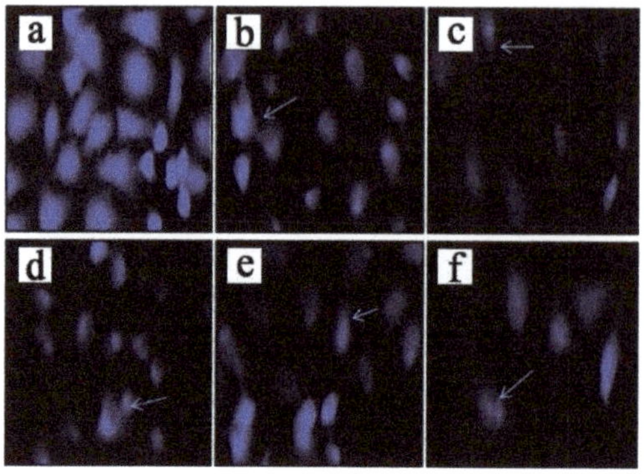

Figure 3. MCF7 cell morphology of the blank group (**a**), cordycepin group (**b**), positive control (**c**), low dose of compound **4a** (**d**), medium dose of compound **4a** (**e**), high dose of compound **4a** (**f**) under fluorescence microscopy (10×40 times).

To further determine the basis of the anti-proliferative action of compound **4a** on the MCF7, we analyzed the cell cycle profile of the treated cells. Compound **4a** treatment at 14, 28 and 56 μM significantly increased the accumulation of G0/G1 cells. Compared with the control group, the percentage of MCF7 cells in the G1 phase significantly increased compared with that in the control group, respectively, indicating that compound **4a** induced MCF7 cell cycle arrest at the G0/G1 phase in a dose-dependent manner (Figure 4). We analyzed the proportion of apoptotic MCF7 cells by flow cytometry. After being treated with compound **4a**, notably, as the compound **4a** concentration increased, the degree of apoptosis also increased. As shown in Figure 4f, at 56 μM compound **4a**, the percentage of apoptotic cells reached $50.422 \pm 1.862\%$. Therefore, these results demonstrate that compound **4a** exerts an apoptotic effect on MCF7 cells.

Previous studies have indicated that in the MCF7 cells, treatment with etoposide leads to shrinkage of both chromatin and whole cells, without the appearance of endonucleosomal cleavage of DNA Typical characteristics of apoptosis [21], including DNA fragmentation, were not evident in cordycepin-treated MCF7 cells [22]. Since apoptosis induction is the mode of action of most anti-cancer drugs [23], we also analyze the potential apoptotic effects of compound **4a** on the MCF7 in subsequent studies. Apoptosis can lead to increased permeability of the cell membrane [24], and previous studies have indicated that cordycepin administration promoted G2/M arrest and apoptosis of MCF7 cells [25]. It is known that inhibition of cell proliferation is often caused by cell cycle arrest [26].

Additionally, to investigate the proapoptotic effect of compound **4a** on breast cancer period of time event, flow cytometry staining with Rhodamine 123 was used to detect the effect of compound **4a** on the mitochondrial membrane potential of MCF7. The effects of different doses of compounds on membrane potential are shown in Figure 5a–e. The results show that, with the increase in dose, the positive expression rate and cell membrane potential decrease gradually, and have a dose–effect relationship. The positive expression rates are shown in Figure 5f. The results are significant and indicate compound **4a** could significantly reduce MCF7 cells. Granular membrane potential induces apoptosis.

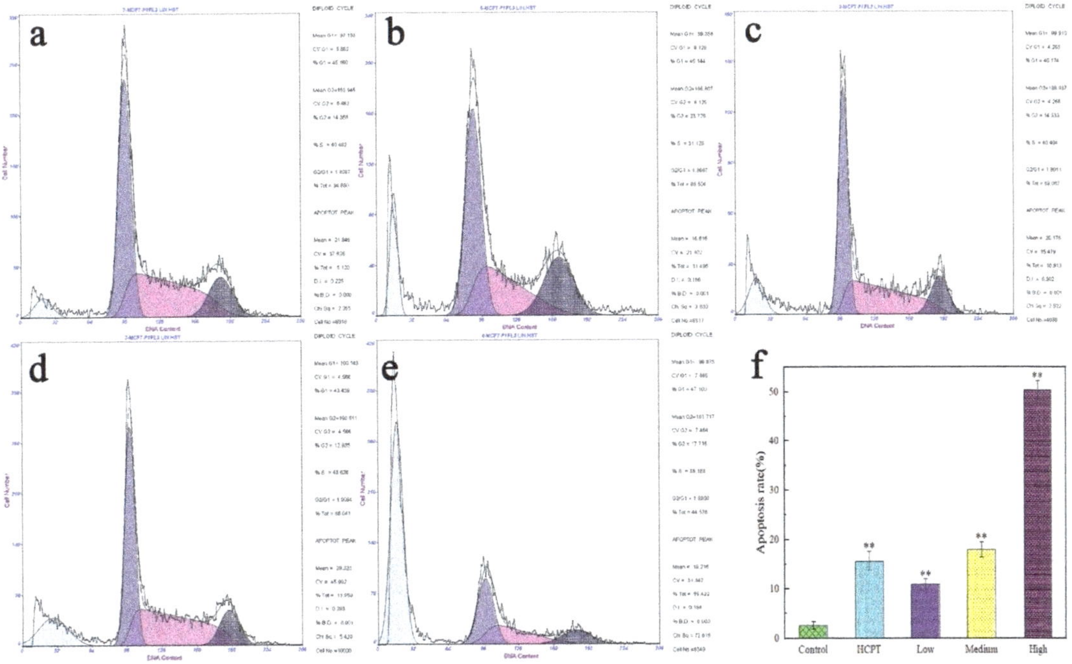

Figure 4. Flow cytometric detection of compound **4a** on the cycle of MCF7 cells (48 h). The blank group (**a**), positive control (**b**), low dose of compound **4a** (**c**), medium dose of compound **4a** (**d**), high dose of compound **4a** (**e**). Apoptosis rate (**f**); ($\overline{x} \pm s$, $n = 3$) compared with control: ** $p < 0.01$.

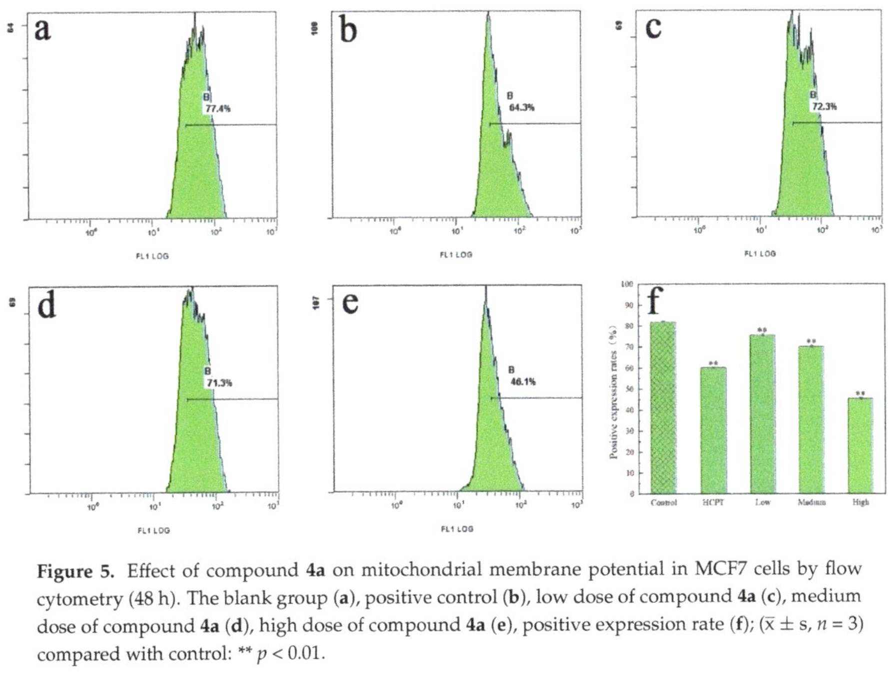

Figure 5. Effect of compound **4a** on mitochondrial membrane potential in MCF7 cells by flow cytometry (48 h). The blank group (**a**), positive control (**b**), low dose of compound **4a** (**c**), medium dose of compound **4a** (**d**), high dose of compound **4a** (**e**), positive expression rate (**f**); ($\overline{x} \pm s$, $n = 3$) compared with control: ** $p < 0.01$.

2.4. Results of Western Blot Analysis

To investigate the mechanism and induction of cordycepin on MCF7 cell activity, after different concentrations of compound **4a** on cells, test results of associated proteins by way of Western Blot assay were shown in Figures 6–9. In this experiment, the blank group, compound **4a** low dose (14 μM), medium dose (28 μM) and high dose (56 μM), Western Blot detected the expression levels of related apoptotic proteins, Caspase-9, Caspase-3, Bcl-2, Bax and p53 in each group. The results showed that compound **4a** could down-regulate the expression of Bcl-2 protein and up-regulate the expression of Caspase-9, Caspase-3, Bax and p53.

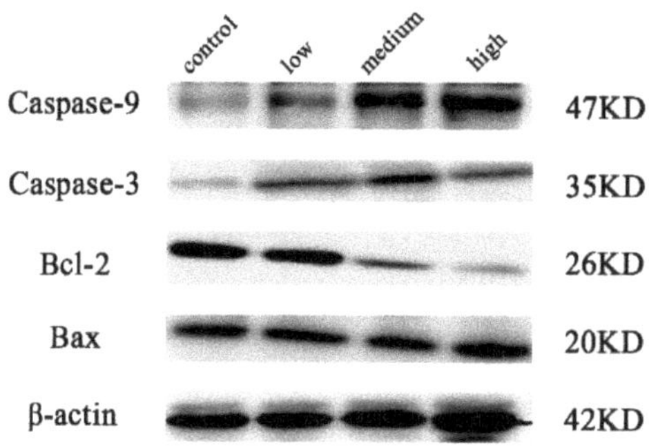

Figure 6. Western Blot method was used to detect the effect of sophocarpine on the expression of Caspase-9, Caspase-3, Bcl-2 and Bax in breast cancer MCF7 cells.

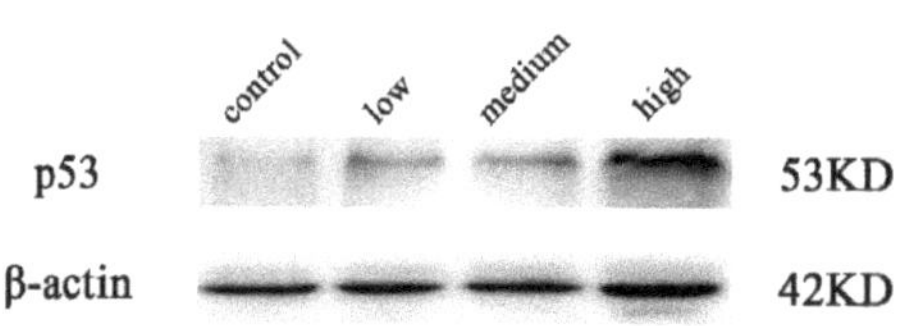

Figure 7. Western Blot method was used to detect the effect of sophocarpine on the expression of p53 in breast cancer MCF7 cells.

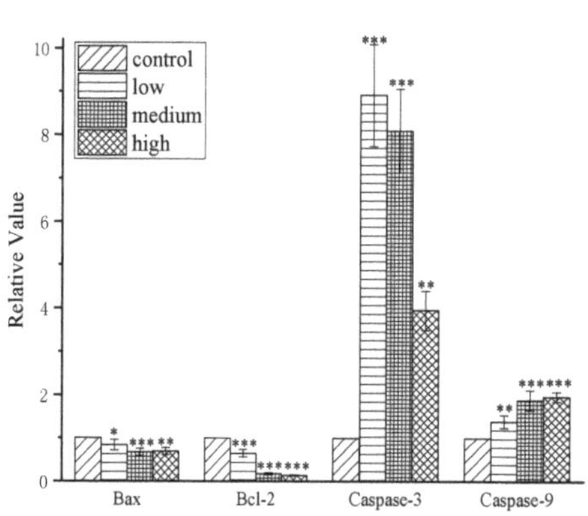

Figure 8. Western Blot detection of Bax, Bcl-2, Caspase-3 and Caspase-9 protein expression ($\bar{x} \pm s$, $n = 3$); compared with control: * $p < 0.05$, ** $p < 0.01$, *** $p < 0.001$.

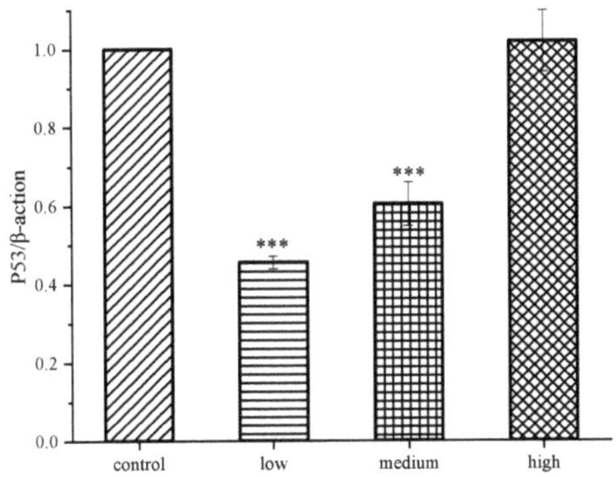

Figure 9. Western Blot detection of p53 protein expression ($\overline{x} \pm s$, $n = 3$); compared with control: *** $p < 0.001$.

There are three confirmed apoptosis pathways: death receptor pathway, mitochondrial pathway and endoplasmic reticulum pathway. After receiving apoptosis stimulation, cyt-C enters the cytoplasm through the mitochondrial membrane and activates caspase-9 and caspase-3. Caspase-3 is in the downstream effector part of various apoptotic pathways and acts as the ultimate executor of the caspase cascade reaction. The Bcl-2 family is an important regulatory protein in the mitochondrial pathway of apoptosis. The pro-apoptotic and anti-apoptotic proteins of the Bcl family remained relatively stable, but when Bax is overexpressed, the number of Bax/Bax homologous dimers increased, and the cell response to the death signal was enhanced, promoting the occurrence of apoptosis. However, when Bcl-2 is highly expressed, the activation of caspase-3 is inhibited, and more heterodimers of Bcl-2/Bax with stable structure are formed, which can resist the induction effect of Bax/Bax on apoptosis. The ratio relationship between Bcl-2 and Bax is critical for cell survival. And p53 functions beyond its ability to trigger cell cycle arrest and programmed cell death and new activities such as regulating metabolism, autophagy and cellular oxidation status. P53 protein can specifically inhibit the expression of Bcl-2, secrete pro-apoptotic proteins (Bax, Bad, Bak) and anti-apoptotic proteins (Bcl-2, Bcl-xl) to induce the activation of pre-apoptotic protein Bax, induce mitochondria to release cyt-C and form apoptotic bodies. Activate the Caspases (Caspase-3, Caspase-9) pathway and cause cell apoptosis through protease hydrolysis. In this experiment, with the increasing dose of compound **4a**, the relative expression levels of BAX, caspase-3 and p53 showed an increasing trend, while the expression of Bcl-2 showed an obvious decreasing trend, which indicated that compound **4a** could promote apoptosis of MCF7, and the promotion effect of high dose on apoptosis was more obvious than that of low dose. We suggest that compound **4a** may induce apoptosis of MCF7 in breast cancer through the mitochondrial pathway.

3. Conclusions

We modified the structure of cordycepin and synthesized new cordycepin derivatives by introducing fat-soluble groups on the N6 groups. The MTT method was used to investigate the bioactivity of the synthesized compounds against MCF7 cells of breast cancer, HepG2 cells of liver cancer and SGC-7901 cells of gastric cancer. Compound **4a** showed significant inhibitory activity against MCF7 cells with an IC_{50} value of 27.57 ± 0.52 μM, which was superior to cordycepin (46.85± 1.62 μM). In the cytotoxicity test, it was found to be less toxic to normal cells MCF-10A, showing high selectivity between breast cancer cells and normal cell lines. Therefore, compound **4a** was further studied to explore the effect of compound **4a** on MCF7 through in vitro cell experiments. The results showed that the number of breast cancer MCF7 cells decreased gradually in a concentration-dependent manner after treatment and spherical dead cells were visible. Compound **4a** can induce apoptosis and arrest the cell cycle in the G0/G1 phase. The related animal experiments need further verification.

In order to clarify the mechanism of compound **4a** affecting MCF7 cells, Western Blot analysis showed that compound **4a** could down-regulate the expression of Bcl-2 protein and up-regulate the expression of p53, Bax, Caspase-3 and Caspase-9 proteins. The mitochondrial membrane potential decreased continuously and the positive expression rate decreased. It was speculated that compound **4a** induced the apoptosis of MCF7 cells through the mitochondrial pathway.

4. Experimental Section

4.1. Chemicals and Reagents

Cordycepin (more than 99% in purity) (TIANBAO Biopharmaceuticals, Canton, China). 2-Thiophenecarbonyl chloride; 2-Furoyl chloride; 6-Chloronicotinoyl chloride, tert-Butyldimethyl chlorosilane (TBDMSCl), imidazole, and tetrabutylammonium fluoride (TBAF) were purchased from Aladdin (Shanghai, China). All reagents were of analytical grade. RPMI-1640 (Gibco, Oakland, CA, USA); fetal bovine serum (Sijiqing, Hangzhou, China); MTT (Solarbio, Beijing, China); HCPT (Shengtai, Harbin, China); DMSO (BASF, Shanghai, China). Mitochondrial membrane potential assay kit, cell cycle apoptosis analysis kit, trypsin, RIPA lysis buffer, and Hoechst 33258 were purchased from Beyotime (Shanghai, China); Infinite F50 Microplate reader (TECAN group, Männedorf, Switzerland); HF90 Incubator (Lishen, Shanghai, China).

4.2. Synthesis

4.2.1. Synthesis of Compound **2**

Cordycepin (**1**) (1.5 g, 6 mmol) and 0.4446 g imidazole (680 mg, 10 mmol) were dissolved in dry DMF (6 mL) under ultrasound-assisted dissolution. TBDMSCl (1.5 g, 10 mmol) was added in batches. The reaction was stirred for 5 h at room temperature and then concentrated in a vacuum. The organic layer was washed with water and ethyl acetate, then afterward dried and concentrated, and then the crude product was purified by silica gel column chromatography with $MeOH/CH_2Cl_2$ (1:15). The compound **2** was obtained.

4.2.2. Synthesis of Compounds **3a**–**3c** (i)

Compound **2** (0.5 g, 1 mmol) was dissolved in dry pyridine (14 mL). 2-Thiophenecarbonyl chloride (0.57 g, 4 mmol), 2-Furoyl chloride (0.59 g, 4 mmol) and 6-Chloronicotinoyl chloride (1.13 g, 4 mmol) were added to the solution under N_2 protection, respectively. The reaction was stirred at room temperature for 8 h and then cooled to 0 °C. NH_4OH (28% aq) (4 mL) was slowly added to the solution. After the mixture was stirred for 30 min, the solvent was removed, and the residue was dissolved in EtOAc. The organic layer was washed with water, $NaHCO_3$ solution and salt water. Organic layers were incorporated together, and then the solvent was removed by vacuum concentration. The residues were purified by silica gel column ($MeOH/CH_2Cl_2$ = 1:25) to obtain target compounds **3a**, **3b** and **3c**, respectively. The products were dried and weighed.

4.2.3. Synthesis of Compound **4a**–**4c** (ii)

Compounds **3a** (0.43 g, 0.4 mmol), **3b** (0.42 g, 0.4 mmol) and **3c** (0.6 g, 0.4 mmol) were dissolved in THF (13 mL) and TBAF (0.94 mL, 1.0 M THF solution) at 0 °C respectively. The reaction was stirred at room temperature for 5 h, and then the solvent was removed. The residues were purified by silica gel column ($MeOH/CH_2Cl_2$ = 1:20) to obtain compounds **4a**, **4b** and **4c** respectively. The products were dried and weighed.

Melting points, optical rotation, TLC, ^{1}H NMR. ^{13}C NMR and MS were used to identify structures of new compounds. Melting points were tested on a YRT-3 melting point apparatus (Tianjin, China). Optical rotation was tested on PRO/P850 Polarimeter (Jinan, China). ^{1}H NMR and ^{13}C NMR were tested with Bruker Avance-500 MHz spectrometer

(^{1}H: 500 MHz, ^{13}C: 125 MHz). Molecular weight was tested with TripIeTOF6600 (AB Sciex, Framingham, MA, USA).

4.3. Characterization

4.3.1. 5′-O-(Tert-butyldimethylsilyl)-3′-deoxyadenosine (**2**)

^{1}H-NMR (500 MHz, $CDCl_3$) δ: δ8.26 (1H, s, 2-H), 8.23 (1H, s, 8-H), 6.08 (2H, s, -NH_2), 5.93 (1H, s, 1′-H), 4.62 (1H, s, 2′-H), 4.54 (1H, s, 4′-H), 3.97–3.66 (2H, ddd, 5′-CH_2), 2.30–2.03 (2H, ddd, 3′-CH_2), 0.80–0.01 (15H, m, -CH_3); ^{13}C-NMR (125 MHz, $CDCl_3$): δ 155.30, 152.30, 148.77, 138.96, 120.14, 92.67, 81.71, 65.64, 63.92, 32.38, 30.61, 25.97, 19.26, 18.51, 13.69. White powder; yield 92.9%; mp 256–258 °C; $[\alpha]^{25}_{D}$ −50 (0.1, MeOH).

4.3.2. 5′-O-(Tert-butyldimethylsilyl)-6-thiophenecarboxamide-3′-deoxya-denosine (**3a**)

^{1}H-NMR (500 MHz, $CDCl_3$) δ: 9.30 (1H, s, -NH), 8.65 (1H, s, 2-H), 8.37 (1H, s, 8-H), 7.76–7.00 (3H, dd, 6-CH=), 6.30 (1H, s, 1′-H), 5.75 (1H, d, 2′-H), 4.50 (1H, dt, 4′-H), 4.00–3.71 (2H, ddd, 5′-CH_2), 2.64–2.18 (2H, ddd, 3′-CH_2), 0.88–0.00 (15H, m, -CH_3); ^{13}C-NMR (125 MHz, $CDCl_3$): δ161.13, 141.39, 134.50, 133.53, 132.62, 132.42, 130.90, 130.10, 128.82, 127.91, 89.27, 81.65, 78.98, 65.57, 63.65, 31.69, 30.53, 26.02, 19.17, 18.52, 13.75. Light yellow oily liquid; yield 92.3%; mp 263–267 °C; $[\alpha]^{25}_{D}$ −60 (0.1, MeOH).

4.3.3. 5′-O-(Tert-butyldimethylsilyl)-6-furancarboxamide-3′-deoxyadenosine (**3b**)

^{1}H-NMR (500 MHz, $CDCl_3$) δ: 9.40 (1H, s, -NH), 8.70 (1H, s, 2-H), 8.40 (1H, s, 8-H), 7.60–7.17 (3H, d, 6-CH=), 6.30 (1H, s, 1′-H), 5.79 (1H, d, 2′-H), 4.51 (1H, dt, 4′-H), 4.00–3.70 (2H, ddd, 5′-CH_2), 2.69–2.18 (2H, ddd, 3′-CH_2), 0.86–0.00 (15H, m, -CH_3);^{13}C-NMR (125 MHz, $CDCl_3$): 157.51, 147.00, 145.28, 143.73, 141.66, 123.20, 117.10, 112.90, 112.19, 89.22, 81.54, 78.67, 63.63, 53.49, 31.67, 30.54, 25.95, 19.15, 18.51, 13.72. Light yellow oily liquid; yield 95.2%; mp 235–236 °C; $[\alpha]^{25}_{D}$ −65 (0.1, MeOH).

4.3.4. 5′-O-(Tert-butyldimethylsilyl)-6-6-chloroformamide-3′-deoxya-denosine (**3c**)

^{1}H-NMR (500 MHz, $CDCl_3$) δ: ^{1}H-NMR (500 MHz, $CDCl_3$): 10.03 (1H, s, -NH), 8.93 (1H, s, 2-H), 8.45 (1H, s, 8-H), 8.18–7.23 (3H, d, 6-CH=), 6.33 (1H, s, 1′-H), 5.81 (1H, s, 2′-H), 4.53 (1H, s, 4′-H), 4.03–3.73 (2H, ddd, 5′-CH_2), 2.66–2.22 (2H, ddd, 3′-CH_2), 0.81–0.01 (15H, m, -CH_3); ^{13}C-NMR (125 MHz, $CDCl_3$): 167.71, 163.43, 156.34, 151.24, 139.71, 132.25, 130.86, 128.80, 124.43, 124.20, 89.11, 81.61, 79.44, 63.61, 53.49, 31.43, 30.57, 25.95, 19.08, 18.49, 13.72. Yellow oily liquid; yield 85.5%; mp 231–233 °C; $[\alpha]^{25}_{D}$ −64 (0.1, MeOH).

4.3.5. 6-Thiophenecarboxamide-3′-deoxyadenosine (**4a**)

IR (KBr) υ_{max}: 3475, 3400, 3145, 2355, 1908, 1714, 1660, 1422, 1286, 1085, 995, 730, 639 cm^{-1}; ^{1}H-NMR (500 MHz, $CDCl_3$) δ: 8.65 (1H, s, 2-H), 8.24 (1H, s, 8-H), 7.80–7.00 (3H, dd, 6-CH=), 6.14 (1H, s, 1′-H), 5.73 (1H, d, 2′-H), 4.58 (1H, t, 4′-H), 4.07–3.66 (2H, ddd, 5′-CH_2), 2.91–2.23 (1H, ddd, 3′-CH_2); ^{13}C-NMR (125 MHz, $CDCl_3$): δ161.38, 152.28, 150.83, 149.64, 142.33, 132.16, 130.46, 128.10, 128.05, 123.50, 91.06, 82.03, 78.89, 62.82, 31.23. MS m/z 362.1094 $[M+H]^+$; white powder; yield 83.5%; mp 267–269 °C; $[\alpha]^{25}_{D}$−71 (0.1, MeOH).

4.3.6. 6-Furancarboxamide-3′-deoxyadenosine (**4b**)

IR (KBr) υ_{max}: 3620, 3118, 2335, 2363, 1915, 1801, 1400, 1155, 1050, 977, 835, 672 cm^{-1}; ^{1}H-NMR (500 MHz, $CDCl_3$) δ: 8.80 (1H, s, 2-H), 8.28 (1H, s, 8-H), 7.62–7.25 (3H, d, 6-CH=), 6.20 (1H, s, 1′-H), 5.82 (1H, dt, 2′-H), 4.67 (1H, t, 4′-H), 3.75–4.18 (2H, ddd, 5′-CH_2), 2.30–3.08 (1H, ddd, 3′-CH_2); ^{13}C-NMR (125 MHz, $CDCl_3$): δ157.8, 154.53, 152.52, 150.72, 149.42, 119.42, 117.44, 113.01, 112.21, 123.65, 91.34, 81.97, 78.70, 63.00, 31.21. MS m/z 346.1324 $[M+H]^+$; white powder; yield 67.2%; mp 213–215 °C; $[\alpha]^{25}_{D}$ −73 (0.1, MeOH).

4.3.7. 6-6-Chloronicotinamide-3′-deoxyadenosine (**4c**)

IR (KBr) υ_{max}: 3400, 3330, 3220, 2949, 1736, 165, 1605, 1298, 1135, 1094, 760, 646, 521 cm^{-1}; ^{1}H-NMR (500 MHz, DMSO-d6): 11.57 (1H, s, -NH$_2$), 9.00 (1H, s, 2-H), 8.77 (1H, s, 8-H), δ 8.77–7.25 (3H, d, 6-CH=), 6.04 (1H, s, 1′-H), 5.78 (1H, d, 2′-OH), 5.10 (1H, t, 5′-OH), 4.65 (1H, s, 2′-H), 4.41 (1H, d, 4′-H), 3.75–3.54 (2H, ddd, 5′-CH$_2$), 1.96–1.93 (1H, ddd, 3′-CH$_2$); ^{13}C-NMR (125 MHz, DMSO-d6): δ163.82, 153.95, 152.35, 152.03, 150.54, 150.00, 143.28, 140.22, 129.13, 125.92, 124.74, 91.38, 81.67, 75.34, 62.8, 34.24. MS m/z 391.1080 [M+H]$^+$; white powder; yield 87.6%; mp 225–227 °C;$[\alpha]_D^{25}$ −74 (0.1, MeOH).

4.4. Cell Culture

MCF7, HepG2, SGC-7901 and MCF-10A were purchased from the National Infrastructure of Cell Line Resources in the Chinese Academy of Sciences (Shanghai Institute of Biochemistry and Cell Biology, Shanghai, China). Cells were maintained using RPMI-1640 medium with 10% fetal bovine serum. The cells were maintained at 37 °C in a 5% CO_2-containing incubator for all treatments. Logarithmic growth phase cells were used in all experiments.

4.5. Cell Proliferation and Cytotoxicity Assays

The effect of compounds on the proliferation of cancer cells was determined with an MTT assay [27–29]. Cells monolayer in exponential growth were seeded in 96-well plates at 5 × 10^3 cells/well. After cultured at 37 °C for 24 h, the culture medium was removed, then compounds (**3a–4c**) (1.25, 2.5, 5, 10, 20, 40, 80 μM) and HCPT (4, 8, 16 μM) were added. After 48 h incubation, 100 μL of MTT was added to each well and then cultured for 4 h in a 37 °C incubator. After that, the medium was aspirated, 150 μL of DMSO was then added to each well, and the 96-well plates were placed on an orbital shaker for 12 min until the formazan dissolved completely. The absorbance was measured at a test wavelength of 490 nm with a microplate reader, the optical density (OD) values were tested and the inhibitory rate and half inhibitory concentration (IC_{50}) were calculated for each administration group.

4.6. Hoechst 33258 Staining Assay

Cells were cultured in the medium for 24 h and treated with compounds, cordycepin and HCPT. Cells were observed under inverted light microscopy (Olympus Corporation Tokyo, Japan). Then, cells were fixed for 1 h at room temperature with 4% paraformaldehyde, and Hoechst 33258 was used to stain the cells. Fluorescence microscopy (ECLIPSE Ti2; Nikon, Tokyo, Japan) was used to observe the apoptotic cells.

4.7. Mitochondrial Membrane Potential Assay

MCF7 cells were cultured in 6-well plates, after treatment with compounds in different concentrations and positive control group (HCPT), cells were harvested, washed with PBS, and then stained with rhodamine 123 (100 μg) for 30 min at 37 °C. The cells were collected by pipetting, washed twice with PBS and analyzed by flow cytometry (BD Biosciences, San Jose, CA, USA).

4.8. Cell Cycle Detection

Cells were cultured in the medium for 24 h and treated with compounds in different concentrations and HCPT incubated for 48 h. The cells were centrifuged at 1000 rpm for 10 min and washed with PBS. Subsequently, the cells were fixed with 70% ethanol for 24 h and then stained with annexin V-FITC (10 μL) and propidium iodide (PI; 10 μL). The proportion of apoptotic cells was then measured by flow cytometry.

After the treatments, the harvested cells were incubated at 70% in ice-ethanol overnight. Subsequently, they were washed twice using PBS and incubated with propidium iodide (PI) and RNase A. Finally, the proportions of cells in different cell cycle phases were determined by flow cytometry.

4.9. Western Blot Assay

Cells were cultured in the medium for 24 h and treated with compounds in different concentrations and HCPT for 48 h. After lysis buffer (RIPA: PMSF = 100:1) was used to lyse cells. The lysate was further sonicated and centrifuged at 12,000 rpm for 15 min, then the supernatant obtained was collected and the denatured protein mixture was subjected to SDS gel electrophoresis and transferred to nitrocellulose membranes, and blocked with containing 5% nonfat milk. Then, the membranes were incubated with the primary antibodies (1:1000, overnight 4 °C) and secondary antibodies (1:5000, 1 h). Then, ECL Solution was used to detect the antigen–antibody complexes. The data were analyzed by NIH Image J 1.53a software.

4.10. Statistical Analysis

Data were presented as mean ± SEM. The statistical significance was tested using Student's *t*-test, one-way ANOVA, or two-way ANOVA with Bonferroni posttests. $p < 0.05$ was considered to be statistically significant. Tested significance is displayed in the figures as * $p < 0.05$; ** $p < 0.01$; *** $p < 0.001$.

Author Contributions: L.C. designed the procedure and wrote the manuscript. L.Z. and T.Y. synthesized the compounds. G.S. analyzed the data. D.Y. and Y.Z. performed the biological assay. B.Y. conceived the project. All authors have read and agreed to the published version of the manuscript.

Funding: This study was supported financially by the Postdoctoral Project of Heilongjiang Province (LBH-Z21066) and training program for young innovative talents in scientific research of Harbin University of Commerce (2023-KYYWF-1025).

Institutional Review Board Statement: Not applicable.

Informed Consent Statement: Not applicable.

Data Availability Statement: Data are contained within the article.

Acknowledgments: All the research was conducted by the authors.

Conflicts of Interest: The authors declare no conflict of interest.

References

1. Cunningham, K.; Manson, W.; Spring, F.; Hutchinson, S. Cordycepin, a metabolic product isolated from cultures of *Cordyceps militaris* (Linn.) Link. *Nature* **1950**, *166*, 949.
2. Tuli, H.S.; Sandhu, S.S.; Sharma, A. Pharmacological and therapeutic potential of Cordyceps with special reference to Cordycepin. *3 Biotech* **2014**, *4*, 1–12. [CrossRef] [PubMed]
3. Yue, K.; Ye, M.; Zhou, Z.; Sun, W.; Lin, X. The genus Cordyceps: A chemical and pharmacological review. *J. Pharm. Pharmacol.* **2013**, *65*, 474–493. [PubMed]
4. Park, J.M.; Lee, J.S.; Lee, K.R.; Ha, S.J.; Hong, E.K. *Cordyceps militaris* extract protects human dermal fibroblasts against oxidative stress-induced apoptosis and premature senescence. *Nutrients* **2014**, *6*, 3711–3726. [CrossRef] [PubMed]
5. Yin, J.; Xin, X.; Weng, Y.; Gui, Z. Transcriptome-wide analysis reveals the progress of *Cordyceps militaris* subculture degeneration. *PLoS ONE* **2017**, *12*, e0186279.
6. Liu, W.; Zhang, L.; Sun, S.; Tang, L.; He, S.; Chen, A.; Yao, L.; Ren, D. Cordycepin inhibits inflammatory responses through suppression of ERK activation in zebrafish. *Dev. Comp. Immunol.* **2021**, *124*, 104178.
7. Zhang, X.; Huang, W.; Tang, P.; Sun, Y.; Zhang, X.; Qiu, L.; Yu, B.; Zhang, X.; Hong, Y.; He, Y. Anti-inflammatory and neuroprotective effects of natural cordycepin in rotenone-induced PD models through inhibiting Drp1-mediated mitochondrial fission. *Neurotoxicology* **2021**, *84*, 1–13. [CrossRef]
8. Liao, X.; Gao, Y.; Zhao, H.; Zhou, M.; Chen, D.; Tao, L.; Guo, W.; Sun, L.; Gu, C.; Chen, H. Cordycepin reverses cisplatin resistance in non-small cell lung cancer by activating AMPK and inhibiting AKT signaling pathway. *Front. Cell Dev. Biol.* **2021**, *8*, 609285.
9. Mai, N.X.D.; Le, U.-C.N.; Nguyen, L.H.T.; Ta, H.T.K.; Van, N.H.; Le, T.M.; Phan, T.B.; Nguyen, L.T.T.; Tamanoi, F.; Doan, T.L.H. Facile synthesis of biodegradable mesoporous functionalized-organosilica nanoparticles for enhancing the anti-cancer efficiency of cordycepin. *Microporous. Mesoporous. Mater.* **2021**, *315*, 110913.
10. Wen, Z.; Du, X.; Meng, N.; Li, Y.; Mi, R.; Li, X.; Sun, Y.; Ma, S.; Li, S. Tussah silkmoth pupae improve anti-tumor properties of *Cordyceps militaris* (L.) Link by increasing the levels of major metabolite cordycepin. *RSC Adv.* **2019**, *9*, 5480–5491. [CrossRef]

11. Lee, C.; Huang, K.; Shaw, J.; Chen, J.; Kuo, W.; Shen, G.; Grumezescu, A.M.; Holban, A.M.; Wang, Y.; Wang, J. Trends in the immunomodulatory effects of *Cordyceps militaris*: Total extracts, polysaccharides and cordycepin. *Front. Pharmacol.* **2020**, *1824*, 575704. [CrossRef] [PubMed]
12. Fei, X.; Zhang, X.; Zhang, G.; Bao, W.; Zhang, Y.; Zhang, M.; Zhou, X. Cordycepin inhibits airway remodeling in a rat model of chronic asthma. *Biomed. Pharmacother.* **2017**, *88*, 335–341. [PubMed]
13. Han, F.; Dou, M.; Wang, Y.; Xu, C.; Li, Y.; Ding, X.; Xue, W.; Zheng, J.; Tian, P.; Ding, C. Cordycepin protects renal ischemia/reperfusion injury through regulating inflammation, apoptosis, and oxidative stress. *Acta Biochim. Biophys. Sin.* **2020**, *52*, 125–132. [CrossRef]
14. Jia, Y.; Li, H.; Bao, H.; Zhang, D.; Feng, L.; Xiao, Y.; Zhu, K.; Hou, Y.; Luo, S.; Zhang, Y. Cordycepin (3′-deoxyadenosine) promotes remyelination via suppression of neuroinflammation in a cuprizone-induced mouse model of demyelination. *Int. Immunopharmacol.* **2019**, *75*, 105777. [CrossRef] [PubMed]
15. Chang, M.M.; Hong, S.Y.; Yang, S.H.; Wu, C.C.; Wang, C.Y.; Huang, B.M. Anti-cancer effect of cordycepin on FGF9-induced testicular tumorigenesis. *Int. J. Mol. Sci.* **2020**, *21*, 8336. [CrossRef] [PubMed]
16. Radhi, M.; Ashraf, S.; Lawrence, S.; Tranholm, A.A.; Wellham, P.A.D.; Hafeez, A.; Khamis, A.S.; Thomas, R.; McWilliams, D.; DeMoor, C.H. A systematic review of the biological effects of cordycepin. *Molecules* **2021**, *26*, 5886. [CrossRef]
17. Yoon, S.Y.; Park, S.J.; Park, Y.J. The anticancer properties of cordycepin and their underlying mechanisms. *Int. J. Mol. Sci.* **2018**, *19*, 3027. [CrossRef]
18. Liu, C.; Qi, M.; Li, L.; Yuan, Y.; Wu, X.; Fu, J. Natural cordycepin induces apoptosis and suppresses metastasis in breast cancer cells by inhibiting the Hedgehog pathway. *Food Funct.* **2020**, *11*, 2107–2116. [CrossRef]
19. Lee, H.J.; Burger, P.; Vogel, M.; Friese, K.; Brüning, A. The nucleoside antagonist cordycepin causes DNA double strand breaks in breast cancer cells. *Investig. New Drugs* **2012**, *30*, 1917–1925.
20. Noh, E.M.; Jung, S.H.; Han, J.H.; Chung, E.Y.; Jung, J.Y.; Kim, B.S.; Lee, S.H.; Lee, Y.R.; Kim, J.S. Cordycepin inhibits TPA-induced matrix metalloproteinase-9 expression by suppressing the MAPK/AP-1 pathway in MCF-7 human breast cancer cells. *Int. J. Mol. Med.* **2010**, *25*, 255–260.
21. Thomadaki, H.; Tsiapalis, C.M.; Scorilas, A. Polyadenylate polymerase modulations in human epithelioid cervix and breast cancer cell lines, treated with etoposide or cordycepin, follow cell cycle rather than apoptosis induction. *Biol. Chem.* **2005**. [CrossRef] [PubMed]
22. Choi, S.; Lim, M.H.; Kim, K.M.; Jeon, B.H.; Song, W.O.; Kim, T.W. Cordycepin-induced apoptosis and autophagy in breast cancer cells are independent of the estrogen receptor. *Toxicol. Appl. Pharmacol.* **2011**, *257*, 165–173. [CrossRef]
23. Bi, Y.; Li, H.; Yi, D.; Sun, Y.; Bai, Y.; Zhong, S.; Song, Y.; Zhao, G.; Chen, Y. Cordycepin augments the chemosensitivity of human glioma cells to temozolomide by activating AMPK and inhibiting the AKT signaling pathway. *Mol. Pharm.* **2018**, *15*, 4912–4925. [CrossRef] [PubMed]
24. Zhou, Y.; Guo, Z.; Meng, Q.; Lu, J.; Wang, N.; Liu, H.; Liang, Q.; Quan, Y.; Wang, D.; Xie, J. Cordycepin affects multiple apoptotic pathways to mediate hepatocellular carcinoma cell death. *Anti-Cancer Agents Med. Chem.* **2017**, *17*, 143–149.
25. Dong, J.; Li, Y.; Xiao, H.; Luo, D.; Zhang, S.; Zhu, C.; Jiang, M.; Cui, M.; Lu, L.; Fan, S. Cordycepin sensitizes breast cancer cells toward irradiation through elevating ROS production involving Nrf2. *Toxicol. Appl. Pharmacol.* **2019**, *364*, 12–21. [CrossRef]
26. Alberts, S.R.; Wagman, L.D. Chemotherapy for colorectal cancer liver metastases. *Oncologist* **2008**, *13*, 1063–1073. [CrossRef]
27. Lu, H.; Li, X.; Zhang, J.; Shi, H.; Zhu, X.; He, H. Effects of cordycepin on HepG2 and EA. hy926 cells: Potential antiproliferative, antimetastatic and anti-angiogenic effects on hepatocellular carcinoma. *Oncol. Lett.* **2014**, *7*, 1556–1562. [CrossRef]
28. Nasser, M.I.; Masood, M.; Wei, W.; Li, X.; Zhou, Y.; Liu, B.; Li, J.; Li, X. Cordycepin induces apoptosis in SGC-7901 cells through mitochondrial extrinsic phosphorylation of PI3K/Akt by generating ROS. *Int. J. Oncol.* **2017**, *50*, 911–919. [CrossRef]
29. Wei, C.; Khan, M.A.; Du, J.; Cheng, J.; Tania, M.; Leung, E.L.H.; Fu, J. Cordycepin inhibits triple-negative breast cancer cell migration and invasion by regulating EMT-TFs SLUG, TWIST1, SNAIL1, and ZEB1. *Front. Oncol.* **2022**, *12*, 898583. [CrossRef]

molecules

MDPI

Article

The Anticancer Effects of Marine Carotenoid Fucoxanthin through Phosphatidylinositol 3-Kinase (PI3K)-AKT Signaling on Triple-Negative Breast Cancer Cells

Shade' A. Ahmed [1], Patricia Mendonca [2,*], Samia S. Messeha [2], Ebenezer T. Oriaku [1] and Karam F. A. Soliman [1,*]

1 Division of Pharmaceutical Sciences, College of Pharmacy and Pharmaceutical Sciences, Institute of Public Health, Florida A&M University, Tallahassee, FL 32307, USA; shade1.ahmed@famu.edu (S.A.A.); ebenezer.oriaku@famu.edu (E.T.O.)

2 Department of Biology, College of Science and Technology, Florida A&M University, Tallahassee, FL 32307, USA; samia.messeha@famu.edu

* Correspondence: patricia.mendonca@famu.edu (P.M.); karam.soliman@famu.edu (K.F.A.S.); Tel.: +1-850-599-3306 (P.M. & K.F.A.S.)

Abstract: Triple-negative breast cancer (TNBC) is an aggressive subtype of breast cancer that lacks specific targets such as estrogen, progesterone, and HER2 receptors. TNBC affects one in eight women in the United States, making up 15–20% of breast cancer cases. Patients with TNBC can develop resistance to chemotherapy over time, leading to treatment failure. Therefore, finding other options like natural products is necessary for treatment. The advantages of using natural products sourced from plants as anticancer agents are that they are less toxic, more affordable, and have fewer side effects. These products can modulate several cellular processes of the tumor microenvironment, such as proliferation, migration, angiogenesis, cell cycle arrest, and apoptosis. The phosphatidyl inositol 3-kinase (PI3K)-AKT signaling pathway is an important pathway that contributes to the survival and growth of the tumor microenvironment and is associated with these cellular processes. This current study examined the anticancer effects of fucoxanthin, a marine carotenoid isolated from brown seaweed, in the MDA-MB-231 and MDA-MB-468 TNBC cell lines. The methods used in this study include a cytotoxic assay, PI3K-AKT signaling pathway PCR arrays, and Wes analysis. Fucoxanthin (6.25 μM) + TNF-α (50 ng/mL) and TNF-α (50 ng/mL) showed no significant effect on cell viability compared to the control in both MDA-MB-231 and MDA-MB-468 cells after a 24 h treatment period. PI3K-AKT signaling pathway PCR array studies showed that in TNF-α-stimulated (50 ng/mL) MDA-MB-231 and MDA-MB-468 cells, fucoxanthin (6.25 μM) modulated the mRNA expression of 12 genes, including *FOXO1*, *RASA1*, *HRAS*, *MAPK3*, *PDK2*, *IRS1*, *EIF4EBP1*, *EIF4B*, *PTK2*, *TIRAP*, *RHOA*, and *ELK1*. Additionally, fucoxanthin significantly downregulated the protein expression of IRS1, EIF4B, and ELK1 in MDA-MB-231 cells, and no change in the protein expression of EIF4B and ELK1 was shown in MDA-MB-468 cells. Fucoxanthin upregulated the protein expression of RHOA in both cell lines. The modulation of the expression of genes and proteins of the PI3K-AKT signaling pathway may elucidate fucoxanthin's effects in cell cycle progression, apoptotic processes, migration, and proliferation, which shows that PI3K-AKT may be the possible molecular mechanism for fucoxanthin's effects. In conclusion, the results obtained in this study elucidate fucoxanthin's molecular mechanisms and indicate that fucoxanthin may be considered a promising candidate for breast cancer-targeted therapy.

Keywords: fucoxanthin; TNBC; PI3K-AKT signaling pathway; tumor microenvironment

1. Introduction

An estimated 297,790 women in the United States will be diagnosed with invasive breast cancer in 2023 [1]. Approximately 13% of American women will be diagnosed with breast cancer throughout their lifetime [2]. Triple-negative breast cancer (TNBC) is an aggressive and metastatic form of breast cancer that is not easily treatable due to its triple-negative status for the progesterone receptor, the estrogen receptor, and human epidermal growth factor receptor-2. TNBC is one of the most aggressive breast cancer subtypes with rapid metastasis and poor prognosis. It accounts for 10–15% of breast cancers, and its mortality rate is 40% within the first five years after diagnosis [3].

Moreover, conventional chemotherapy is not highly effective in treating TNBC [4]. since TNBC patients can develop resistance to chemotherapy over time, leading to treatment failure. Therefore, finding other options like natural products is necessary for treatment. Natural compounds that can be used as anticancer agents are less toxic and affordable, offering strategies to overcome chemoresistance and present anti-inflammatory, anti-proliferative, anti-angiogenic, and antioxidant properties [5].

Many natural products, such as carotenoids, taxanes, and flavonoids, effectively display antitumor activity and have minimal toxicity on normal healthy cells and tissues [6]. Fucoxanthin (3′-acetoxy-5, 6-epoxy-3, 5′-dihydroxy-6′, 7′-dinero-5,6,7,8,5′,6′-hexahydro-β,β-carotene-8-one) is a marine xanthophyll which contributes to more than 10% of the carotenoids isolated from brown seaweeds such as *Fucus vesiculosus* [7], *Hijikia fusiformis* [8], *Undaria pinnatifida* [9], *Petalonia binghamiae* [10], and *Laminaria japponica* [11].

Fucoxanthin is metabolized into fucoxanthinol by digestive enzymes in the gastrointestinal tract, such as esterase and lipase [12]. Fucoxanthinol is further metabolized into amarouciaxanthin A through dehydrogenation in the liver. Fucoxanthin is a safe carotenoid with no side effects at 0.5% w/v in human skin and 20–2000 mg/kg body weight in rodents. Toxicity studies conducted on rats for 4 weeks, with oral dosing of fucoxanthin, showed no harmful adverse effects after daily treatments. Additionally, in vivo investigations with fucoxanthinol showed no significant adverse effects [13].

Previous studies suggest that fucoxanthin exhibits several anticancer effects in various cancer cell lines. These effects include anti-proliferation, anti-migration, anti-angiogenic, cell cycle arrest, apoptotic, and tumor inhibition [14,15]. Fucoxanthin's anticancer potential has also been demonstrated through the inhibition of metastasis and by decreased cell growth, angiogenesis, and the induction of cell cycle arrest and apoptosis, as described in our previous literature review and studies [16,17].

Fucoxanthin was found to induce cell cycle arrest and apoptosis in various cancer cell lines such as human colon carcinoma WiDr cells, hepatocarcinoma HepG2 cells, gastric cancer adenocarcinoma MGC-803 cells, and T-cell leukemia cells [18]. In prostate cancer DU145 cells, human cervical cancer HeLa cells, and breast cancer MCF-7 cells, fucoxanthin exhibited anti-proliferative effects [19–21]. Fucoxanthin also displayed anti-angiogenic effects in human lymphatic endothelial cells (HLEC) and was found to downregulate the expression of VEGF-C, NF-κB, phospho-AKT, and phospho-PI3K [22]. In human glioma cancer cell lines, fucoxanthin at concentrations of 25 and 50 µM was shown to decrease the expression of phosphorylated AKT and mTOR, showing that fucoxanthin inhibits the PI3K/AKT/mTOR pathway [23].

Several signaling pathways, such as phosphatidylinositol 3-kinase (PI3K), AKT, mitogen protein kinase (MAPK), and nuclear factor kappa B (NF-κB), are associated with TNBC progression. These pathways are involved in cancer cell proliferation, the cell cycle, apoptosis, migration, invasion, and cell survival [24]. The PI3K-AKT signaling pathway, in particular, performs a vital role in selecting cellular processes, including proliferation, motility, cell growth, apoptosis, angiogenesis, and tumor survival [25]. The PI3K-AKT signaling pathway is important in regulating inflammatory pathways, which can be modulated by tumor necrosis factor-α (TNF-α) [26]. TNF-α is an important inflammatory cytokine that is a major contributor to tumor progression and is secreted into the tumor microenvironment [27]. The tumor microenvironment in TNBC is essential in tumor progression and inflammatory promotion. It is also involved in all stages of breast cancer development, affecting tumor growth, proliferation, survival, and recurrence [28]. TNF-α binds to TNF-receptor 1 (TNFR1) and can activate NF-κB, which promotes cyclin D1 and cell proliferation, to induce the PI3K-AKT signaling pathway [29]. TNF-α has also been found to increase CXCL10 expression, which plays a role in cell growth, apoptosis, and angiogenesis effects through the PI3K-AKT pathway [30,31].

The molecular genetic profiles of Caucasian and African American TNBC cell line populations are poorly characterized [32]. The genomic profiles of each TNBC cell line have been shown to differ, showing altered gene expression of the *AKT*, *TP53*, and *NFB1* pathways amongst Caucasian and African American TNBC cell lines [33]. A data study of TCGA with 90 Caucasian and 52 African American patients revealed that African American (28.42%) patients had a higher TNBC incidence than Caucasian American (11.89%) patients. A MAPK signaling gene set showed upregulation of *GADD45C*, *MAPK12*, *TRADD*, *NTRK1*, and *FGFR2* genes in African American patients compared to Caucasian patients, which could contribute to African American TNBC patients having the worst prognosis [34].

To elucidate fucoxanthin's molecular mechanisms in proliferation, migration, angiogenesis, and cell cycle arrest in these genetically different TNBC cell lines, the current study aimed to examine fucoxanthin's modulatory effects through TNF-α on the PI3K-AKT signaling pathway in the MDA-MB-231 and MDA-MB-468 TNBC cell lines.

2. Results

2.1. Fucoxanthin Inhibited Cell Viability in TNF-α-Stimulated MDA-MB-231 and MDA-MB-468 Cell Lines

The effect of fucoxanthin on cell viability in TNBC was investigated in the MDA-MB-231 and MDA-MB-468 cell lines after 24 h of treatment (Figure 1). The results show that both cell lines produced cell viability values over 75% in all the different treatment groups (control, fucoxanthin (6.25 μM), TNF-α, and fucoxanthin (6.25 μM) + TNF-α (50 ng/mL)).

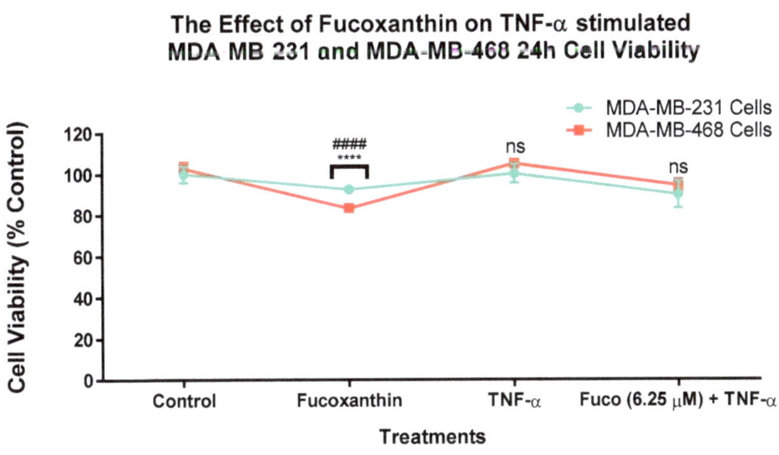

Figure 1. The effect of fucoxanthin on cell viability in MDA-MB-231 and MDA-MB-468 TNBC cells. The effect of fucoxanthin was tested at 6.25 μM on cell viability. The control cells were treated with DMSO (<0.1%). Each experiment was performed 3 times with n = 5. The cytotoxic effect was measured at 24 h of treatment. The data are presented as the mean ± SEM. Statistically significant differences between control and treatments were evaluated by a one-way ANOVA, followed by Dunnett's multiple comparison tests. **** $p < 0.0001$ (MDA-MB-231 cells), #### $p < 0.0001$ (MDA-MB-468 cells), ns = $p > 0.05$.

2.2. Fucoxanthin Modulation of the PI3K-AKT Signaling Pathway in MDA-MB-231 and MDA-MB-468 TNBC Cells

The mRNA expression of 89 genes mediating the PI3K-AKT signaling pathway was investigated in the MDA-MB-231 and MDA-MB-468 cell lines. Both cell lines were exposed to treatment groups consisting of control (cells + DMSO), fucoxanthin (6.25 μM), TNF-α (50 ng/mL), and fucoxanthin (6.25 μM) + TNF-α (50 ng/mL). The expression of

12 genes was modulated by fucoxanthin in the MDA-MB-231 and MDA-MB-468 cell lines (Figure 2). In MDA-MB-231 cells, the expression of three tumor suppressor genes was significantly upregulated by fucoxanthin compared to TNF-α alone (Figure 3). These genes include forkhead box 1 (*FOXO1*) (Figure 3A), Rho family protein A (*RHOA*) (Figure 3B), and Ras P21 protein activator (*RASA1*) (Figure 3C). In MDA-MB-231 cells, the expression of nine genes was significantly downregulated by fucoxanthin compared to TNF-α treatment only (Figure 4). These genes include ETS transcription factor (*ELK1*) (Figure 4A), HRas Proto-Oncogene, GTPase (*HRAS*) (Figure 4B), mitogen-activated protein kinase-3 (*MAPK3*) (Figure 4C), pyruvate dehydrogenase kinase 2 (*PDK2*) (Figure 4D), insulin receptor substrate 1 (*IRS1*) (Figure 4E), eukaryotic transition initiation factor 4E-binding protein 1 (*EIF4EBP1*) (Figure 4F), eukaryotic translation initiation factor 4b (*EIF4B*) (Figure 4G), protein tyrosine kinase 2 (*PTK2*) (Figure 4H), and TIR domain-containing adaptor protein (*TIRAP*) (Figure 4I). In MDA-MB-468 cells, the expression of *EIF4EBP1* (Figure 5) was significantly upregulated by fucoxanthin compared to TNF-α alone. Also, in the MDA-MB-468 cells, the expression of 11 genes was significantly downregulated compared to TNF-α alone (Figure 6). These genes include *HRAS* (Figure 6A), *FOXO1* (Figure 6B), *MAPK3* (Figure 6C), *PDK2* (Figure 6D), *RASA1* (Figure 6E), *IRS1* (Figure 6F), *EIF4B* (Figure 6G), *PTK2* (Figure 6H), *TIRAP* (Figure 6I), *RHOA* (Figure 6J), and *ELK1* (Figure 6K). In MDA-MB-231 cells, the fold changes (FCs) for significantly upregulated genes were shown to be *FOXO1* (+2.66), *RASA1* (+3.24), and *RHOA* (+1.89). The FCs for downregulated genes in MDA-MB-231 cells were *HRAS* (−1.97), *MAPK3* (−1.90), *PDK2* (−2.46), *IRS1* (−3.51), *EIF4EBP1* (−1.23), *EIF4B* (−3.83), *PTK2* (−2.64), *TIRAP* (−2.14), and *ELK1* (−3.91) (Table 1). In MDA-MB-468 cells, the FC for the significantly upregulated gene was *EIF4EBP1* (+2.16). The FCs for MDA-MB-468 downregulated genes were shown to be *FOXO1* (−1.21), *HRAS* (−1.34), *MAPK3* (−1.22), *PDK2* (−1.14), *RASA1* (−1.19), *IRS1* (−1.49), *EIF4B* (−1.51), *PTK2* (−1.51), *TIRAP* (−1.61), *RHOA* (−1.38), and *ELK1* (−1.44) (Table 2).

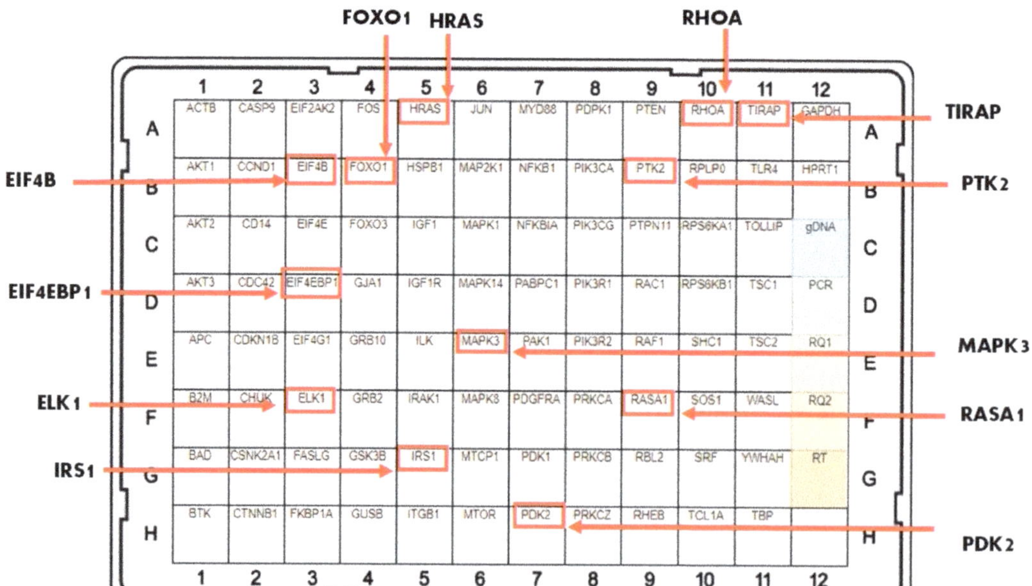

Figure 2. The mRNA expression of 12 genes in the PI3K-AKT signaling pathway was modulated (indicated by red boxes) by fucoxanthin (6.25 μM) + TNF-α (50 ng/mL) compared to TNF-α treatment in the MDA-MB-231 and MDA-MB-468 cell lines.

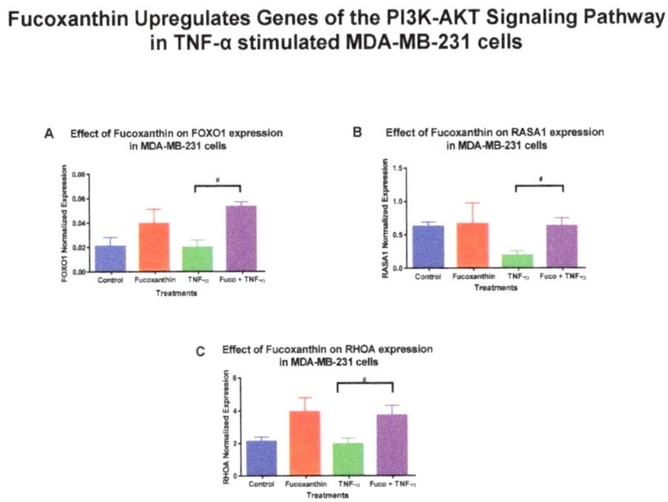

Figure 3. Fucoxanthin upregulated the mRNA expression of PI3K-AKT signaling pathway genes in TNF-α-stimulated MDA-MB-231 cells. Fucoxanthin upregulated the expression of 3 genes in the TNF-α-stimulated PI3K-AKT signaling pathway in MDA-MB-231 cells. These genes include forkhead box 1 (*FOXO1*) (**A**), Ras P21 protein activator (*RASA1*) (**B**), and Rho family protein A (*RHOA*) (**C**). Each data point represents the mean ± SEM, representing 4 treatments: control (cells + DMSO), fucoxanthin (6.25 μM), TNF-α (50 ng/mL), and fucoxanthin (6.25 μM) + TNF-α (50 ng/mL). A *t*-test was used to evaluate statistically significant differences between TNF-α and fucoxanthin + TNF-α. ns = $p > 0.05$, # $p < 0.05$.

Table 1. Data show FCs for 3 genes that were significantly upregulated in MDA-MB-231 cells, which include *FOXO1* (+2.66), *RASA1* (+3.24), and *RHOA* (+1.89), and FCs for 10 genes that were significantly downregulated in MDA-MB-231 cells, which include *HRAS* (−1.97), *MAPK3* (−1.90), *PDK2* (−2.46), *IRS1* (−3.51), *EIF4EBP1* (−1.23), *EIF4B* (−3.83), *PTK2* (−2.64), *TIRAP* (−2.14), and *ELK1* (−3.91).

Gene Name	Description	Fold Change (Downregulated or Upregulated)	Function
RASA1	Ras P21 Protein Activator 1	+3.24	Provides instruction for protein p120-RasGap and helps regulate Ras-MAPK [35].
FOXO1	Forkhead Box 1	+2.66	Tumor suppressor that induces cell cycle arrest and apoptosis [36].
RHOA	Rho Family Protein A	+1.89	Contributes to cell cycle progression, adhesion, and microtubule and actin cytoskeleton organization [37].
ELK1	ETS Transcription Factor 1	−3.91	Regulates a network of genes related to actin/migration [38].
EIF4B	Eukaryotic Translation Initiation Factor 4B	−3.83	Regulated by both Ras-MAPK and PI3K/mTOR and plays a role in cell viability, cell growth, and survival [39].
IRS1	Insulin Receptor Substrate 1	−3.51	Regulates cell motility, proliferation, and cell survival [40].
PTK2	Protein Tyrosine Kinase 2	−2.64	Stimulates tumor cell invasiveness and adhesion [41].
PDK2	Pyruvate Dehydrogenase Kinase 2	−2.46	Associated with proliferation, anti-apoptosis, tumor aggressiveness, and therapy resistance [42].
TIRAP	TIR Domain-Containing Adaptor Protein	−2.14	Activates NF-κB, MAPK3, and JNK signaling, resulting in the inflammatory response and cytokine release [43].

Table 1. *Cont.*

Gene Name	Description	Fold Change (Downregulated or Upregulated)	Function
HRAS	HRas Proto-Oncogene, GTPase	−1.97	Regulates adhesion, cytoskeleton rearrangements, proliferation, cell survival, and cell motility [44].
MAPK3	Mitogen-Activated Protein Kinase 3	−1.90	Regulates inflammation, proliferation, apoptosis, and differentiation [45].
EIF4EBP1	Eukaryotic Transition Initiation Factor 4E-Binding Protein 1	−1.23	Increases metastasis, invasion, and tumor formation [46].

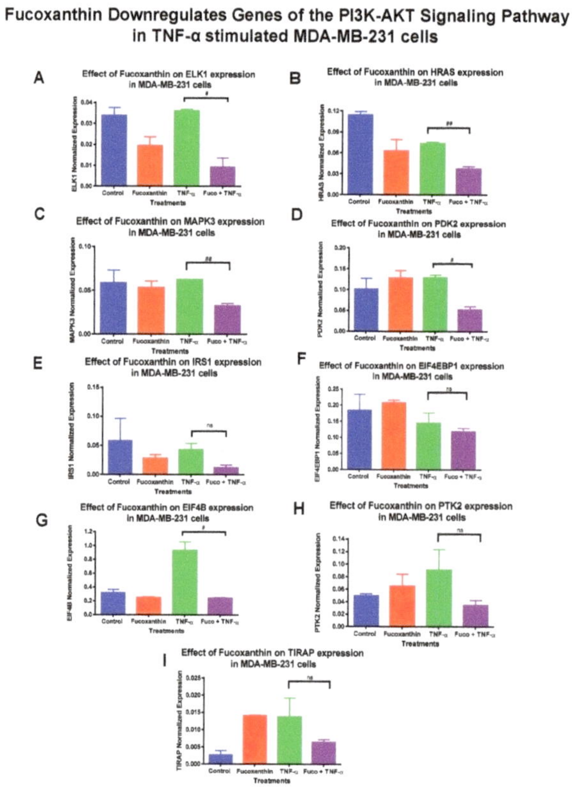

Figure 4. Fucoxanthin downregulated the mRNA expression of PI3K-AKT signaling pathway genes in TNF-α-stimulated MDA-MB-231 cells. The data display the effects of fucoxanthin on the *mRNA* gene expression of genes associated with the PI3K-AKT signaling pathway. Fucoxanthin downregulated the expression of 10 genes in TNF-α-stimulated MDA-MB-231 cells. These genes include ETS transcription factor (*ELK1*) (**A**), HRas Proto-Oncogene, GTPase (*HRAS*) (**B**), mitogen-activated protein kinase 3 (*MAPK3*) (**C**), pyruvate dehydrogenase kinase 2 (*PDK2*) (**D**), insulin receptor substrate 1(*IRS1*) (**E**), eukaryotic transition initiation factor 4E-binding protein 1 (*EIF4EBP1*) (**F**), eukaryotic translation initiation factor 4b (*EIF4B*) (**G**), protein tyrosine kinase 2 (*PTK2*) (**H**), and TIR domain-containing adaptor protein (*TIRAP*) (**I**). Each datapoint represents the mean ± SEM, representing 4 treatments: control (cells + DMSO), fucoxanthin (6.25 μM), TNF-α (50 ng/mL), and fucoxanthin (6.25 μM) + TNF-α (50 ng/mL). A *t*-test evaluated statistically significant differences between TNF-α and fucoxanthin + TNF-α (#). ns = $p > 0.05$, # $p < 0.05$, ## $p < 0.01$.

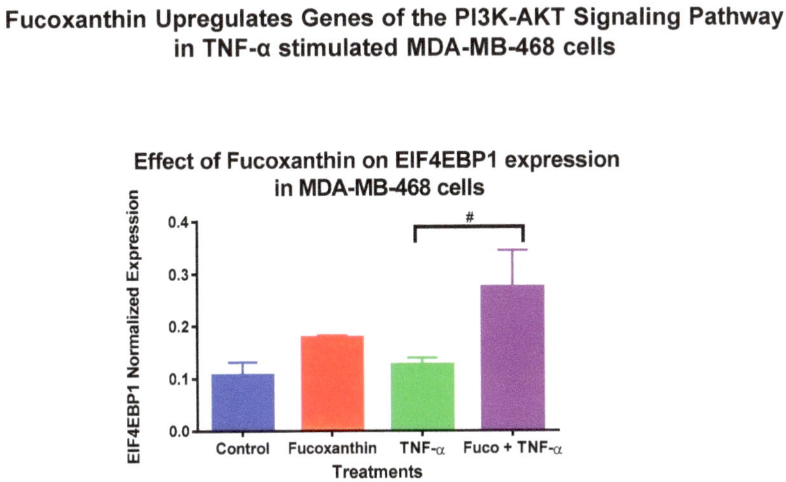

Figure 5. Fucoxanthin upregulated the mRNA expression of the PI3K-AKT signaling pathway EIF4EBP1 in TNF-α-stimulated MDA-MB-468 cells. The data display the effects of fucoxanthin on mRNA gene expression in the PI3K-AKT signaling pathway of TNF-α-stimulated MDA-MB-468 cells. Fucoxanthin upregulated the expression of eukaryotic transition initiation factor 4E-binding protein 1 (*EIF4EBP1*) in TNF-α-stimulated MDA-MB-468 cells. Each data point represents the mean ± SEM, representing 4 treatments: control (cells + DMSO), fucoxanthin (6.25 μM), TNF-α (50 ng/mL), and fucoxanthin (6.25 μM) + TNF-α (50 ng/mL). A *t*-test evaluated statistically significant differences between TNF-α and fucoxanthin + TNF-α (#). # $p < 0.05$.

Table 2. Data show FC for 1 gene significantly upregulated in MDA-MB-468 cells: *EIF4EBP1* (+2.16). Also shown are FCs for 11 genes that were significantly downregulated in MDA-MB-468 cells, which include *FOXO1* (−1.21), *HRAS* (−1.34), *MAPK3* (−1.22), *PDK2* (−1.14), *RASA1* (−1.19), *IRS1* (−1.49), *EIF4B* (−1.51), *PTK2* (−1.51), *TIRAP* (−1.61), *RHOA* (−1.38), and *ELK1* (−1.44).

Gene Name	Description	Fold Change (Downregulated or Upregulated)	Function
EIF4EBP1	Eukaryotic Transition Initiation Factor 4E-Binding Protein 1	+2.16	Increases metastasis, invasion, and tumor formation [46].
TIRAP	TIR Domain-Containing Adaptor Protein	−1.61	Activates NF-κB, MAPK3, and JNK signaling, resulting in the inflammatory response and cytokine release [43].
EIF4B	Eukaryotic Translation Initiation Factor 4B	−1.51	Regulated by both Ras-MAPK and PI3K/mTOR, it plays a role in cell viability, growth, and survival [39].
PTK2	Protein Tyrosine Kinase 2	−1.51	Stimulates tumor cell invasiveness and adhesion [41].
IRS1	Insulin Receptor Substrate 1	−1.49	Regulates cell motility, proliferation, and cell survival [40].
ELK1	ETS Transcription Factor 1	−1.44	Regulates a network of genes related to actin/migration [38].
RHOA	Rho Family Protein A	−1.38	Contributes to cell cycle progression, adhesion, and microtubule and actin cytoskeleton organization [37].
HRAS	HRas Proto-Oncogene, GTPase	−1.34	Regulates adhesion, cytoskeleton rearrangements, proliferation, cell survival, and cell motility [44].
MAPK3	Mitogen-Activated Protein Kinase 3	−1.22	Regulates inflammation, proliferation, apoptosis, and differentiation [45].

Table 2. *Cont.*

Gene Name	Description	Fold Change (Downregulated or Upregulated)	Function
FOXO1	Forkhead Box 1	−1.21	Tumor suppressor that induces cell cycle arrest and apoptosis [36].
RASA1	Ras P21 Protein Activator 1	−1.19	Provides instruction for protein p120-RasGap and helps regulate Ras-MAPK [35].
PDK2	Pyruvate Dehydrogenase Kinase 2	−1.14	Associated with proliferation, anti-apoptosis, tumor aggressiveness, and therapy resistance [42].

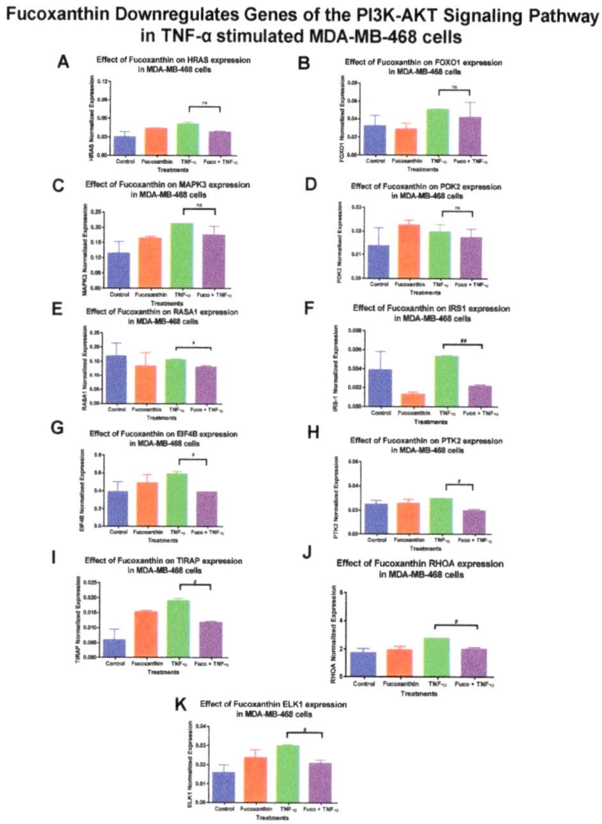

Figure 6. Fucoxanthin downregulated the mRNA expression of PI3K-AKT signaling pathway genes in TNF-α-stimulated MDA-MB-468 cells. The data display the effects of fucoxanthin on *mRNA* gene expression in the PI3K-AKT signaling pathway of TNF-α-stimulated MDA-MB-468 cells. Fucoxanthin upregulated the expression of 11 genes in TNF-α-stimulated MDA-MB-468 cells. These genes include HRas Proto-Oncogene, GTPase (*HRAS*) (**A**) forkhead box 1 (*FOXO1*) (**B**), mitogen-activated protein kinase 3 (*MAPK3*) (**C**), pyruvate dehydrogenase kinase 2 (*PDK2*) (**D**), Ras P21 protein activator (*RASA1*) (**E**), insulin receptor substrate 1(*IRS1*) (**F**), eukaryotic translation initiation factor 4b (*EIF4B*) (**G**), protein tyrosine kinase 2 (*PTK2*) (**H**), TIR domain-containing adaptor protein (*TIRAP*) (**I**), Rho family protein A (*RHOA*) (**J**), and ETS transcription factor (*ELK1*) (**K**). Each data point represents the mean ± SEM, representing 4 treatments: control (cells + DMSO), fucoxanthin (6.25 μM), TNF-α (50 ng/mL), and fucoxanthin (6.25 μM) + TNF-α (50 ng/mL). A *t*-test evaluated statistically significant differences between TNF-α and fucoxanthin + TNF-α (#). ns = $p > 0.05$, # $p < 0.05$, ## $p < 0.01$.

2.3. Fucoxanthin Modulated PI3K-AKT Signaling Pathway Protein Expression in MDA-MB-231 and MDA-MB-468 Cells

The protein levels of the PI3K-AKT PCR array results were verified using Wes analysis, which was performed using total proteins and specific antibodies for IRS-1 (Figure 7A,C), EIF4B (Figure 7B,D), ELK-1 (Figure 7E,F), and RHOA (Figure 8A,B) in MDA-MB-231 cells, and specific antibodies for EIF4B (Figure 9A,C), ELK1 (Figure 9B,D), and RHOA (Figure 10A,B) in MDA-MB-468 cells. The stimulation of cells with TNF-α upregulated IRS-1, EIF4B, and ELK-1 and downregulated RHOA protein expression in MDA-MB-231 cells. However, co-treatment with fucoxanthin + TNF-α downregulated IRS-1, EIF4B, and ELK-1 and upregulated RHOA protein expression in MDA-MB-231 cells (Figures 7 and 8). In MDA-MB-468 cells, stimulation with TNF-α downregulated RHOA, and no change was shown for ELK-1 and EIF4B protein expression. Co-treatment (fucoxanthin + TNF-α) upregulated RHOA protein expression, and no expression change was shown for ELK-1 and EIF4B protein expression in MDA-MB-468 cells (Figures 9 and 10).

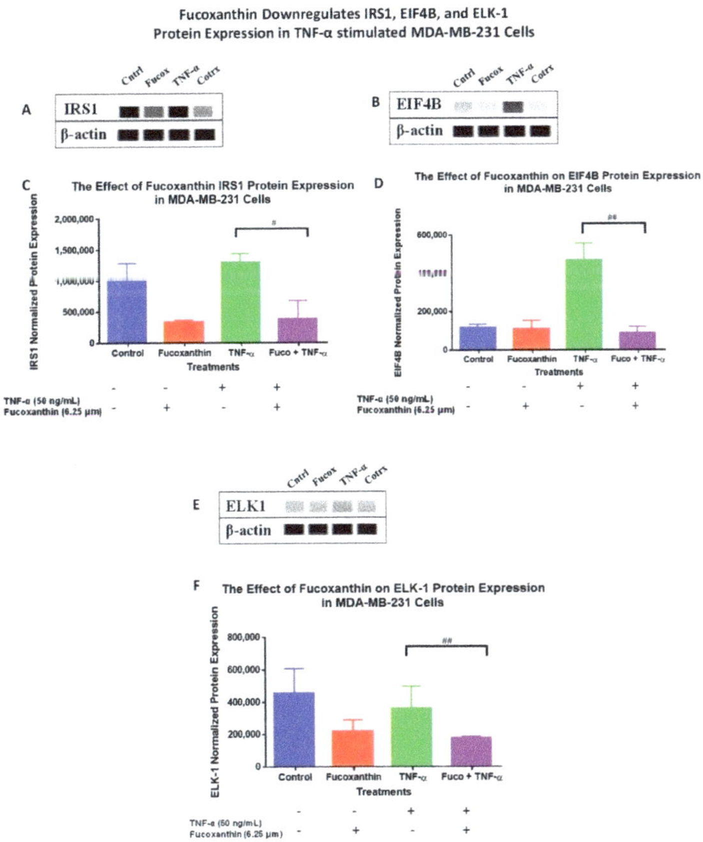

Figure 7. The data display the effects of fucoxanthin on IRS1 (**A**,**C**), EIF4B (**B**,**D**), and ELK1 (**E**,**F**) protein expression with quantitative normalized protein expression in TNF-α-stimulated MDA-MB-231 cells. Immunodetection of bands was achieved via Wes analysis from ProteinSimple www.bio-techne.com/brands/proteinsimple (accessed on 6 May 2023). Each data point represents the mean ± SEM, representing 4 treatments: control (cells + DMSO), fucoxanthin (6.25 μM), TNF-α (50 ng/mL), and fucoxanthin (6.25 μM) + TNF-α (50 ng/mL). A *t*-test was used to evaluate statistically significant differences between TNF-α and fucoxanthin + TNF-α (#). # $p < 0.05$, ## $p < 0.01$.

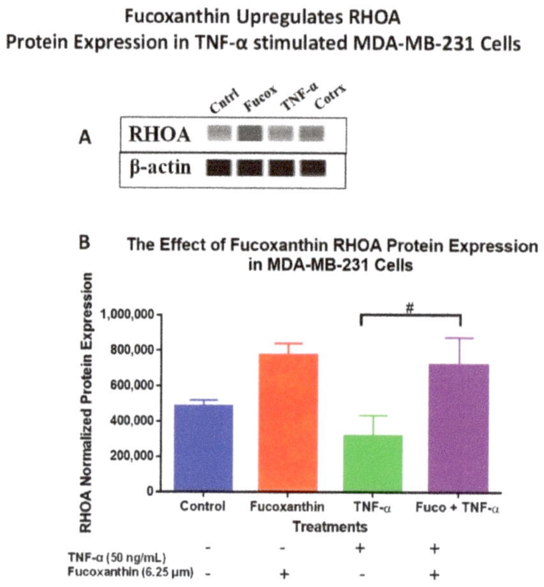

Figure 8. The data display the effects of fucoxanthin on RHOA (**A**,**B**) protein expression and quantitative normalized protein expression in TNF-α-stimulated MDA-MB-231 cells. Immunodetection of bands was achieved via Wes analysis using ProteinSimple. Each data point represents the mean ± SEM, representing 4 treatments: control (cells + DMSO), fucoxanthin (6.25 μM), TNF-α (50 ng/mL), and fucoxanthin (6.25 μM) + TNF-α (50 ng/mL). A *t*-test was used to evaluate statistically significant differences between TNF-α and fucoxanthin + TNF-α (#). # $p < 0.05$.

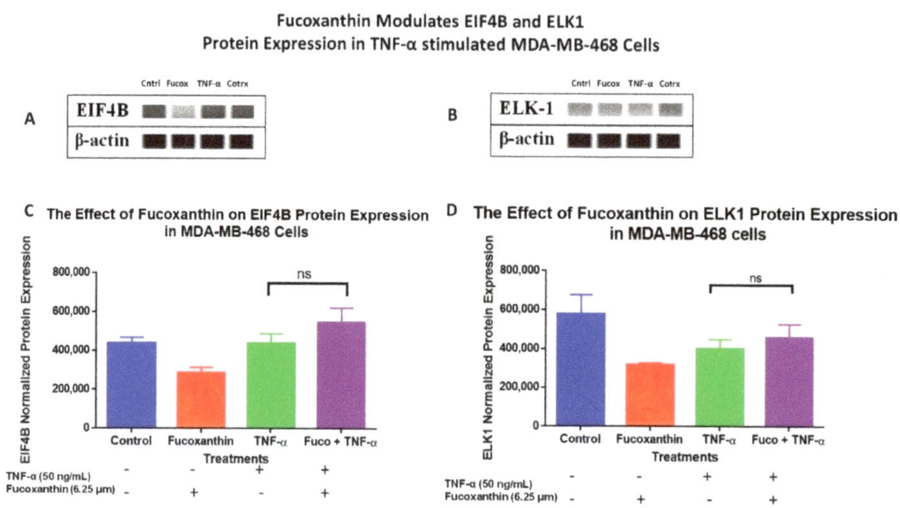

Figure 9. The data display the effects of fucoxanthin on EIF4B (**A**,**C**) and ELK1 (**B**,**D**) protein expression and quantitative normalized expression in TNF-α-stimulated MDA-MB-468 cells. Immunodetection of bands was achieved via Wes analysis using ProteinSimple. Each datapoint represents the mean ± SEM, representing 4 treatments: control (cells + DMSO), fucoxanthin (6.25 μM), TNF-α (50 ng/mL), and fucoxanthin (6.25 μM) + TNF-α (50 ng/mL). A *t*-test was used to evaluate statistically significant differences between TNF-α and fucoxanthin + TNF-α (#). ns = $p > 0.05$.

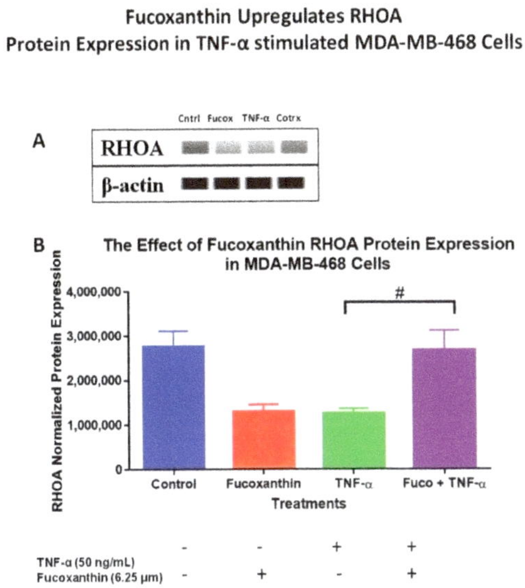

Figure 10. The data display the effects of fucoxanthin on RHOA (**A**,**B**) protein expression and quantitative normalized protein expression in TNF-α-stimulated MDA-MB-468 cells. Immunodetection of bands was achieved via Wes analysis using ProteinSimple. Each datapoint represents the mean ± SEM, representing 4 treatments: control (cells + DMSO), fucoxanthin (6.25 μM), TNF-α (50 ng/mL), and fucoxanthin (6.25 μM) + TNF-α (50 ng/mL). A *t*-test evaluated statistically significant differences between TNF-α and fucoxanthin + TNF-α (#). # $p < 0.05$.

2.4. Activation of the PI3K-AKT Signaling Pathway through Stimulation of TNF-α Using KEGG Analysis Database

The KEGG (Kyoto Encyclopedia of Genes and Genomes) database was used to show PI3K-AKT signaling activation in this investigation. The KEGG pathway displays PI3K-AKT pathway activation by TNF signaling (Figure 11), where TNF binds to the TNF receptor 2 (TNFR2) and mediates intracellular signaling, activating many genes that induce cell survival and apoptosis. TNFR2 stimulates PI3K, which then activates AKT and induces tumor cell survival [47].

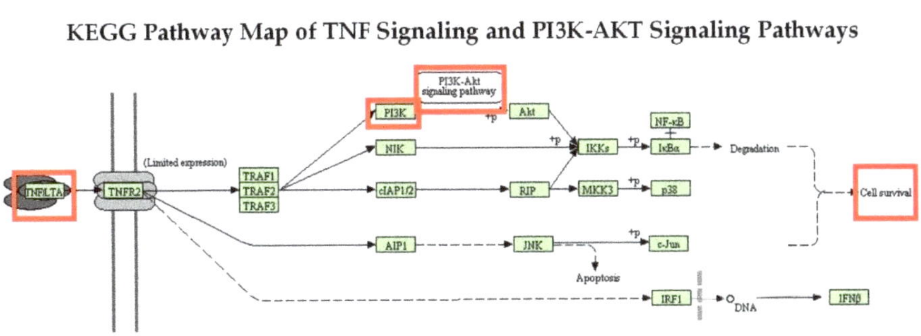

Figure 11. KEGG TNF-α signaling and its association with the PI3K-AKT signaling pathway. The KEGG pathway map displays the PI3K-AKT pathway activated by TNF signaling. TNF binds to TNF receptor 2 (TNFR2) and mediates intracellular signaling, which activates many genes, leading to cell survival and apoptosis. TNFR2 leads to the stimulation of PI3K, which then activates AKT and cell survival. Red boxes indicate TNF activation of the PI3K-AKT signaling pathway.

2.5. STRING, Functional Enrichment Analysis of Fucoxanthin's Effects on Proteins of the PI3K-AKT Signaling Pathway

Another bioinformatics tool used in this study was the STRING (search tool for retrieving interacting genes/proteins). It is a proteomic database that contains computational prediction methods, experimental data, and text collections that allow for the study of interactions of genes and proteins. The STRING functional enrichment analysis database was used to integrate and study the protein–protein analysis of six proteins of the PI3K-AKT signaling pathway, including FOXO1, PDK2, IRS1, EIF4B, RHOA, and ELK1 [48]. The network displayed the direct interaction between FOXO1, PDK2, IRS1, EIF4B, and RHOA with predicted functional partner proteins (Figure 12). The phosphatidylinositol-4, 5-bisphosphate 3 kinase (PIK3CA) gene encodes subunit p110α, which plays a role in activating PI3K-AKT signaling [49]. The interaction of FOXO1, PDK2, IRS1, EIF4B, and RHOA was also shown with the predicted functional partner, PIK3CA. However, ELK1 had no interactions with the other proteins or the associated predicted functional partner proteins. Identifying proteins that are close to neighboring proteins aids in the identification of similar biological functions within a protein network. The thickness of the edges showed protein–protein association confidence levels from low to highest (0.150–0.900). The analysis showed a high confidence interaction between FOXO1 and IRS1, a medium confidence interaction between PDK2 and IRS1 and between PDK2 and FOXO1, and a low confidence interaction between RHOA and IRS1, RHOA and FOXO1, RHOA and PDK2, and EIF4B and all proteins (Table 3). The PPI enrichment value shows the significance of interactions between all proteins. The lower the significance value, the lower the amounts of protein interactions.

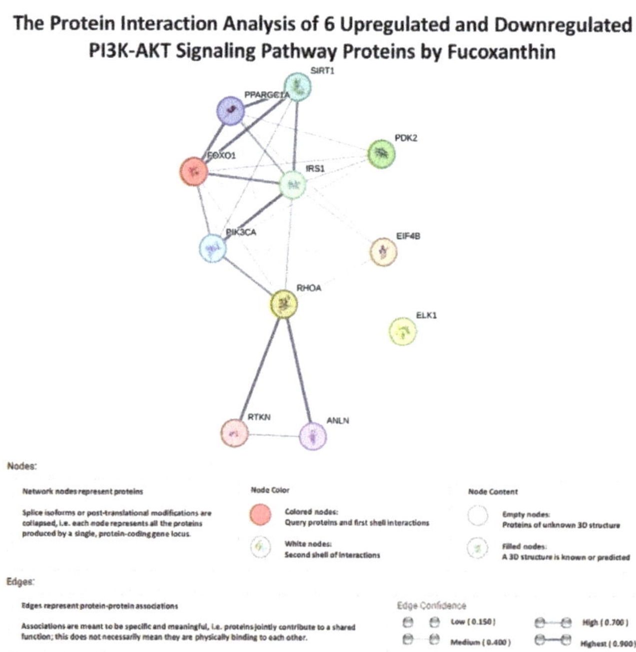

Figure 12. STRING, functional enrichment analysis of fucoxanthin's effects on 6 proteins of the PI3K-AKT signaling pathway in TNF-α-stimulated MDA-MB-231 and MDA-MB-468 cells. STRING protein–protein interaction network functional enrichment analysis of PI3K-AKT signaling pathway in TNF-α-stimulated MDA-MB-231 and MDA-MB-468 cells. The network included 6 proteins with predicted functional partners, including FOXO1, RHOA, ELK1, EIF4B, IRS1, and PDK2. The legend displays the node network, color and content, and edge protein–protein associations and confidence.

Table 3. This table displays interaction scores of proteins in the PI3K-AKT signaling pathway.

Node 1	Node 2	Node 1 Accession	Node 2 Accession	Score
RHOA	PDK2	ENSPO0000400175	ENSP00000420927	0.184
RHOA	IRS1	ENSPO0000400175	ENSP0000030489S	0.386
RHOA	FOXO1	ENSP00000400175	ENSP00000368880	0.364
RHOA	EIF4B	ENSP00000400175	ENSP00000388806	0.324
PDK2	RHOA	ENSP00000420927	ENSPO0000400175	0.184
PDK2	IRS1	ENSPO0000420927	ENSPO0000304895	0.436
PDK2	FOXO1	ENSP00000420927	ENSP00000368880	0.413
PDK2	EIF4B	ENSP00000420927	ENSP00000388806	0.179
IRS1	RHOA	ENSP00000304895	ENSPO0000400175	0.386
IRS1	PDK2	ENSP00000304895	ENSPO0000420927	0.436
IRS1	FOXO1	ENSP00000304895	ENSP00000368880	0.854
IRS1	EIF4B	ENSPO0000304895	ENSP00000388806	0.357
FOXO1	RHOA	ENSP00000368880	ENSPO0000400175	0.364
FOXO1	PDK2	ENSP00000368880	ENSP00000420927	0.413
FOXO1	IRS1	ENSP00000368880	ENSP00000304895	0.854
FOXO1	EIF4B	ENSP00000368880	ENSPO0000388806	0.244
EIF4B	RHOA	ENSP00000388806	ENSP00000400175	0.324
EIF4B	PDK2	ENSPO0000388806	ENSPO0000420927	0.179
EIF4B	IRS1	ENSPO0000388806	ENSPO0000304895	0.357
EIF4B	FOXO1	ENSPO0000388806	ENSPO0000368880	0.244

3. Discussion

Several natural products have been studied for their potential to treat or prevent breast cancer [50]. Specifically, marine carotenoids were investigated for their efficacy, potency, and low toxicity against breast cancer. Carotenoids target signaling pathways such as the PI3K-AKT, MAPK, and NF-κβ signaling pathways [51]. Fucoxanthin is a marine carotenoid isolated from brown seaweed and has many anticancer effects, such as anti-proliferative, anti-metastasis, cell cycle growth arrest, and apoptotic effects [52]. The PI3K-AKT signaling pathway mediates several of these cellular processes, contributing to the progression of tumors [53]. The current study examined fucoxanthin's inductive and modulatory effects on the PI3K-AKT signaling pathway in MDA-MB-231 and MDA-MB-468 cells.

Normal breast cells have an equilibrium between proliferation, cell cycle arrest, and apoptosis to maintain homeostasis [54]. Both apoptosis and cellular proliferation are linked by cell cycle regulators and apoptotic stimulation, which affect both processes [55]. The apoptotic process occurs when cell cycle checkpoints are interrupted, and the involved proteins cause cellular death, cell cycle arrest, or cell proliferation, thereby contributing to tumor progression. The PI3K-AKT signaling pathway has been associated with many proteins contributing to the induction of apoptosis [56], and in breast cancer cells, it can be triggered by inflammatory cytokines like TNF-α, which can induce uncontrolled cell death and disrupt the balance of anti-apoptotic and pro-apoptotic pathways [57]. The PI3K-AKT signaling pathway has been found to favor the cell cycle and regulate apoptosis. The ability to induce cell cycle arrest and inhibit cell proliferation has also been associated with inhibiting the PI3K-AKT signaling pathway [58]. It has been reported that the PI3K-AKT signaling pathway is also associated with inflammation during breast cancer progression [59].

The PI3K-AKT signaling pathway is a functional and critical pathway that consists of several genes that interconnect and are involved in cellular processes such as cell survival, cell proliferation, cell growth, cell motility, cell cycle arrest, and apoptosis in TNBC [60,61]. It has been confirmed that exposure to the inflammatory cytokine TNF-α can activate the PI3K-AKT signaling pathway in TNBC cell lines [62]. Furthermore, studies have shown that TNF-α stimulates NF-κB, J.N.K., and the PI3K-AKT signaling pathway in human breast cancer cells [63]. Several genes, such as *MAPK*, phosphatase and tensin homolog (*PTEN*), forkhead box (*FOXO*), caspase 3 (*CASP3*), x-linked inhibitor of apoptosis (*XIAP*), focal adhesion kinase (*FAK*), and nuclear factor-kappa-B (*NF-κB*), are involved in the PI3K-

AKT signaling pathway [64]. Meanwhile, PI3K is an enzyme that plays a crucial role in regulating cellular processes such as cell growth, survival, and metabolism. When PI3K is phosphorylated (phospho-PI3K), it becomes activated and initiates a cascade of events that can promote cell proliferation and survival [65]. Fucoxanthin has been shown to act on the PI3K-AKT signaling pathway, thereby inducing cell cycle arrest and apoptosis and inhibiting proliferation, angiogenesis, and migration [66,67].Through the downregulation of PI3K/AKT/mTOR signaling, fucoxanthin has been proven to inhibit proliferation and stimulate apoptosis in ovarian cancer cells [68]. Fucoxanthin has also been shown to inhibit PI3K-AKT signaling by suppressing cell motility, migration, invasion, adhesion, and survival through the downregulation of p-PI3K, p-AKT, AKT1/2/3, and the focal adhesion proteins FAK/paxillin in MCF-7 cells and mouse lung tumors [69]. Studies showed that phospho-PI3K and phospho-Akt protein expressions were decreased after treatment with fucoxanthin in human lymphatic endothelial cells (HLEC). These results also showed that fucoxanthin inhibits tube migration and the formation of HLEC through suppression of the PI3K-AKT signaling pathway [70]. In glioma cells, fucoxanthin has been shown to cause the inhibition of PI3K-AKT and MAPKs in a time-dependent manner [71]. In vivo studies show that fucoxanthin inhibits PI3K/AKT/mTOR in BALB/C mice injected with U251 and glioma U87 cells [72]. Fucoxanthin has also been shown to inactivate the PI3K-AKT signaling pathway, thereby preventing metastasis in MCF-7 breast cancer cells [73]. Fucoxanthin has also been shown to inhibit PI3K-AKT signaling in several in vitro models; however, no effects of fucoxanthin on PI3K-AKT signaling have been demonstrated in MDA-MB-468 cells [74].

The present study used a PI3K-AKT signaling PCR array to identify the expression of 89 genes to attain an all-inclusive insight into signaling pathway molecular gene profiles. Through the stimulation of TNF-α, our results show that the mRNA expression of 12 genes was modulated by fucoxanthin in the MDA-MB-231 and MDA-MB-468 cell lines. Fucoxanthin was found to downregulate the expression of several genes in the PI3K-AKT signaling pathway in both the TNF-α-stimulated MDA-MB-231 and MDA-MB-468 cell lines, including *HRAS*, *MAPK3*, *PDK2*, *IRS1*, *EIF4B*, *PTK2*, *TIRAP*, and *ELK1*. Many of these genes induce several processes, such as cell cycle progression, apoptosis, cellular adhesion, mitosis, cell motility, and cytoskeleton remodeling [75,76]

Several targets of AKT signaling, such as insulin receptor substrate-1 (IRS1) and MAPK3, are involved in protein synthesis, leading to cell growth and cell cycle progression [77]. MAPKs play a role in inflammation, proliferation, apoptosis, differentiation, progression, and the development of breast cancer [78]. IRS1 is a mediator of the PI3K-AKT signaling pathway known to promote tumorigeneses. IRS proteins have been found to play an important role in breast cancer survival by regulating motility, proliferation, and cell survival [79]. IRS1 has been identified to promote PTK2 (protein tyrosine kinase 2), also known as FAK (focal adhesion kinase). PTK2 or FAK has been shown to assist in and stimulate tumor cell invasiveness. FAK is linked to inducing cell features such as cell adhesion and cellular migration, leading to metastasis and cellular invasion, and the overexpression of FAK leads to the inhibition of apoptosis [80]. Cancer cells have been found to avoid apoptosis and choose cellular pathways associated with proliferation, motility, adhesion, and migration, where the overexpression of PTK2 and IRS1 proteins plays a role in survival. Together, IRS1 and PTK2 may provide an explanation as to why MDA-MB-468 cells evaded apoptosis [81].

Harvey rat viral oncogene homolog (HRAS) is a member of the RAS family, which are GTP-binding proteins and play a crucial role in the regulation of adhesion, cytoskeleton rearrangements, proliferation, cell survival, and cell motility [82]. When activated, Ras proteins act as a cellular switch that forms an active GTP-bound form of RAS, which activates various cellular processes [83]. In breast epithelial cells, activated HRAS induces invasion through the recruitment of p38. FAK is required for PI3K-dependent breast tumorigenesis and activated Ras [84]. Pyruvate dehydrogenase kinase 2 (PDK2) is associated with proliferation, anti-apoptosis, tumor aggressiveness, and therapy resistance [85]. PDK2 has a co-responsibility with PDK1 to phosphorylate and activate AKT [86]. AKT binds to PIP3 at the plasma membrane and becomes phosphorylated by PDK1 and PDK2 [87]. PDK1 carries out partial activation of AKT, and full activation is carried out by PDK2, which acts to gain control of AKT and regulates its activity [88].

Cancer cells have a high demand for proteins that sustain an increased metabolic rate and proliferation for cancer progression. Eukaryotic translation initiation factor 4B (EIF4B) is regulated by both the Ras-MAPK and PI3K/mTOR proto-oncogenic signaling pathways [89]. The MAPK and mTOR signaling pathways converge, and EIF4B is one of the focal points between both pathways, where survival and proliferation are increased. EIF4B has been shown to induce cell transformation, is considered an oncogene, and leads to further cancer risk [90]. Decreases in the phosphorylation of EIF4B lead to a reduction in viability, growth, and survival in TNBC [91]. Moreover, Toll/Interleukin-1-receptor domain-containing adaptor protein (TIRAP) plays a role in immune signaling and inflammatory signaling pathways. TIRAP activates NF-κB, MAPK3, and JNK signaling, resulting in the inflammatory response and cytokine release [92]. ETS-like (ELK) acts as an integration point for MAP kinase pathways and regulates a network of genes related to actin/migration [93]. The knockdown of ELK1 has reduced pancreatic cancer, breast cancer, and glioblastoma cells [94].

Fucoxanthin +TNF-α was found to downregulate the gene expression of RHOA, with an FC of −1.38, in TNF-α-stimulated MDA-MB-468 cells, and upregulate the expression of *RHOA*, with an FC of +1.89, in MDA-MB-231 cells in the PI3K-AKT signaling pathway. For the function of FAK-mediated cellular migration to occur, actin and cytoskeleton remodeling by small GTPases such as G protein Rho family protein A (RHOA) is required [95]. Guanine nucleotide exchange factor proteins facilitate converting the GDP-bound (inactive) to the GTP-bound (active) form, and GTPases facilitate converting from the active to the inactive form. RHOA GTPases are found to cycle between GDP and GTP-bound states and contribute to various processes such as cell polarity, cell morphogenesis, cell cycle progression, gene expression, vesicle trafficking, and the organization of microtubule and actin cytoskeletons [96].

In the present work, fucoxanthin was found to upregulate the expression of two genes, including *FOXO1* and *RASA1*, with FCs of +2.66 and +3.24, respectively, in TNF-α-stimulated MDA-MB-231 cells, and was found to upregulate EIF4EBP1, with an FC of +2.16, in TNF-α-stimulated MDA-MB-468 cells. Fucoxanthin downregulated the expression of EIF4EBP1, with an FC of −1.23, in MDA-MB-231, and downregulated the expression of two genes, including *FOXO1* and *RASA1*, with FCs of −1.21 and −1.19, respectively in TNF-α-stimulated MDA-MB-468, the PI3K-AKT signaling pathway of TNF-α-stimulated MDA-MB-231 cells. AKT has been found to inactivate tumor suppressor forkhead families of transcription factors such as FOXO1, which has been found to inactivate pro-caspase proteins such as Fas-L and Bad [97]. AKT directly phosphorylates FOXO, which promotes the nuclear elimination of FOXO-dependent transcription and suppresses FOXO activity [98]. The depletion of PI3K and activation of FOXO1 has led to cell cycle arrest and apoptosis in MCF-7 breast cancer cells [99].

Ras p21 protein activator 1 or Ras GTPase-activating protein (RASA1) expression is prevalent in TNBC tumors and is a tumor suppressor. RASA1 provides instruction for protein p120-RasGap and helps regulate Ras-MAPK [100]. Studies show that low expression of RASA1 is associated with TNBC and activates PI3K-AKT signaling. Loss of RASA1 expression is also involved in angiogenesis, metastasis, tumorigenesis, and invasion [101]. In breast cancer, EIF4EBP1 is highly overexpressed and frequently amplified, and is connected to increased metastasis, invasion, and tumor formation [102].

According to our findings, fucoxanthin has been shown to modulate several genes of the PI3K-AKT signaling pathway. These genes are involved in metastasis, invasion, cell growth, the cell cycle, proliferation, apoptosis inhibition, cytoskeleton and motility regulation, and microtubule and actin organization. To validate the results at the protein level, Wes analysis was performed. Fucoxanthin was shown to downregulate IRS-1,

EIF4B, and ELK-1 and upregulate RHOA protein expression in TNF-α MDA-MB-231 cells. Fucoxanthin + TNF-α showed no change in the protein expression of EIF4B and ELK1 and upregulated RHOA protein expression in MDA-MB-468 cells. Low expression of this protein is known to initiate the actin cytoskeleton, permit cellular migration, encode small GTPases, and increase breast cancer metastasis. The cells' actin cytoskeleton changes during mitosis, and actin cytoskeleton disruption leads to cell cycle arrest at the S phase [103]. These studies show that RHOA protein expression is associated with cell cycle arrest and cellular migration.

This study has certain limitations, one of which is the genetic variation between both cell lines. The molecular genetic profiles of Caucasian and African American TNBC cell line populations have not been well characterized [104]. Several studies have also revealed that the genomic profiles of TNBC cell lines vary, with altered gene expression observed in both Caucasian and African American TNBC cell lines [105].

As previously stated, several studies examined fucoxanthin's effects on the PI3K-AKT signaling pathway in MDA-MB-231 cells; however, there was a lack of studies on MDA-MB-468 cells. The present study fills this gap, demonstrating that fucoxanthin modulated genes and proteins associated with the PI3K-AKT signaling pathway in both cell lines but showing different responses after fucoxanthin treatment. The PI3K-AKT signaling pathway is linked to cellular processes such as tumor cell migration, proliferation, angiogenesis, cell cycle progression, apoptosis, and cell adhesion. By showing the modulatory effects of fucoxanthin in genes and proteins of this pathway, this investigation indicates a possible molecular mechanism for fucoxanthin's effects in both cell lines and shows fucoxanthin's potential in halting cancer progression.

4. Materials and Methods

4.1. Cell Lines, Chemicals, and Reagents

MDA-MB-231 (Caucasian American) and MDA-MB-468 (African American) TNBC cells were purchased from the American Type Culture Collection (ATCC). Heat-inactivated fetal bovine serum (HI-FBS), Dulbecco's modified Eagle's medium (DMEM) high glucose, penicillin/streptomycin, and phosphate-buffered saline (PBS) were obtained from Genesee Scientific (San Diego, CA, USA). Alamar Blue®, fucoxanthin (95% purity), dimethyl sulfoxide (DMSO), chloroform, and isopropyl alcohol were purchased from Sigma-Aldrich Co. (St. Louis, MO, USA). Human Angiogenesis Antibody Array (Cat# AAH-ANG-1000-4) and TNF-alpha (Cat# ELH-TNFα) were purchased from RayBiotech (Norcross, GA, USA). TRIzol was purchased from Thermo Fischer Scientific (Wilmington, DE, USA). PrimePCR_PI3K-AKT signaling pathway (SABTarget list) H96 arrays, SYBR Green, and iScript advanced reverse transcriptase kit were purchased from Bio-Rad (Hercules, CA, USA). A propidium iodide DNA staining kit was purchased from Abcam USA (Cambridge, MA, USA). The Wes kit and reagents used were obtained from ProteinSimple (San Jose, CA, USA). The antibodies used were obtained from Cell Signaling Technology (Danvers, MA, USA) and ThermoFisher (Waltham, MA, USA).

4.2. Cell Culture

TNBC cells (MDA-MB-231 and MDA-MB-468) were cultured in DMEM supplemented with 1% penicillin (100 U/mL)/streptomycin (0.1 mg/mL) and 10% heat-inactivated fetal bovine serum (HI-FBS). Cells were incubated at 5% CO_2 and 37 °C, subcultured in T-75 flasks, and then, grown to 90% confluency before setting the cells for each assay. The plating used was composed of a DMEM experimental medium supplemented with 2.5% HI-FBS and no penicillin/streptomycin for each experiment.

4.3. Cell Viability

Cells were seeded at a density of 3×10^4 cells/100 μL/well in 96-well plates and incubated for attachment overnight in experimental media. After overnight incubation, cells were treated with control (DMSO + media), fucoxanthin (6.25 μM), TNF-α (50 ng/mL), and fucoxanthin (6.25 μM) + TNF-α (50 ng/mL) and incubated for another 24 h. In our previous study, we tested various concentrations of fucoxanthin (1.56–300 μM) alone at 24 h (Ahmed et al., 2023). In the current study, we selected fucoxanthin (6.25 μM) at 24 h because this concentration shows cell viability at more than 75% in both cell lines. The compound fucoxanthin was dissolved in DMSO and then in a medium, with the final concentration of DMSO not exceeding 0.1%. A volume of 100 μL of each treatment was added to the plates containing cells, with a final volume of 200 μL. Alamar Blue® (Resazurin) assay was used to assess cell viability, and 20 μL of Alamar Blue® solution (0.5 mg/mL) was added to each plate and incubated for a total of 4 h. An Infinite M200 microplate reader (Tecan Trading AG) was used to read the fluorescence of excitation/emission at 550/580 nm wavelengths for quantification. Fluorescence changes were observed as resazurin was reduced to resorufin by the viable cells. The fluorescent signal was proportional to the number of viable cells in the sample, and the data were displayed as a percentage of live, untreated controls.

4.4. cDNA Synthesis and RT-PCR Array (PI3K-AKT Signaling Pathway)

4.4.1. Extraction of RNA

Cells were exposed to different treatments for 24 h, and treatments included control (cells + DMSO), fucoxanthin treatment (6.25 μM), TNF-α stimulation (50 ng/mL), and co-treatment with fucoxanthin (6.25 μM) + TNF-α (50 ng/mL). After 24 h, cells were harvested, and cell pellets were collected. An amount of 1 mL of TRIzol was added to each homogenized pellet. Next, 200 μL of chloroform was added to each lysed sample. For 3–5 min, the samples sat at room temperature on a rack, were vortexed for 15–30 s, and then centrifuged at 11,000 rpm for 12 min at 2–8 °C. New tubes containing 500 μL of isopropyl alcohol were used to collect the supernatant of each sample. Tubes were inverted 3 times and, for 10 min, were incubated at room temperature. For RNA precipitation, samples were centrifuged at 11,000 rpm for 12 min at 2–8 °C, and the sample supernatant was removed. Pellets were washed with 75% ethanol (inverted carefully) and centrifuged at 9000 rpm for 5 min at 2–8 °C. The ethanol was aspirated, and for 15 min, tubes sat at room temperature until ethanol droplets were evaporated. The pellets were dissolved in 35 μL of RNase-free water and incubated for 30 min on ice. The NanoDrop Spectrophotometer (Thermo Fischer Scientific, Wilmington, DE, USA) was used for each sample to determine RNA purity and quantification.

4.4.2. cDNA Conversion and RT-PCR

The cDNA strands were synthesized from mRNA using iScript advanced transcriptase from Bio-Rad. A combination of 4 μL of the 5X iScript advanced reaction mix, 1 μL of reverse transcriptase, 7.5 μL of the sample (1.5 μg/reaction), and 7.5 μL of water was combined in a 0.2 mL tube to make a total volume of 20 μL. Reverse transcription was performed using the thermocycler program, and the transcription steps included 46 °C for 20 min and 95 °C for 1 min. Next, for RT-PCR array studies, a combination of 5 μL of the sample (200 ng cDNA/reaction) and 1045 μL of nuclease-free water was mixed and vortexed to make a cDNA dilution. Then, 30 μL from the cDNA dilution and 3 μL of the PCR control assay were mixed and vortexed in a small tube to prepare the PCR control and added to the positive control well. Then, 10 μL of the cDNA dilution and 10 μL of SYBR Green were combined and added to each well except the control well. For the positive control well, a combination of 10 μL of the PCR control assay dilution and 10 μL of SYBR Green were combined. RT-PCR was performed at 95 °C for 2 min, and denaturation was carried out at 95 °C for 10 s, followed by 39 cycles of 60 °C for 30 s (annealing/extension), and 65–95 °C for 5 s/step (Melting curve) using the Bio-Rad CFX96 Real-Time System (Hercules, CA, USA). The PCR arrays (PI3K-AKT signaling pathway SAB Target List H96—Bio-Rad Hercules, CA, USA) were specific to genes that are involved in the PI3K-AKT signaling pathway.

4.5. Wes Analysis

A Wes machine (ProteinSimple, San Jose, CA, USA) was utilized to determine total proteins. ProteinSimple provided reagents, and analysis was performed following the user manual. Protocols were adjusted for each antibody and protein loading. Cells were exposed to the following treatments: control (media + DMSO), fucoxanthin (6.25 μM), TNF-α (50 ng/mL), and a combination of fucoxanthin (6.25 μM) + TNF-α (50 ng/mL). After a 24-h treatment, cells were washed and pelleted, and a lysis buffer comprising a protease inhibitor cocktail was added to each pellet. Protein concentrations of 0.25 mg/mL, 0.50 mg/mL, and 1 mg/mL were used. Proteins were integrated with a master mix to obtain final concentrations of 0.25 mg/mL to 1 mg/mL total protein, 1x sample buffer, 1x fluorescent molecular weight markers, and 40 mM dithiothreitol. For 5 min, samples were heated to 95 °C. Blocking solution, samples, primary antibodies (dilution: 1:5 to 1:125), a chemiluminescent substrate, horseradish peroxidase-conjugated secondary antibodies, and separation and stacking matrices were inserted into assigned wells on a microplate. Target protein identification was analyzed using a primary antibody specific to each protein, an HRP-conjugated secondary antibody, and a chemiluminescent substrate. The microplate was loaded and inserted in the instrument where fully automated electrophoresis and immunodetection occurred following the manufacturer's protocol (ProteinSimple). The reaction occurred within the capillary system utilizing exclusive antibodies; the proteins were detected, the chemiluminescence reaction was elucidated, and digital blot images were captured. A charge-coupled-device camera captured chemiluminescence, and the digital image was quantified and analyzed using ProteinSimple Compass software (Wes-WS2709). Protein expression was normalized using beta-actin as the loading control, and normalization was also calculated using ProteinSimple Compass Software. The selected antibodies specific to each protein from Cell Signaling Technology® were as follows: IRS-1 (59G8) Rabbit mAb #2390S, EIF4B Antibody #3592S, RhoA (67B9) Rabbit mAb #2117S, Elk-1 Antibody #9182S, and Beta-Actin (13E5) Rabbit mAb #4970S.

4.6. Data Analysis

Statistical analysis was performed using GraphPad Prism (version 6.07). All data were expressed as mean ± SEM from at least 3 independent experiments, and the significance of the difference between the groups was assessed using Student's *t*-test and a one-way ANOVA, followed by Dunnett's multiple comparisons test. * $p < 0.05$, ** $p < 0.01$, *** $p < 0.001$, and **** $p < 0.0001$, and ns = $p > 0.05$. KEGG (Kyoto Encyclopedia of Genes and Genomes: https://www.genome.jp/kegg/ accessed on 12 June 2023) and STRING (Search Tool for Retrieving Interacting Genes/proteins: https://string-db.org/ accessed on 15 June 2023) databases were utilized to display gene, protein, and signaling pathway interactions.

5. Conclusions

In conclusion, fucoxanthin modulated several genes of the PI3K-AKT signaling pathway that are differentially expressed in each of the MDA-MB-231 and MDA-MB-468 cell lines. The differential expression of the identified genes supports the abnormal activation of the signaling pathway that leads to mitotic spindle formation, a dysregulated cell cycle, continued division, and angiogenesis. Fucoxanthin has been shown to be effective at modulating several genes and proteins in the PI3K-AKT signaling pathway that aid in inducing cell cycle arrest and apoptosis or inhibiting mitotic spindle formation and cellular division (Figure 13), indicating that this pathway may be the molecular signaling involved in fucoxanthin's anticancer effects. This study shows that fucoxanthin may be a promising chemotherapeutic for breast cancer therapy and could be combined with chemotherapy to enhance synergistic therapeutic efficacy in TNBC.

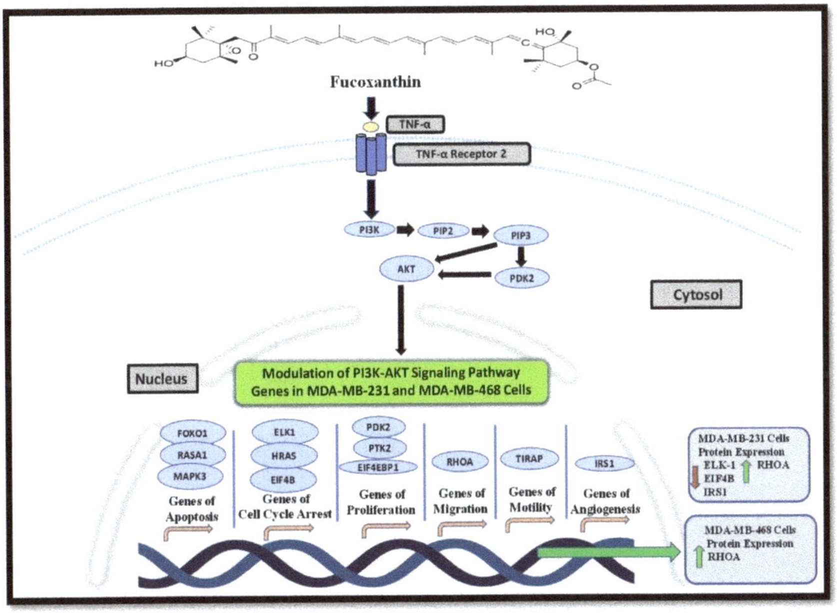

Figure 13. Proposed molecular effects of fucoxanthin through modulation of genes and proteins of the PI3K-AKT signaling pathway in TNF-α-stimulated MDA-MB-231 and MDA-MB-468 cells. The black arrow shows that fucoxanthin modulated the PI3K-AKT signaling pathway genes in MDA-MB-231 and MDA-MB-468 cells. Red arrows indicate inhibition by fucoxanthin, and green arrows show activation by fucoxanthin. The graph was created utilizing Microsoft PowerPoint version 1808.

Author Contributions: Conceptualization: S.A.A., P.M., S.S.M. and K.F.A.S.; funding acquisition: K.F.A.S.; project administration: K.F.A.S.; investigation: S.A.A., P.M., S.S.M. and K.F.A.S.; resources: K.F.A.S.; supervision: K.F.A.S.; writing—original draft: S.A.A., P.M. and S.S.M.; writing—review and editing: S.A.A., P.M., S.S.M., E.T.O. and K.F.A.S. All authors have read and agreed to the published version of the manuscript.

Funding: This research was supported by grants from the National Institute of Minority Health and Health Disparities of the National Institutes of Health under Award Number U54 MD007582.

Institutional Review Board Statement: Not applicable.

Informed Consent Statement: Not applicable.

Data Availability Statement: All data from this study are included in this published article.

Conflicts of Interest: The authors declare no conflict of interest.

References

1. Breast Cancer Research Foundation. *Breast Cancer Statistics and Resources*; Breast Cancer Research Foundation: New York, NY, USA, 2023; Available online: https://www.bcrf.org/breast-cancer-statistics-and-resources/ (accessed on 18 June 2023).
2. National Cancer Institute. *Cancer Stat Facts: Female Breast Cancer*; National Cancer Institute: Bethesda, MD, USA, 2023. Available online: https://seer.cancer.gov/statfacts/html/breast.html (accessed on 16 June 2023).
3. Aysola, K.; Desai, A.; Welch, C.; Xu, J.; Qin, Y.; Reddy, V.; Matthews, R.; Owens, C.; Okoli, J.; Beech, D.J.; et al. Triple Negative Breast Cancer—An Overview. *Hered. Genet.* **2013**, *2013* (Suppl. S2), 1. [CrossRef]
4. Wahba, H.A.; El-Hadaad, H.A. Current approaches in the treatment of triple-negative breast cancer. *Cancer Biol. Med.* **2015**, *12*, 106–116. [CrossRef] [PubMed]
5. Dewi, C.; Fristiohady, A.; Amalia, R.; Khairul Ikram, N.K.; Ibrahim, S.; Muchtaridi, M. Signaling Pathways and Natural Compounds in Triple-Negative Breast Cancer Cell Line. *Molecules* **2022**, *27*, 3661. [CrossRef] [PubMed]

6. Hashem, S.; Ali, T.A.; Akhtar, S.; Nisar, S.; Sageena, G.; Ali, S.; Al-Mannai, S.; Therachiyil, L.; Mir, R.; Elfaki, I.; et al. Targeting cancer signaling pathways by natural products: Exploring promising anticancer agents. *Biomed. Pharmacother.* **2022**, *150*, 113054. [CrossRef] [PubMed]
7. Zaragozá, M.C.; López, D.; Sáiz, M.P.; Poquet, M.; Pérez, J.; Puig-Parellada, P.; Màrmol, F.; Simonetti, P.; Gardana, C.; Lerat, Y.; et al. Toxicity and antioxidant activity in vitro and in vivo of two Fucus vesiculosus extracts. *J. Agric. Food Chem.* **2008**, *56*, 7773–7780. [CrossRef] [PubMed]
8. Nishino, H. Cancer prevention by carotenoids. *Mutat. Res./Fundam. Mol. Mech. Mutagen.* **1998**, *402*, 159–163. [CrossRef]
9. Airanthi, M.K.; Sasaki, N.; Iwasaki, S.; Baba, N.; Abe, M.; Hosokawa, M.; Miyashita, K. Effect of brown seaweed lipids on fatty acid composition and lipid hydroperoxide levels of mouse liver. *J. Agric. Food Chem.* **2011**, *59*, 4156–4163. [CrossRef]
10. Murakami, C.; Takemura, M.; Sugiyama, Y.; Kamisuki, S.; Asahara, H.; Kawasaki, M.; Ishidoh, T.; Linn, S.; Yoshida, S.; Sugawara, F.; et al. Vitamin A-related compounds, all-trans-retinal and retinoic acids, selectively inhibit the activities of mammalian replicative DNA polymerases. *Biochim. Biophys. Acta* **2002**, *1574*, 85–92. [CrossRef]
11. Ming, J.X.; Wang, Z.C.; Huang, Y.; Ohishi, H.; Wu, R.J.; Shao, Y.; Wang, H.; Qin, M.Y.; Wu, Z.L.; Li, Y.Y.; et al. Fucoxanthin extracted from Laminaria Japonica inhibits metastasis and enhances the sensitivity of lung cancer to Gefitinib. *J. Ethnopharmacol.* **2021**, *265*, 113302. [CrossRef]
12. Mohibbullah, M.; Haque, M.N.; Sohag, A.A.M.; Hossain, M.T.; Zahan, M.S.; Uddin, M.J.; Hannan, M.A.; Moon, I.S.; Choi, J.S. A Systematic Review on Marine Algae-Derived Fucoxanthin: An Update of Pharmacological Insights. *Mar. Drugs* **2022**, *20*, 279. [CrossRef]
13. Zhang, H.; Tang, Y.; Zhang, Y.; Zhang, S.; Qu, J.; Wang, X.; Kong, R.; Han, C.; Liu, Z. Fucoxanthin: A Promising Medicinal and Nutritional Ingredient. *Evid. Based Complement. Alternat. Med.* **2015**, *2015*, 723515. [CrossRef] [PubMed]
14. Kumar, S.R.; Hosokawa, M.; Miyashita, K. Fucoxanthin: A marine carotenoid exerting anticancer effects by affecting multiple mechanisms. *Mar. Drugs* **2013**, *11*, 5130–5147. [CrossRef] [PubMed]
15. Martin, L.J. Fucoxanthin and Its Metabolite Fucoxanthinol in Cancer Prevention and Treatment. *Mar. Drugs* **2015**, *13*, 4784–4798. [CrossRef]
16. Ahmed, S.A.; Mendonca, P.; Elhag, R.; Soliman, K.F.A. Anticancer Effects of Fucoxanthin through Cell Cycle Arrest, Apoptosis Induction, Angiogenesis Inhibition, and Autophagy Modulation. *Int. J. Mol. Sci.* **2022**, *23*, 16091. [CrossRef] [PubMed]
17. Ahmed, S.A.; Mendonca, P.; Messeha, S.S.; Soliman, K.F.A. Anticancer Effects of Fucoxanthin through Cell Cycle Arrest, Apoptosis Induction, and Angiogenesis Inhibition in Triple-Negative Breast Cancer Cells. *Molecules* **2023**, *28*, 6536. [CrossRef]
18. Yoshiko, S. Antitumor and Cancer-preventative Function of Fucoxanthin: A Marine Carotenoid. *Anticancer Res.* **2017**, *37*, 1557. Available online: http://ar.iiarjournals.org/content/37/4/1557.abstract (accessed on 7 July 2023).
19. Ye, G.; Wang, L.; Yang, K.; Wang, C. Fucoxanthin may inhibit cervical cancer cell proliferation via downregulation of HIST1H3D. *J. Int. Med. Res.* **2020**, *48*, 0300060520964011. [CrossRef]
20. Kotake-Nara, E.; Kushiro, M.; Zhang, H.; Sugawara, T.; Miyashita, K.; Nagao, A. Carotenoids affect proliferation of human prostate cancer cells. *J. Nutr.* **2001**, *131*, 3303–3306. [CrossRef]
21. Lau, T.Y.; Kwan, H.Y. Fucoxanthin Is a Potential Therapeutic Agent for the Treatment of Breast Cancer. *Mar. Drugs* **2022**, *20*, 370. [CrossRef]
22.Méresse, S.; Fodil, M.; Fleury, F.; Chénais, B. Fucoxanthin, a Marine-Derived Carotenoid from Brown Seaweeds and Microalgae: A Promising Bioactive Compound for Cancer Therapy. *Int. J. Mol. Sci.* **2020**, *21*, 9273. [CrossRef]
23. Liu, Y.; Zheng, J.; Zhang, Y.; Wang, Z.; Yang, Y.; Bai, M.; Dai, Y. Fucoxanthin Activates Apoptosis via Inhibition of PI3K/Akt/mTOR Pathway and Suppresses Invasion and Migration by Restriction of p38-MMP-2/9 Pathway in Human Glioblastoma Cells. *Neurochem. Res.* **2016**, *41*, 2728–2751. [CrossRef] [PubMed]
24. He, Y.; Sun, M.M.; Zhang, G.G.; Yang, J.; Chen, K.S.; Xu, W.W.; Li, B. Targeting PI3K/Akt signal transduction for cancer therapy. *Signal Transduct. Target. Ther.* **2021**, *6*, 425. [CrossRef] [PubMed]
25. Porta, C.; Paglino, C.; Mosca, A. Targeting PI3K/Akt/mTOR Signaling in Cancer. *Front. Oncol.* **2014**, *4*, 64. [CrossRef] [PubMed]
26. Xu, C.; Feng, C.; Huang, P.; Li, Y.; Liu, R.; Liu, C.; Han, Y.; Chen, L.; Ding, Y.; Shao, C.; et al. TNFα and IFNγ rapidly activate PI3K-AKT signaling to drive glycolysis that confers mesenchymal stem cells enhanced anti-inflammatory properties. *Stem Cell Res. Ther.* **2022**, *13*, 491. [CrossRef] [PubMed]
27. Laha, D.; Grant, R.; Mishra, P.; Nilubol, N. The Role of Tumor Necrosis Factor in Manipulating the Immunological Response of Tumor Microenvironment. *Front. Immunol.* **2021**, *12*, 656908. [CrossRef]
28. Martínez-Reza, I.; Díaz, L.; García-Becerra, R. Preclinical and clinical aspects of TNF-α and its receptors TNFR1 and TNFR2 in breast cancer. *J. Biomed. Sci.* **2017**, *24*, 90. [CrossRef]
29. Mercogliano, M.F.; Bruni, S.; Elizalde, P.V.; Schillaci, R. Tumor Necrosis Factor α Blockade: An Opportunity to Tackle Breast Cancer. *Front. Oncol.* **2020**, *10*, 584. [CrossRef]
30. Wang, Z.; Ao, X.; Shen, Z.; Ao, L.; Wu, X.; Pu, C.; Guo, W.; Xing, W.; He, M.; Yuan, H.; et al. TNF-α augments CXCL10/CXCR3 axis activity to induce Epithelial-Mesenchymal Transition in colon cancer cell. *Int. J. Biol. Sci.* **2021**, *17*, 2683–2702. [CrossRef]
31. Liu, M.; Guo, S.; Stiles, J.K. The emerging role of CXCL10 in cancer (Review). *Oncol. Lett.* **2011**, *2*, 583–589. [CrossRef]
32. Saleh, M.; Chandrashekar, D.S.; Shahin, S.; Agarwal, S.; Kim, H.-G.; Behring, M.; Shaikh, A.J.; Moloo, Z.; Eltoum, I.-E.A.; Yates, C.; et al. Comparative analysis of triple-negative breast cancer transcriptomics of Kenyan, African American and Caucasian Women. *Transl. Oncol.* **2021**, *14*, 101086. [CrossRef]

33. Davis, M.; Martini, R.; Newman, L.; Elemento, O.; White, J.; Verma, A.; Datta, I.; Adrianto, I.; Chen, Y.; Gardner, K.; et al. Identification of Distinct Heterogenic Subtypes and Molecular Signatures Associated with African Ancestry in Triple Negative Breast Cancer Using Quantified Genetic Ancestry Models in Admixed Race Populations. *Cancers* **2020**, *12*, 1220. [CrossRef] [PubMed]
34. Matias, A.T.; Jacinta-Fernandes, A.; Maia, A.-T.; Braga, S.; Jacinto, A.; Cabral, M.G.; Brito, P.H. Differential expression between African-ancestry and White patients diagnosed with Triple-Negative Breast Cancer: EGFR, Myc, Bcl2 and β-Catenin as ancestry-associated markers. *bioRxiv* **2020**. [CrossRef]
35. Zhang, Y.; Li, Y.; Wang, Q.; Su, B.; Xu, H.; Sun, Y.; Sun, P.; Li, R.; Peng, X.; Cai, J. Role of RASA1 in cancer: A review and update. *Oncol. Rep.* **2020**, *44*, 2386–2396. [CrossRef] [PubMed]
36. Reagan-Shaw, S.; Ahmad, N. RNA interference-mediated depletion of phosphoinositide 3-kinase activates forkhead box class O transcription factors and induces cell cycle arrest and apoptosis in breast carcinoma cells. *Cancer Res.* **2006**, *66*, 1062–1069. [CrossRef] [PubMed]
37. Etienne-Manneville, S.; Hall, A. Rho GTPases in cell biology. *Nature* **2002**, *420*, 629–635. [CrossRef] [PubMed]
38. Odrowaz, Z.; Sharrocks, A.D. ELK1 Uses Different DNA Binding Modes to Regulate Functionally Distinct Classes of Target Genes. *PLoS Genet.* **2012**, *8*, e1002694. [CrossRef]
39. Chen, B.; Chen, Y.; Rai, K.R.; Wang, X.; Liu, S.; Li, Y.; Xiao, M.; Ma, Y.; Wang, G.; Guo, G.; et al. Deficiency of eIF4B Increases Mouse Mortality and Impairs Antiviral Immunity. *Front. Immunol.* **2021**, *12*, 723885. [CrossRef]
40. Porter, H.A.; Perry, A.; Kingsley, C.; Tran, N.L.; Keegan, A.D. IRS1 is highly expressed in localized breast tumors and regulates the sensitivity of breast cancer cells to chemotherapy, while IRS2 is highly expressed in invasive breast tumors. *Cancer Lett.* **2013**, *338*, 239–248. [CrossRef]
41. Rigiracciolo, D.C.; Santolla, M.F.; Lappano, R.; Vivacqua, A.; Cirillo, F.; Galli, G.R.; Talia, M.; Muglia, L.; Pellegrino, M.; Nohata, N.; et al. Focal adhesion kinase (FAK) activation by estrogens involves GPER in triple-negative breast cancer cells. *J. Exp. Clin. Cancer Res.* **2019**, *38*, 58. [CrossRef]
42. Wang, X.; Shen, X.; Yan, Y.; Li, H. Pyruvate dehydrogenase kinases (PDKs): An overview toward clinical applications. *Biosci. Rep.* **2021**, *41*, BSR20204402. [CrossRef]
43. Bin, L.-H.; Xu, L.-G.; Shu, H.-B. TIRP, a Novel Toll/Interleukin-1 receptor (TIR) Domain-containing Adapter Protein Involved in TIR signaling*. *J. Biol. Chem.* **2003**, *278*, 24526–24532. [CrossRef] [PubMed]
44. Murugan, A.K.; Grieco, M.; Tsuchida, N. RAS mutations in human cancers: Roles in precision medicine. *Semin. Cancer Biol.* **2019**, *59*, 23–35. [CrossRef] [PubMed]
45. Zhang, W.; Liu, H.T. MAPK signal pathways in the regulation of cell proliferation in mammalian cells. *Cell Res.* **2002**, *12*, 9–18. [CrossRef] [PubMed]
46. Chen, Q.; Yang, B.; Nass, N.; Schatz, C.; Haybaeck, J. Impact of Eukaryotic Translation Initiation Factors on Breast Cancer: Still Much to Investigate. *Cancers* **2020**, *12*, 1984. Available online: https://www.mdpi.com/2072-6694/12/7/1984 (accessed on 19 September 2023). [CrossRef] [PubMed]
47. KEGG TNF Signaling Pathway—Homo Sapiens (Human). 2019. Available online: https://www.genome.jp/pathway/hsa04668 (accessed on 14 November 2022).
48. STRING. STRING CONSORTIUM 2023. 2023. Available online: https://string-db.org/ (accessed on 15 June 2023).
49. Asati, V.; Bharti, S.K.; Mahapatra, D.K.; Asati, V.; Budhwani, A.K. Triggering PIK3CA Mutations in PI3K/Akt/mTOR Axis: Exploration of Newer Inhibitors and Rational Preventive Strategies. *Curr. Pharm. Des.* **2016**, *22*, 6039–6054. [CrossRef] [PubMed]
50. Li, Y.; Li, S.; Meng, X.; Gan, R.Y.; Zhang, J.J.; Li, H.B. Dietary Natural Products for Prevention and Treatment of Breast Cancer. *Nutrients* **2017**, *9*, 728. [CrossRef] [PubMed]
51. Metibemu, D.S.; Ogungbe, I.V. Carotenoids in Drug Discovery and Medicine: Pathways and Molecular Targets Implicated in Human Diseases. *Molecules* **2022**, *27*, 6005. [CrossRef] [PubMed]
52. Doolaanea, A.A.; Alfatama, M.; Alkhatib, H.; Mawazi, S.M. Fucoxanthin. In *Handbook of Food Bioactive Ingredients: Properties and Applications*; Jafari, S.M., Rashidinejad, A., Simal-Gandara, J., Eds.; Springer International Publishing: Berlin/Heidelberg, Germany, 2023; pp. 729–755. [CrossRef]
53. Hoxhaj, G.; Manning, B.D. The PI3K-AKT network at the interface of oncogenic signaling and cancer metabolism. *Nat. Rev. Cancer* **2020**, *20*, 74–88. [CrossRef]
54. Pucci, B.; Kasten, M.; Giordano, A. Cell cycle and apoptosis. *Neoplasia* **2000**, *2*, 291–299. [CrossRef]
55. Alenzi, F.Q. Links between apoptosis, proliferation and the cell cycle. *Br. J. Biomed. Sci.* **2004**, *61*, 99–102. [CrossRef]
56. Yuan, L.; Cai, Y.; Zhang, L.; Liu, S.; Li, P.; Li, X. Promoting Apoptosis, a Promising Way to Treat Breast Cancer with Natural Products: A Comprehensive Review. *Front. Pharmacol.* **2022**, *12*, 801662. [CrossRef] [PubMed]
57. Wang, L.; Wu, J.; Lu, J.; Ma, R.; Sun, D.; Tang, J. Regulation of the cell cycle and PI3K/Akt/mTOR signaling pathway by tanshinone I in human breast cancer cell lines Retraction in /10.3892/mmr.2021.12080. *Mol. Med. Rep.* **2015**, *11*, 931–939. [CrossRef] [PubMed]
58. Khan, M.A.; Jain, V.K.; Rizwanullah, M.; Ahmad, J.; Jain, K. PI3K/AKT/mTOR pathway inhibitors in triple-negative breast cancer: A review on drug discovery and future challenges. *Drug Discov. Today* **2019**, *24*, 2181–2191. [CrossRef] [PubMed]
59. Nitulescu, G.M.; Van De Venter, M.; Nitulescu, G.; Ungurianu, A.; Juzenas, P.; Peng, Q.; Olaru, O.T.; Grădinaru, D.; Tsatsakis, A.; Tsoukalas, D.; et al. The Akt pathway in oncology therapy and beyond. *Int. J. Oncol.* **2018**, *53*, 2319–2331. [CrossRef] [PubMed]

60. Ortega, M.A.; Fraile-Martínez, O.; Asúnsolo, Á.; Buján, J.; García-Honduvilla, N.; Coca, S. Signal Transduction Pathways in Breast Cancer: The Important Role of PI3K/Akt/mTOR. *J. Oncol.* **2020**, *2020*, 9258396. [CrossRef] [PubMed]
61. Qiao, Y.; Shiue, C.N.; Zhu, J.; Zhuang, T.; Jonsson, P.; Wright, A.P.; Zhao, C.; Dahlman-Wright, K. AP-1-mediated chromatin looping regulates ZEB2 transcription: New insights into TNFα-induced epithelial-mesenchymal transition in triple-negative breast cancer. *Oncotarget* **2015**, *6*, 7804–7814. [CrossRef]
62. Rivas, M.A.; Carnevale, R.P.; Proietti, C.J.; Rosemblit, C.; Beguelin, W.; Salatino, M.; Charreau, E.H.; Frahm, I.; Sapia, S.; Brouckaert, P.; et al. TNF alpha acting on TNFR1 promotes breast cancer growth via p42/P44 MAPK, JNK, Akt and NF-kappa B-dependent pathways. *Exp. Cell Res.* **2008**, *314*, 509–529. [CrossRef]
63. Catasus, L.; D'Angelo, E.; Pons, C.; Espinosa, I.; Prat, J. Expression profiling of 22 genes involved in the PI3K–AKT pathway identifies two subgroups of high-grade endometrial carcinomas with different molecular alterations. *Mod. Pathol.* **2010**, *23*, 694–702. [CrossRef]
64. Fang, X.; Zhu, Y.; Zhang, T.; Li, Q.; Fan, L.; Li, X.; Jiang, D.; Lin, J.; Zou, L.; Ren, J.; et al. Fucoxanthin Inactivates the PI3K/Akt Signaling Pathway to Mediate Malignant Biological Behaviors of Non-Small Cell Lung Cancer. *Nutr. Cancer* **2022**, *74*, 3747–3760. [CrossRef]
65. Cuesta, C.; Arévalo-Alameda, C.; Castellano, E. The Importance of Being PI3K in the RAS Signaling Network. *Genes* **2021**, *12*, 1094. [CrossRef]
66. Li, Y.; Leisi, T.; Bao, L.; Arunachalam, C.; Alharbi, S.; Cui, J. Fucoxanthin inhibits cell proliferation and stimulates apoptosis through downregulation of PI3K/AKT/mTOR signaling pathway in human ovarian cancer cells. *Pharmacogn. Mag.* **2020**, *16*, 311–316. [CrossRef]
67. Wang, W.; Fu, C.; Lin, M.; Lu, Y.; Lian, S.; Xie, X.; Zhou, G.; Li, W.; Zhang, Y.; Jia, L.; et al. Fucoxanthin prevents breast cancer metastasis by interrupting circulating tumor cells adhesion and transendothelial migration. *Front. Pharmacol.* **2022**, *13*, 960375. [CrossRef] [PubMed]
68. Wang, J.; Ma, Y.; Yang, J.; Jin, L.; Gao, Z.; Xue, L.; Hou, L.; Sui, L.; Liu, J.; Zou, X. Fucoxanthin inhibits tumour-related lymphangiogenesis and growth of breast cancer. *J. Cell. Mol. Med.* **2019**, *23*, 2219–2229. [CrossRef] [PubMed]
69. Wu, H.L.; Fu, X.Y.; Cao, W.Q.; Xiang, W.Z.; Hou, Y.J.; Ma, J.K.; Wang, Y.; Fan, C.D. Induction of Apoptosis in Human Glioma Cells by Fucoxanthin via Triggering of ROS-Mediated Oxidative Damage and Regulation of MAPKs and PI3K-AKT Pathways. *J. Agric. Food Chem.* **2019**, *67*, 2212–2219. [CrossRef] [PubMed]
70. Ismylatifa, D.; Diah, P.; Nita, P.; HerinSetianingsih. The Effect of Brown Seaweed (*Sargassum* sp.) Extract on Apoptosis Process in Breast Cancer—A Literature Review. *Indian J. Forensic Med. Toxicol.* **2021**, *15*, 3247–3251. [CrossRef]
71. Miricescu, D.; Totan, A.; Stanescu-Spinu, I.-I.; Badoiu, S.C.; Stefani, C.; Greabu, M. PI3K/AKT/mTOR Signaling Pathway in Breast Cancer: From Molecular Landscape to Clinical Aspects. *Int. J. Mol. Sci.* **2020**, *22*, 173. [CrossRef]
72. Fresno Vara, J.A.; Casado, E.; de Castro, J.; Cejas, P.; Belda-Iniesta, C.; González-Barón, M. PI3K/Akt signalling pathway and cancer. *Cancer Treat. Rev.* **2004**, *30*, 193–204. [CrossRef]
73. Guan, B.; Chen, K.; Tong, Z.; Chen, L.; Chen, Q.; Su, J. Advances in Fucoxanthin Research for the Prevention and Treatment of Inflammation-Related Diseases. *Nutrients* **2022**, *14*, 4768. [CrossRef]
74. Li, D.; Zhang, Y.; Zhang, H.; Zhan, C.; Li, X.; Ba, T.; Qiu, Z.; E, F.; Lv, G.; Zou, C.; et al. CADM2, as a new target of miR-10b, promotes tumor metastasis through the FAK/AKT pathway in hepatocellular carcinoma. *J. Exp. Clin. Cancer Res.* **2018**, *37*, 46. [CrossRef]
75. Bianchi, N.; Brugnoli, F.; Grassilli, S.; Bourgeois, K.; Keillor, J.W.; Bergamini, C.M.; Aguiari, G.; Volinia, S.; Bertagnolo, V. The Motility and Mesenchymal Features of Breast Cancer Cells Correlate with the Levels and Intracellular Localization of Transglutaminase Type 2. *Cells* **2021**, *10*, 3059. [CrossRef]
76. Yoeli-Lerner, M.; Yiu, G.K.; Rabinovitz, I.; Erhardt, P.; Jauliac, S.; Toker, A. Akt Blocks Breast Cancer Cell Motility and Invasion through the Transcription Factor NFAT. *Mol. Cell* **2005**, *20*, 539–550. [CrossRef] [PubMed]
77. Song, M.; Finley, S.D. Mechanistic insight into activation of MAPK signaling by pro-angiogenic factors. *BMC Syst. Biol.* **2018**, *12*, 145. [CrossRef] [PubMed]
78. Lee, J.-S.; Tocheny, C.E.; Shaw, L.M. The Insulin-like Growth Factor Signaling Pathway in Breast Cancer: An Elusive Therapeutic Target. *Life* **2022**, *12*, 1992. [CrossRef] [PubMed]
79. Timbrell, S.; Aglan, H.; Cramer, A.; Foden, P.; Weaver, D.; Pachter, J.; Kilgallon, A.; Clarke, R.B.; Farnie, G.; Bundred, N.J. FAK inhibition alone or in combination with adjuvant therapies reduces cancer stem cell activity. *NPJ Breast Cancer* **2021**, *7*, 65. [CrossRef] [PubMed]
80. Fernald, K.; Kurokawa, M. Evading apoptosis in cancer. *Trends Cell Biol.* **2013**, *23*, 620–633. [CrossRef]
81. Banys-Paluchowski, M.; Milde-Langosch, K.; Fehm, T.; Witzel, I.; Oliveira-Ferrer, L.; Schmalfeldt, B.; Müller, V. Clinical relevance of H-RAS, K-RAS, and N-RAS mRNA expression in primary breast cancer patients. *Breast Cancer Res. Treat.* **2020**, *179*, 403–414. [CrossRef]
82. Pązik, M.; Michalska, K.; Żebrowska-Nawrocka, M.; Zawadzka, I.; Łochowski, M.; Balcerczak, E. Clinical significance of HRAS and KRAS genes expression in patients with non–small-cell lung cancer—Preliminary findings. *BMC Cancer* **2021**, *21*, 130. [CrossRef]
83. Galiè, M. RAS as Supporting Actor in Breast Cancer. *Front. Oncol.* **2019**, *9*, 1199. [CrossRef]

84. Wu, C.-C.; Pan, M.-R.; Shih, S.-L.; Shiau, J.-P.; Wu, C.-C.; Chang, S.-J.; Kao, C.-N.; Chen, F.-M.; Hou, M.-F.; Luo, C.-W. Combination of FAK inhibitor and cytokine-induced killer cell therapy: An alternative therapeutic strategy for patients with triple-negative breast cancer. *Biomed. Pharmacother.* **2023**, *163*, 114732. [CrossRef]
85. Atas, E.; Oberhuber, M.; Kenner, L. The Implications of PDK1–4 on Tumor Energy Metabolism, Aggressiveness and Therapy Resistance. *Front. Oncol.* **2020**, *10*, 583217. [CrossRef]
86. Chan, T.O.; Tsichlis, P.N. PDK2: A complex tail in one Akt. *Sci. STKE* **2001**, *2001*, pe1. [CrossRef] [PubMed]
87. Hemmings, B.A.; Restuccia, D.F. PI3K-PKB/Akt pathway. *Cold Spring Harb. Perspect. Biol.* **2012**, *4*, a011189. [CrossRef] [PubMed]
88. Sophie, P.; Nicolas, C.; Jérôme, T.; Valérie, B.; Pascale, C.-L.; Lise, W.; Alexa, G.; Patrick, M.; Catherine, L.; Didier, B. Role of the PI3K/AKT and mTOR signaling pathways in acute myeloid leukemia. *Haematologica* **2010**, *95*, 819–828. [CrossRef]
89. Shahbazian, D.; Parsyan, A.; Petroulakis, E.; Hershey, J.; Sonenberg, N. eIF4B controls survival and proliferation and is regulated by proto-oncogenic signaling pathways. *Cell Cycle* **2010**, *9*, 4106–4109. [CrossRef] [PubMed]
90. Lazaris-Karatzas, A.; Montine, K.S.; Sonenberg, N. Malignant transformation by a eukaryotic initiation factor subunit that binds to mRNA 5′ cap. *Nature* **1990**, *345*, 544–547. [CrossRef] [PubMed]
91. Madden, J.M.; Mueller, K.L.; Bollig-Fischer, A.; Stemmer, P.; Mattingly, R.R.; Boerner, J.L. Abrogating phosphorylation of eIF4B is required for EGFR and mTOR inhibitor synergy in triple-negative breast cancer. *Breast Cancer Res. Treat.* **2014**, *147*, 283–293. [CrossRef] [PubMed]
92. Fitzgerald, K.A.; Rowe, D.C.; Barnes, B.J.; Caffrey, D.R.; Visintin, A.; Latz, E.; Monks, B.; Pitha, P.M.; Golenbock, D.T. LPS-TLR4 Signaling to IRF-3/7 and NF-κB Involves the Toll Adapters TRAM and TRIF. *J. Exp. Med.* **2003**, *198*, 1043–1055. [CrossRef] [PubMed]
93. Yang, S.H.; Bumpass, D.C.; Perkins, N.D.; Sharrocks, A.D. The ETS domain transcription factor Elk-1 contains a novel class of repression domain. *Mol. Cell. Biol.* **2002**, *22*, 5036–5046. [CrossRef]
94. Yan, Q.; Ni, C.; Lin, Y.; Sun, X.; Shen, Z.; Zhang, M.; Han, S.; Shi, J.; Mao, J.; Yang, Z.; et al. ELK1 Enhances Pancreatic Cancer Progression Via LGMN and Correlates with Poor Prognosis. *Front. Mol. Biosci.* **2021**, *8*, 764900. [CrossRef]
95. Chen, B.H.; Tzen, J.T.; Bresnick, A.R.; Chen, H.C. Roles of Rho-associated kinase and myosin light chain kinase in morphological and migratory defects of focal adhesion kinase-null cells. *J. Biol. Chem.* **2002**, *277*, 33857–33863. [CrossRef]
96. Phuyal, S.; Farhan, H. Multifaceted Rho GTPase Signaling at the Endomembranes. *Front. Cell Dev. Biol.* **2019**, *7*, 127. [CrossRef] [PubMed]
97. Fu, Z.; Tindall, D.J. FOXOs, cancer and regulation of apoptosis. *Oncogene* **2008**, *27*, 2312–2319. [CrossRef] [PubMed]
98. Zanella, F.; Link, W.; Carnero, A. Understanding FOXO, new views on old transcription factors. *Curr. Cancer Drug Targets* **2010**, *10*, 135–146. [CrossRef] [PubMed]
99. Yao, S.; Fan, L.Y.; Lam, E.W. The FOXO3-FOXM1 axis: A key cancer drug target and a modulator of cancer drug resistance. *Semin. Cancer Biol.* **2018**, *50*, 77–89. [CrossRef] [PubMed]
100. Chau, J.E.; Vish, K.J.; Boggon, T.J.; Stiegler, A.L. SH3 domain regulation of RhoGAP activity: Crosstalk between p120RasGAP and DLC1 RhoGAP. *Nat. Commun.* **2022**, *13*, 4788. [CrossRef] [PubMed]
101. Xiao, W.; Zheng, S.; Zou, Y.; Yang, A.; Xie, X.; Tang, H.; Xie, X. CircAHNAK1 inhibits proliferation and metastasis of triple-negative breast cancer by modulating miR-421 and RASA1. *Aging* **2019**, *11*, 12043–12056. [CrossRef] [PubMed]
102. Rutkovsky, A.C.; Yeh, E.S.; Guest, S.T.; Findlay, V.J.; Muise-Helmericks, R.C.; Armeson, K.; Ethier, S.P. Eukaryotic initiation factor 4E-binding protein as an oncogene in breast cancer. *BMC Cancer* **2019**, *19*, 491. [CrossRef] [PubMed]
103. Assoian, R.K.; Zhu, X. Cell anchorage and the cytoskeleton as partners in growth factor-dependent cell cycle progression. *Curr. Opin. Cell Biol.* **1997**, *9*, 93–98. [CrossRef]
104. O'Meara, T.; Safonov, A.; Casadevall, D.; Qing, T.; Silber, A.; Killelea, B.; Hatzis, C.; Pusztai, L. Immune microenvironment of triple-negative breast cancer in African-American and Caucasian women. *Breast Cancer Res. Treat.* **2019**, *175*, 247–259. [CrossRef]
105. Yin, L.; Duan, J.J.; Bian, X.W.; Yu, S.C. Triple-negative breast cancer molecular subtyping and treatment progress. *Breast Cancer Res. BCR* **2020**, *22*, 61. [CrossRef]

molecules

MDPI

Review

Gamma-Tocopherol: A Comprehensive Review of Its Antioxidant, Anti-Inflammatory, and Anticancer Properties

Basma Es-Sai [1,†], Hicham Wahnou [2,†], Salma Benayad [1], Soufiane Rabbaa [1], Yassir Laaziouez [1], Riad El Kebbaj [1], Youness Limami [1] and Raphaël Emmanuel Duval [3,*]

[1] Sciences and Engineering of Biomedicals, Biophysics and Health Laboratory, Higher Institute of Health Sciences, Hassan First University, Settat 26000, Morocco; essaibassma1999pfe@gmail.com (B.E.-S.); salmabenayad6@gmail.com (S.B.); rabbaasoufian@gmail.com (S.R.); y.laaziouez@gmail.com (Y.L.); elkebbajriad@gmail.com (R.E.K.); youness.limami@uhp.ac.ma (Y.L.)

[2] Laboratory of Immunology and Biodiversity, Faculty of Sciences Ain Chock, Hassan II University, B.P. 2693, Maarif, Casablanca 20100, Morocco; hwwahnou@gmail.com

[3] Université de Lorraine, F-54000 Nancy, France

* Correspondence: raphael.duval@univ-lorraine.fr

† These authors contributed equally to this work.

Abstract: Gamma-tocopherol (γ-tocopherol), a major isoform of vitamin E, exhibits potent antioxidant, anti-inflammatory, and anticancer properties, making it a promising therapeutic candidate for treating oxidative stress-related diseases. Unlike other tocopherol isoforms, γ-tocopherol effectively neutralizes reactive oxygen species (ROS) and reactive nitrogen species (RNS), providing robust cellular protection against oxidative damage and lipid peroxidation. Its anti-inflammatory effects are mediated through the modulation of pathways involving cyclooxygenase-2 (COX-2) and tumor necrosis factor-alpha (TNF-α), reducing chronic inflammation and its associated risks. In cancer therapy, γ-tocopherol demonstrates multifaceted activity, including the inhibition of tumor growth, induction of apoptosis, and suppression of angiogenesis, with significant efficacy observed in cancers such as prostate, lung, and colon. Preclinical and clinical studies support its efficacy in mitigating oxidative stress, inflammation, and cancer progression, with excellent tolerance at physiological levels. However, high doses necessitate careful evaluation to minimize adverse effects. This review consolidates current knowledge on γ-tocopherol's biological activities and clinical implications, underscoring its importance as a natural compound for managing inflammation, oxidative stress, and cancer. As a perspective, advancements in nanoformulation technology could enhance γ-tocopherol's bioavailability, stability, and targeted delivery, offering the potential to optimize its therapeutic application in the future.

Keywords: γ-tocopherol; antioxidant; anti-inflammatory; anticancer; clinical trials; nanoformulations

1. Introduction

Gamma-tocopherol (γ-tocopherol) is one of the eight naturally occurring isoforms of vitamin E, a fat-soluble compound recognized for its essential role in maintaining cellular health and combating oxidative damage [1]. Unlike alpha-tocopherol (α-tocopherol), the most studied and abundant form of vitamin E, γ-tocopherol possesses distinct chemical and biological properties that make it uniquely effective in certain physiological contexts [1]. Found abundantly in dietary sources such as nuts, seeds, and vegetable oils (Figure 1A), γ-tocopherol is distinguished by its ability to neutralize reactive nitrogen species, particularly peroxynitrite, a feature less prominent in other tocopherol isoforms [1,2].

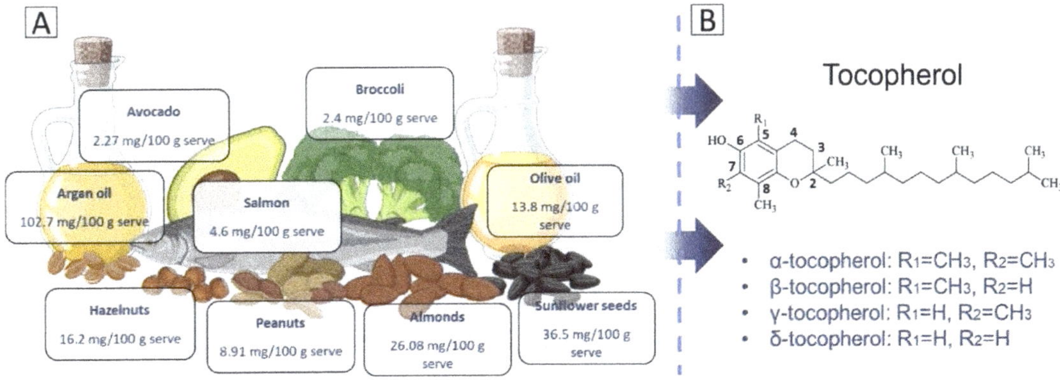

Figure 1. Dietary sources and structural variants of tocopherols: (**A**) key dietary sources of tocopherols, including argan oil, avocado, broccoli, salmon, hazelnuts, almonds, sunflower seeds, peanuts, and olive oil, with their respective tocopherol content per serving. (**B**) The chemical structure of tocopherol, highlighting the differences in methyl group positions that distinguish alpha (α), beta (β), gamma (γ), and delta (δ) tocopherols.

Oxidative stress, characterized by an imbalance between reactive oxygen and nitrogen species (ROS and RNS) and the body's antioxidant defenses, is a key driver of cellular damage and dysfunction [3]. This pathological state contributes to the progression of numerous chronic diseases, including cardiovascular disorders, neurodegenerative conditions, and cancer [4,5]. Closely linked to oxidative stress, chronic inflammation acts as a mediator in many of these diseases. Persistent inflammatory signals promote tissue damage and creates a microenvironment conducive to the development and progression of disorders such as arthritis, atherosclerosis, and malignancies [6]. In cancer, oxidative stress and inflammation synergistically drive tumorigenesis by promoting DNA damage, genetic instability, and immune evasion [7]. Despite advancements in conventional therapies, such as nonsteroidal anti-inflammatory drugs (NSAIDs) and chemotherapeutic agents, these treatments are often associated with significant adverse effects [8]. Long-term use of NSAIDs, for instance, is linked to gastrointestinal and cardiovascular risks, while many anticancer agents cause severe systemic toxicity, leading to compromised quality of life and treatment discontinuation [8–10]. These limitations underscore the pressing need for safer, more targeted therapeutic strategies that mitigate disease progression without imposing undue harm on patients.

γ-tocopherol has emerged as a promising candidate for addressing these challenges due to its multifaceted biological properties [2]. Its potent antioxidant activity not only neutralizes harmful ROS and RNS but also interrupts the vicious cycle of oxidative stress and inflammation [11]. Furthermore, its ability to modulate inflammatory pathways and disrupt cancer cell proliferation positions it as a natural compound with therapeutic potential across a spectrum of oxidative stress and inflammation-driven diseases [12]. Clinical studies have begun to reveal its promise in reducing disease progression and improving patient outcomes in various pathological contexts [13]. The emergence of nanoformulation technologies further enhances the therapeutic potential of γ-tocopherol. By improving its bioavailability, stability, and targeted delivery, nanoformulations offer a means to maximize its efficacy while minimizing potential side effects [14]. This review examines the broad therapeutic spectrum of γ-tocopherol, consolidates evidence from clinical studies, and explores the innovative role of nanoformulations in advancing its application as a modern therapeutic agent.

2. Structure and Properties

Tocopherols are a group of lipid-soluble compounds classified under vitamin E [15]. They are recognized for their potent antioxidant properties, which play a crucial role in protecting cell membranes from the oxidative damage caused by free radicals. Structurally, tocopherols are characterized by a chromanol ring attached to a long hydrophobic phytyl tail, making them amphipathic molecules [16]. This allows tocopherols to embed within biological membranes, where they stabilize lipid bilayers and prevent lipid peroxidation. There are four main types of tocopherols [17], alpha (α), beta (β), gamma (γ), and delta (δ), which differ in terms of the number and position of methyl groups on the chromanol ring (Figure 1B).

- α-tocopherol: the chromanol ring has methyl groups at positions R1 = CH_3 and R2 = CH_3 . This fully methylated structure allows α-tocopherol to have the highest antioxidant activity among tocopherols and makes it the most bioavailable form.
- β-tocopherol: structurally, β-tocopherol has one methyl group located at R1 = CH_3 and one hydrogen atom in R2 = H. Its activity is intermediate, between that of α-tocopherol and γ-tocopherol, due to this arrangement.
- γ-tocopherol: in this variant, the hydrogen atom is located at R1 = H, while the methyl group is located at R2 = CH_3. The absence of a methyl group at R1 alters its antioxidant properties. While its radical-scavenging activity is slightly lower than α-tocopherol, γ-tocopherol has a unique ability to trap RNS, such as peroxynitrite, which may confer additional protective effects.
- δ-tocopherol: the chromanol ring in δ-tocopherol has hydrogen atoms in both substituents, which are positioned at R1 = H and R2 = H. This structure results in δ-tocopherol having the highest ability to neutralize ROS among the tocopherol isomers.

γ-tocopherol is distinguished by its unique chemical structure and physical properties, which together underpin its essential biological roles [1]. Its structure features a chromanol ring as the aromatic head, containing a hydroxyl group (-OH) at the 6th position (Figure 1B). This hydroxyl group is critical for its antioxidant activity, as it donates hydrogen atoms to neutralize free radicals and prevent oxidative damage [18]. γ-tocopherol's substituent pattern, with methyl groups at the 7th and 8th positions, differs from α-tocopherol, which also has a methyl group at the 5th position (Figure 1B). This structural variation influences γ-tocopherol's antioxidant function and biological activity. Additionally, the molecule possesses a long hydrophobic phytyl tail attached to the chromanol ring [19]. This tail anchors the molecule within lipid bilayers of cell membranes, where it protects membrane lipids from peroxidation and maintains cellular integrity [19]. Its molecular formula is $C_{28}H_{48}O_2$, with a molecular weight of 416.68 g/mol. γ-tocopherol is lipophilic, making it soluble in fats and oils but insoluble in water [20].

Physically, γ-tocopherol appears as a viscous oil with a yellow to amber color. It remains stable under normal physiological conditions but can decompose at high temperatures. While it is insoluble in water, it dissolves readily in organic solvents such as ethanol, acetone, and lipids, facilitating its use in lipid-based applications [11]. γ-tocopherol is relatively stable when exposed to heat and light but is prone to degradation upon prolonged exposure to ultraviolet radiation or oxygen [21,22]. This susceptibility can diminish its antioxidant efficacy over time.

3. Antioxidant Effects

Oxidative stress reflects an imbalance between the systemic manifestation of the reactivity of oxygen species and the ability of biological systems to detoxify the body from free radicals, which leads to the accumulation of ROS in cells, resulting cellular toxicity and subsequently leading to tissue damage [3]. Antioxidants play a crucial role in neutralizing ROS and, among them, γ-tocopherol stands out due to its potent radical-scavenging capabilities [23,24]. Its antioxidant activity is attributed to its unique chemical structure, particularly the chromanol ring, which facilitates the donation of hydrogen atoms and stabilization of the resulting radical species [23,24].

Density functional theory (DFT) has been employed to explore the thermodynamic parameters governing the antioxidant activity of γ-tocopherol [25], suggesting that γ-tocopherol possesses a lower band dissociation than α-tocopherol [26]. This property may enhance the efficiency of γ-tocopherol in donating a hydrogen atom from its phenolic hydroxyl group, which can facilitate free radical neutralization. In addition, the electronic configuration of γ-tocopherol characterized by one less methyl group compared to α-tocopherol, enables it to trap more effectively with RNS, such as peroxynitrite (ONOO-) [27]. Furthermore, key parameters include the bond dissociation enthalpy (BDE), which measures the ease of hydrogen donation, with a low BDE indicating high reactivity towards free radicals [25,28]. The ionization potential (IP) reflects the compound's electron-donating capacity, with lower IP values enhancing radical scavenging via electron transfer [29].

Theoretical studies also highlight two dominant mechanisms of action: (i) hydrogen atom transfer (HAT), where the phenolic hydrogen atom neutralizes free radicals [30], and (ii) sequential proton loss electron transfer (SPLET), which involves deprotonation followed by electron transfer [31]. These mechanisms underscore γ-tocopherol's versatility and effectiveness in mitigating oxidative stress and preventing cellular damage.

Various studies have demonstrated, experimentally, the protective effects of γ-tocopherol against oxidative stress through its impact on key markers of this biological process involved in the apparition of several pathologies (Table 1).

Table 1. Antioxidant actions and mechanisms across different animal models.

Model	Antioxidant Action	Mechanisms	References
Human (Plasma) Mice (Kidney, Liver) Rat (Plasma) Sheep (Lung) Chicken (Plasma, Liver)	Anti-lipid peroxidation	↓ MDA ↓ MDA+HNE ↓ Lipid peroxide ↓ 8-isoprostane	[32–36]
Mice (Colon tumor and tissue adjacent, Colon) Rat (Mammary gland)	Anti-DNA damage	↓ 8-oxo-dG ↓ γH2AX	[35,37–39]
Mice (Kidney, Liver, Prostate, mammary tumor) Rat (Plasma, Mammary gland, aorta) PBM Cells	Defense antioxidant	↑ mRNA SOD1 ↑ SOD activity ↓ Superoxide generation ↓ ROS ↔/↑ Catalase ↔/↑ GPx ↔ HO-1 ↑ NQO1 ↑ GCLM ↑ HO-1 ↔/↑ Nrf2 Inhibition CpG hypermethylation in Nrf2 promoter	[32,35,40–43]

MDA: malondialdehyde, MDA+HNE: malondialdehyde + 4-hydroxy-2-nonenal, lipid peroxide: lipid peroxidation products, 8-isoprostane: 8-iso-prostaglandin F2α, 8-oxo-dG: 8-Oxo-2′-deoxyguanosine, γH2AX: phosphorylated H2A.X histone variant, mRNA SOD1: messenger RNA for superoxide dismutase 1, SOD activity: superoxide dismutase activity, superoxide generation: generation of superoxide anions ($O_2•−$), ROS: reactive oxygen species, catalase: catalase enzyme activity, GPx: glutathione peroxidase, HO-1: heme oxygenase-1, NQO1: NAD(P)H quinone dehydrogenase 1, GCLM: glutamate-cysteine ligase modifier subunit, Nrf2: nuclear factor erythroid 2-related factor 2, and PBM cells: peripheral blood mononuclear cells; ↑: increase, ↓: decrease, and ↔: standardization.

3.1. Effects Against Lipid Peroxidation

ROS cause lipid peroxidation, which is a complex process, renowned for the rearrangement of doubles bonds in polyunsaturated fatty acids, particularly arachidonic acids. This reaction generates an inactivation of receptors, and membrane permeability and fluidity, which leads to apoptosis [44,45].

γ-tocopherol exhibits remarkable antioxidant properties through its ability to modulate lipid peroxidation markers, particularly malondialdehyde (MDA) levels, in diverse biological systems. In diabetic-afflicted mice kidneys, γ-tocopherol was shown to decrease MDA concentrations [32]. Similarly, in normal mice stimulated by lipopolysaccharide (LPS), γ-tocopherol attenuated lipid peroxidation in the liver [33], highlighting its preventive and protective effect across different tissues. In sheep exposed to burn and smoke inhalation, γ-tocopherol nebulization decreased MDA levels in the lungs [34], reflecting its impact in modulating lipid peroxidation even under environmental toxin exposure. Interestingly, when γ-tocopherol is combined with α-tocopherol, MDA levels decreased in the plasma and livers of chicken, suggesting that this interaction might influence specific antioxidant pathways [46]. In light of this, natural sources such as argan oil, abundant in antioxidant compounds like γ-tocopherol [47], are recognized for their restorative properties in cases of oxidative stress [48,49]. Previous studies demonstrate that argan oil can normalize depleted MDA levels in the livers and the brains of mice under oxidative stress challenges, reaffirming the role of diet-derived antioxidants in hepatoprotection [49]. In addition, it was demonstrated that γ-tocopherol attenuated the serum levels of 8-isoprostane in rats implanted with 17β-estradiol (E2), with impacts varying based on the concentration and the period of treatment [33,35] (Figures 2 and 3). Clinically, γ-tocopherol has been explored in conditions like metabolic syndrome, where it regulates lipid peroxidation by decreasing plasmatic MDA+HNE (4-hydroxy 2-nonenal) and lipid peroxides levels [36]. Given the relationship between γ-tocopherol and lipid peroxidation, it has been well-explained how this component can limit this process. One key mechanism involves the formation of metabolites from its oxidation, such as tocopherylquinone, which can scavenge lipid peroxyl radicals (ROO·) and stop lipid peroxidation [2].

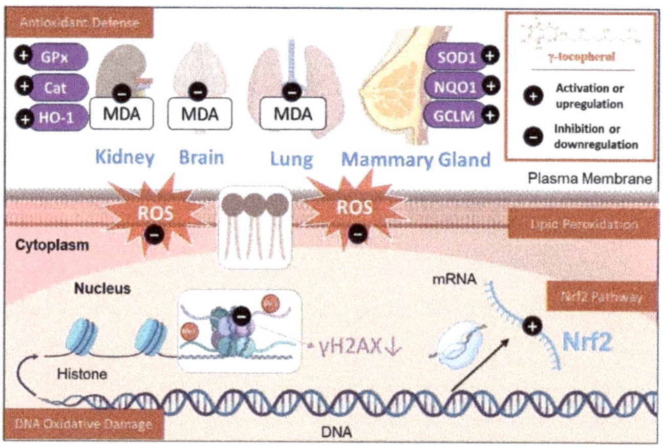

Figure 2. γ-tocopherol's therapeutic effects in oxidative stress: γ-tocopherol's role in reducing lipid peroxidation (via suppression of MDA and ROS production), mitigating DNA oxidative damage (indicated by reduced γH2AX, 8-oxo-dG, and PHIP-induced damage), enhancing antioxidant defense systems (elevating levels of GPx, SOD1, Cat, NQO1, HO-1, and GCLM), and its involvement in the Nrf2 pathway (inducing Nrf2 activation and antioxidant gene expression).

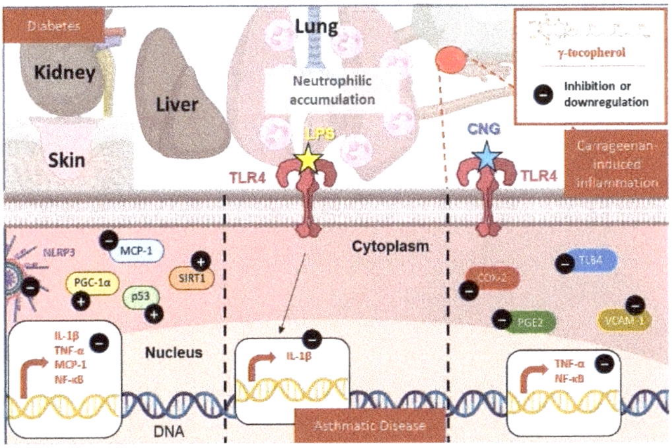

Figure 3. γ-tocopherol's therapeutic effects in inflammatory and metabolic diseases: γ-tocopherol modulates inflammation and oxidative stress in diabetes (by reducing CRP, MCP-1, IL-1β, TNF-α, and NLRP3, while enhancing p53, PGC-1α, and SIRT1), carrageenan-induced inflammation (by lowering LTB4, TNF-α, PGE2, COX-2, VCAM-1, and NF-κB), and asthmatic disease (by decreasing neutrophilic accumulation and IL-1β levels in LPS-induced inflammation).

This suggests that γ-tocopherol successfully helps to attenuate oxidative stress by regulating lipid peroxidation through reducing and normalizing the markers involved in this process, offering a potential beneficial therapeutic for oxidative stress-related disorders.

3.2. Effects Against DNA Oxidative Damage

ROS, particularly hydroxyl radical (OH•), has the potential to induce DNA damage, which encompasses processes such as base mutation, single- and double-strand breaks, and the formation of DNA adducts [37].

γ-tocopherol has been evaluated in preclinical models, demonstrating its effectiveness in protecting against DNA damage induced by oxidative stress. For instance, in the mammary gland of rats, γ-tocopherol administration significantly attenuated the E2-induced 8-oxo-2′-deoxyguanosine (8-oxo-dG) level, which is a marker of DNA oxidative damage and a product of nitrogenous base oxidation [37], by 72% and 67% at 7 and 14 days of treatments, respectively [35]. Similarly, Chen et al. [38] found that γ-tocopherol lowered the 8-oxo-dG level by 35% in colon tumors and by 42% in the adjacent tissues. On the other hand, the level of 8-oxo-dG was markedly decreased in the wild-type and nuclear factor erythroid 2–related factor 2 (Nrf2) (−/−) epithelial cells of mice, which suggests that the antioxidant potential of γ-tocopherol is independent of Nrf2-regulated antioxidative enzymes [39] (Figure 2). Additionally, it was demonstrated that the supplementation of γ-tocopherol significantly attenuated the levels of the marker of DNA damage phosphorylated histone γH2AX in the colon of 2-amino-1-methyl-6-phenylimidazo(4,5-b) pyridine (PHIP) -treated mice [38]. According to these available studies, γ-tocopherol exerts a protective effect against DNA oxidative damage, which is primarily demonstrated in vivo. These studies suggest an indirect effect of γ-tocopherol is to preserve DNA integrity and stability. However, the direct interactions between γ-tocopherol and DNA have not been conclusively demonstrated. Its role as an antioxidant highlights its preventive effect in maintaining DNA structure. Further research is needed to elucidate the direct interaction between γ-tocopherol and DNA to gain a deeper understanding of the associated biological pathways (Figure 2).

3.3. *Effect on Antioxidant Defense*

The body has a complex system of antioxidant defenses to protect against the harmful effects of ROS and RNS. There are two distinct levels of antioxidant effects: one is provided by nutrition through vegetable and fruits rich in vitamins C, E, carotenoids, flavonoids, or glutathione, whose role is to scavenging free radicals; the other is endogenous and consists of antioxidant enzymes such as superoxide dismutase (SOD), catalase (Cat), glutathione peroxidase (GPx), and heme oxygenase (HO-1), whose role is to degrade free radicals [50–52].

γ-tocopherol treatment is quite successful in improving the oxidative state in the kidneys of diabetic mice by normalizing the protein expression of antioxidant enzymes (catalase, GPx, and HO-1) [32]. At the transcriptional level, Das Gupta et al. found that the supplementation of γ-tocopherol significantly upregulating the mRNA expression of antioxidant enzymes (SOD1, GPx, and catalase) in the mammary glands of rats treated with E2, which can lead to degradation of the ROS produced by E2 [35]. Moreover, an in vitro study revealed that γ-tocopherol neutralizes ROS levels in PBMCs stimulated by LPS [40]. Similarly, the arterial superoxide anion generation, a well-established substrate of SOD enzyme, was significantly decreased in the γ-tocopherol-fed rats [53], suggesting that its degradation is facilitated by the enhancement of this enzyme. In addition, it was demonstrated that there was a notable rise in SOD activity in aortic homogenates as well as in the plasma of rats administered by γ-tocopherol [53]. Furthermore, γ-tocopherol significantly increases the proteins levels of NQO1 (NAD(P)H quinone dehydrogenase 1), GCLM (glutamate-cysteine ligase modifier subunit), and HO-1 in mammary tumors in mice subjected to E2-induced alterations, thereby mitigating oxidative stress, which helps to prevents cellular homeostasis [35] (Figure 2).

3.4. *Effect on Nrf2 Pathway*

Nrf2 is a transcription factor that is key for the preservation of cellular redox homeostasis and the elimination of reactive and carcinogenic mediators. The connection between Nrf2 and oxidative stress has been thoroughly elucidated by El Kebbaj et al. [54], who highlighted how natural compounds like fatty acids, tocopherols (α, β, γ, and δ), phytosterols (particularly schottenol and spinasterol), and ferulic acid can regulate Nrf2 expression during the oxidative stress process.

In this regard, it was demonstrated that γ-tocopherol-enhanced mammary gland Nrf2 mRNA levels in rats treated with E2 are correlated by increasing the mRNA expression of antioxidant enzyme [35]. Furthermore, a diet enriched with γ-tocopherol mixtures was shown to increase Nrf2 at both the mRNA and protein levels in the livers of mice [41,42]. However, it was also evaluated that Nrf2 did not exhibit a direct correlation with the modulation of MDA levels in the livers of mice [42], suggesting that the mechanism governing these pathways may be independent or may involve additional mediators indirectly linked to Nrf2 modulation. To better understand how γ-tocopherol influences Nrf2, a mechanistic study revealed that a tocopherol-rich mixture containing γ-tocopherol maintains the mRNA and protein levels of Nrf2 by inhibiting CpG hypermethylation in the Nrf2 promoter during prostate carcinogenesis in mice and in TRAMP-C1 cells [43] (Figure 2). These studies suggest that γ-tocopherol inhibits oxidative stress through modulating the Nrf2 signaling pathway to upregulate antioxidant proteins involved in free radical neutralization.

4. Anti-Inflammatory Activity

Inflammation is a key immune response that helps repair tissues after injury, but chronic inflammation can increase the risk of various diseases. γ-tocopherol, a powerful antioxidant, has shown potential in reducing inflammation by limiting neutrophils infiltration and cytokines' production [55,56]. Several studies have evaluated its effects, revealing that γ-tocopherol may help modulate inflammatory responses, offering therapeutic benefits in conditions associated with acute inflammation and also with chronic inflammation such as diabetes and asthma.

4.1. Carrageenan-Induced Inflammation

γ-tocopherol's effects in acute inflammation models, where the inflammatory response is more immediate and transient, have been thoroughly evaluated. In the carrageenan-induced inflammation model in the air pouch, γ-tocopherol significantly reduced the pro-inflammatory eicosanoids prostaglandin E2 (PGE2) (46%) and Leukotriene B4 (LTB4) (70%), the latter being a potent chemotactic factor produced by 5-lipoxygenase in neutrophils [57]. At a higher dose (i.e., 100 mg/kg), γ-tocopherol exhibited similar inhibitory effects on PGE2 (51%) and LTB4. Additionally, γ-tocopherol's effects on TNF-α, a key inflammation mediator, were investigated. At the lower dose (i.e., 33 mg/kg), γ-tocopherol did not significantly inhibit TNF-α (51%), while at the higher dose, γ-tocopherol reduced TNF-α by 65%. These results suggest that γ-tocopherol modulates multiple components of the inflammatory response, offering potential therapeutic benefits in acute inflammation settings (Figure 3).

In a carrageenan-induced inflammation model in rats, a combination of aspirin and γ-tocopherol exhibited enhanced anti-inflammatory effects compared to aspirin alone [58]. The combination reduced PGE2 levels by 40% and decreased exudate volume by 15%, suggesting a prolonged anti-inflammatory action [58]. However, no significant effects were observed on immune cell infiltration. Of note, the combination of aspirin and γ-tocopherol also partially alleviated aspirin-induced gastric lesions and spared the reduction in food intake typically caused by aspirin. Additionally, aspirin and γ-tocopherol reduced gastric PGE2 depletion and decreased 8-isoprostane levels, indicating a protective effect against oxidative stress in the stomach [58]. These findings suggest that γ-tocopherol may improve the therapeutic profile of aspirin by enhancing its anti-inflammatory efficacy and reducing its adverse gastrointestinal effects.

In the IFN-γ/PMA-stimulated Caco-2 intestinal cell model system, γ-tocopherol exhibits significant anti-inflammatory effects, particularly by reducing IL-8 at both mRNA and protein levels [59]. In fact, at concentrations of 10 and 100 μM, γ-tocopherol significantly decreases IFN-γ/PMA-induced IL-8 production without inducing cytotoxicity. Additionally, γ-tocopherol downregulates IL-8 mRNA expression at 8 h, although this effect diminishes by 24 h [59]. Regarding NF-κB activation, γ-tocopherol is an inhibitor with moderate effectiveness, and δ-tocopherol the least effective [59]. Globally, these results highlight γ-tocopherol as a potent anti-inflammatory tocopherol for modulating inflammatory signaling pathways in this model.

In murine RAW264.7 macrophages treated with LPS and human A549 epithelial cancer cells treated with IL-1β, γ-tocopherol and its metabolite gamma-carboxyethyl hydroxychroman (γ-CEHC) effectively inhibit PGE2 synthesis in a dose-dependent manner, with γ-tocopherol demonstrating greater potency (IC_{50} values: 7.5 ± 2 μM in macrophages and 4 ± 1 μM in A549 cells) [60]. γ-tocopherol also reduces prostaglandin D2 (PGD2) levels in LPS-stimulated macrophages but is slightly less effective than it is on PGE2. While α-tocopherol has minimal or no effect on PGE2 and PGD2 synthesis, γ-CEHC directly inhibits COX-2 activity after short exposures, while γ-tocopherol requires prolonged exposure [60]. However, none of the tocopherols significantly reduce COX-2 protein expression or mRNA levels, despite inflammatory stimulation (Figure 3).

TNF-α stimulation increased Vascular Cell Adhesion Molecule 1 (VCAM-1) expression in HMEC-1 cells, which was significantly reduced by all tocopherols at 20 and 40 μM, with γ-tocopherol showing the strongest effect. At 10 μM, none of the tocopherols affected VCAM-1 expression. Moreover, in lymphatic endothelial cells, tocopherols did not reduce VCAM-1 expression after TNF-α exposure [61] (Figure 4).

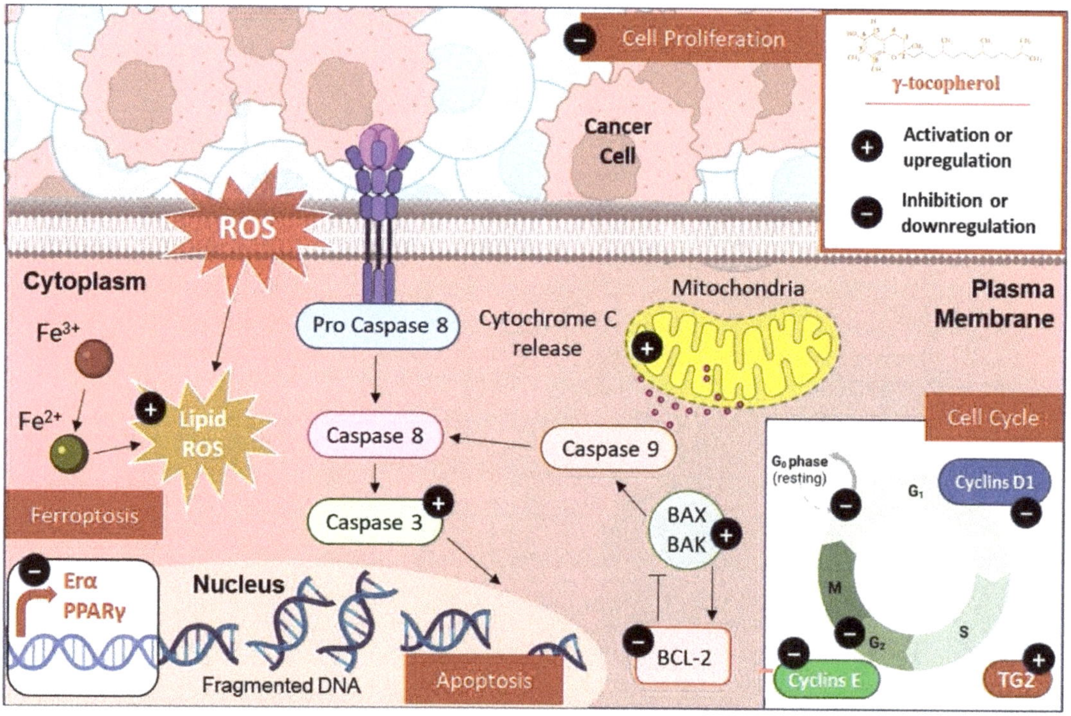

Figure 4. γ-tocopherol's therapeutic effects in cancer: γ-tocopherol promotes cell death through apoptosis and ferroptosis, evidenced by interactions of Bcl-2, Bax proteins, and activated Caspase 3. It also highlights its impact on the cell cycle, notably inhibiting progression via cyclins D1 and E, which affects the resting (G0) phase and other stages. Lastly, γ-tocopherol influences cell proliferation by downregulating estrogen receptor α (Erα) and PPARγ, thereby reducing cancer cell growth in experimental models.

γ-tocopherol exhibits potent anti-inflammatory effects by modulating key inflammatory markers such as IL-8 and VCAM-1, with significant inhibition of NF-κB activation and COX-2 activity. The mechanism of action of γ-tocopherol seems to involve both direct inhibition of inflammatory mediators and modulation of the endothelial barrier function, highlighting its therapeutic potential in inflammatory diseases and angiogenesis-related conditions [62]. Its effects were more pronounced than those of α-tocopherol, making γ-tocopherol a promising candidate for further exploration in inflammatory and vascular health management.

4.2. Diabetes-Induced Inflammation

γ-tocopherol demonstrates significant anti-inflammatory effects in diabetic models.

The studies collectively demonstrate that γ-tocopherol exerts anti-inflammatory effects by targeting key molecular pathways, particularly in severe diabetic conditions. γ-tocopherol consistently reduced NF-κB activity across tissues, leading to decreased levels of pro-inflammatory cytokines like IL-1β, TNF-α, and Monocyte Chemoattractant Protein-1 (MCP-1) [32,63]. It also elevated Sirtuin 1 (SIRT1), which inhibits NF-κB signaling and enhances antioxidant and repair mechanisms. The reduction in inflammasome components, such as NOD-like Receptor Pyrin Domain Containing 3 (NLRP3) and caspase-1 in hepatic tissue, further supports γ-tocopherol's role in mitigating severe inflammation. Additionally, γ-tocopherol's ability to increase nuclear peroxisome proliferator-activated receptor gamma coactivator 1 alpha (PGC-1α) and p53 in wound healing suggests it improves mitochondrial function and cellular stress responses [63].

4.3. *Asthmatic Disease*

Asthma is one of the most prevalent chronic childhood illnesses, characterized as a heterogeneous and multifactorial disorder influenced by a complex interplay of genetic and environmental factors [64]. Many in vivo and clinical trials have been conducted to assess the potential of γ-tocopherol in improving this disease.

4.3.1. Non-Allergic Asthma Induced by LPS

A combined animal and clinical study evaluated γ-tocopherol's effects on LPS-induced airway inflammation [65]. In rats, LPS challenge-induced significant neutrophils accumulation in bronchoalveolar lavage fluid (BALF) and airway tissues, which were mitigated by 4 days of γ-tocopherol supplementation (30 mg/kg/day) [65]. A clinical trial using a double-blind, placebo-controlled crossover design involved 13 non-asthmatic individuals receiving γ-tocopherol supplementation (830 mg/day) or placebo before an LPS challenge. γ-tocopherol significantly reduced neutrophils and eosinophils in sputum and prevented LPS-induced increases in IL-1β levels, although it did not significantly alter IL-6, TNF-α, or markers such as TLR4 and CD11b [65]. These findings underline γ-tocopherol's potential to reduce neutrophilic and eosinophilic inflammation in both animal and human models (Figure 3). Furthermore, a randomized study examined γ-tocopherol supplementation (1200 mg/day for 14 days) vs. placebo in participants undergoing an LPS challenge [66]. γ-tocopherol significantly reduced post-treatment sputum eosinophils, total mucins, and MUC5AC content when compared to the placebo group. It also attenuated LPS-induced neutrophilia at 6 h and 24 h post-challenge. While mucociliary clearance (MCC) slowed significantly during placebo treatment, no such slowing was observed during γ-tocopherol treatment [66]. Furthermore, γ-tocopherol mitigated LPS-induced increases in IL-1β and IL-8 but did not affect IL-6 or other TH1 cytokines [66]. These results underscore γ-tocopherol's ability to reduce airway inflammation, neutrophils infiltration, and mucin production while potentially preserving MCC during LPS-induced airway challenges (Figure 3). Additionally, in a phase I open-label study, 16 volunteers (8 asthmatics and 8 healthy controls) were treated with a γ-tocopherol-rich supplement: one gel tab per day for one week, followed by two gel tabs per day for another week, with a washout period between dosing cycles. Blood samples collected before and after each dosing period showed no significant changes in serum cytokines (e.g., IL-1α, IL-1β, IL-12p40, granulocyte-macrophage colony-stimulating factor (GM-CSF), TNF, eotaxin, Rantes, macrophage inflammatory protein-1 alpha (MIP1α), and interferon gamma-inducible protein 10 (IP-10)) [40]. PBMCs collected from 14 participants and exposed to LPS exhibited significantly reduced secretion of pro-inflammatory cytokines (IL-1β, IL-6, and TNF-α) and chemokines (MCP1 and MIP1α) after γ-tocopherol supplementation, without affecting anti-inflammatory cytokines (Interleukin-1 receptor antagonist (IL-1RA) and Interleukin-10 (IL-10)) [40]. No differences in cytokine responses were observed between asthmatic and non-asthmatic participants.

4.3.2. Other Asthma Inducers

In a pilot randomized clinical trial [67], volunteers aged 18–45 underwent a controlled 2 h exposure to 500 mg/m^3 of wood smoke particulates (WSPs) to assess γ-tocopherol supplementation's effects on airway inflammation. Treatment included four doses of 1253 mg over 2 days or daily doses of 1154–1253 mg over 7 days. WSP exposure increased sputum neutrophils at 6 and 24 h, with no significant differences between γ-tocopherol and placebo groups [67]. Unlike its effects on endotoxin-induced inflammation, γ-tocopherol did not mitigate WSP or O_3-induced neutrophilia, suggesting it may be more effective against endotoxin-induced airway responses [67].

4.4. Nanoformulation

Nanoformulation strategies have emerged as a promising solution to enhance γ-tocopherol pharmacokinetics and therapeutic efficacy [68]. By incorporating γ-tocopherol into nanoparticles, liposomes, or micelles, its stability, solubility, and targeted delivery to specific tissues are significantly improved [69]. Although studies on the biological effects of γ-tocopherol nanoformulations remain limited, existing research highlights their significant potential. For instance, a study by Kuo et al. demonstrated that γ-tocopherol nanoemulsions enhance anti-inflammatory efficacy by facilitating targeted delivery to inflamed tissues. This targeted approach allows for more efficient suppression of pro-inflammatory cytokines, such as TNF-α and IL-1β. In preclinical models of inflammation induced by 2% croton oil, γ-tocopherol nanoemulsions showed a marked reduction in inflammatory markers, underscoring their therapeutic promise for managing inflammatory conditions [14].

5. Anticancer Activity

The anti-tumoral potential of γ-tocopherol has been extensively studied through a variety of in vitro and in vivo experimental models. The underlying mechanisms by which the molecule exerts its anticancer effects are multifaceted and involve modulation of key signaling pathways and molecular targets that are crucial for tumorigenesis and cell proliferation. In this section, we will discuss the detailed mechanistic role of γ-tocopherol in cancer chemoprevention and destruction.

5.1. Modulation of Cell Death

The discovery of novel cell death pathways has significantly expanded our understanding of how cells respond to various forms of cellular damage. Apoptosis and ferroptosis are two critical forms of cell death that are intricately linked in cancer treatment [70,71]. Apoptosis, a regulated process, is often dysregulated in cancer, leading to uncontrolled cell proliferation; therapies that activate apoptotic pathways or inhibit anti-apoptotic proteins hold promise for inducing cancer cell death [72]. Ferroptosis, on the other hand, is a distinct form of programmed cell death that is characterized by the accumulation of lipid hydroperoxides to lethal levels, leading to membrane damage and cell death [73]. It is characterized by iron and lipidic ROS/peroxides' accumulation due to the Fenton reaction and via the loss of balance in ROS production and cell glutathione (GSH)-dependent antioxidants, which protect cells from lipid peroxidation [74]. Emerging evidence suggests that ferroptosis may have a tumor-suppressive function by removing cells that lack access to critical nutrients or have been compromised by infection or environmental stress [75]. Jiang et al. showed that γ-tocopherol treatment induced apoptosis in prostate cancer cells [76] (Figure 4). γ-tocotrienol was more effective than γ-tocopherol to inhibit prostate tumor growth in vivo [77]. Furthermore, according to Gopalan et al., γ-tocopherol triggers apoptosis and enhances the expression of death receptor-5 in human breast cancer cells [78]. Mechanistic analysis revealed that treatment with γ-tocopherol leads to the intracellular buildup of dihydroceramides and dihydrosphingosine in prostate cancer cells [79], as well as ceramides and dihydroceramides in breast cancer cells [78]. The use of chemical inhibitors to block de novo ceramide biosynthesis significantly reduced γ-tocopherol's apoptotic effects [78,79]. These results highlight that γ-tocopherol's anticancer properties are mediated through its influence on the de novo synthesis of sphingolipids. Regarding its metabolites, while γ-tocopherol -13′-COOH has not been specifically investigated, similar metabolites such as δT-13′-COOH and δTE-13′-COOH have been shown to inhibit the proliferation of colon cancer cells, while having minimal effect on normal cells [80].

Moreover, Nrf2 plays a crucial role in protecting against ferroptosis. In fact, Nrf2 upregulate genes involved in antioxidant defense, iron metabolism, and NADPH synthesis. Through this regulation, Nrf2 enhances glutathione synthesis, supporting glutathione peroxidase 4 (GPX4) activity, the primary defense against lipid peroxidation. Nrf2 also regulates iron metabolism by increasing ferroportin expression, which reduces intracellular iron levels and limits ROS generation. Furthermore, it promotes NADPH regeneration via NQO1, an essential process for maintaining GPX4 function [81]. As the downregulation of Nrf2 is generally associated with increased susceptibility to ferroptosis, the importance of maintaining Nrf2 activity becomes apparent. In this regard, it has been demonstrated that a γ-tocopherol-rich mixture of tocopherols (γ-TmT) suppresses prostate tumor development in transgenic adenocarcinoma of the mouse prostate (TRAMP) by epigenetically maintaining Nrf2 expression. Specifically, γ-TmT inhibits DNA methyltransferases (DNMT1, DNMT3A, and DNMT3B), preventing hypermethylation of the Nrf2 promoter and, thus, preserving Nrf2 activity. This sustained Nrf2 expression leads to increased levels of downstream antioxidant and detoxification enzymes, ultimately protecting against prostate tumorigenesis [43] (Figure 4).

In addition to ferroptosis, γ-tocopherols have also been observed to modulate apoptosis, which is particularly relevant in the context of cancer. Apoptosis, a tightly regulated form of programmed cell death, plays a crucial role in maintaining tissue homeostasis by eliminating damaged or unwanted cells. Evasion of apoptosis is recognized as a hallmark of cancer, enabling uncontrolled cell proliferation and tumor development [82]. Mechanistic studies have revealed that the pro-apoptotic effect of γ-tocopherol is mediated through its ability to disrupt the balance between pro- and anti-apoptotic proteins, such as Bcl-2 and Bax [83]. In this study, it has been demonstrated that the combination of γ-tocopherol and 5-fluorouracil (5-FU) modulated the expression of the apoptosis-related genes as it increased the expression of Bax, a pro-apoptotic protein, while decreasing the expression of Bcl-2, an anti-apoptotic protein [83]. This disruption leads to the release of cytochrome c from the mitochondria, triggering the activation of the intrinsic mitochondrial apoptotic pathway. In fact, combining γ-tocopherol and lovastatin in HT-29 cells led to a loss of mitochondrial membrane potential and cytochrome c release, indicating the involvement of the intrinsic apoptotic pathway. Caspase-3 activation, a key executioner of apoptosis, was also observed [84]. In breast cancer cells, it has been demonstrated that γ-tocopherol induces apoptosis by activating the JNK/CHOP/DR5 signaling pathway. In fact, γ-tocopherol treatment activates JNK, a stress-activated protein kinase critical for apoptosis. JNK upregulates CHOP (GADD153), a transcription factor induced by cellular stress, which promotes apoptosis by regulating pro-apoptotic genes. CHOP also increases DR5 expression, a TNF receptor family death receptor, triggering the extrinsic apoptotic pathway and linking intrinsic and extrinsic apoptotic cascades [78].

5.2. Modulation of Cell Cycle

γ-tocopherol has also been shown to modulate the expression and activity of key proteins involved in cell-cycle regulation and survival. A study by Gysin et al. revealed that γ-tocopherol induces cell-cycle arrest at the G0/G1 phase, halting progression to the S phase and subsequent division. This transition, controlled by cyclins D and E, is significantly impaired as γ-tocopherol reduces these proteins' levels. This cycle arrest is observed in various cancer cell lines, including prostate (DU145 and LNCaP), colon, and bone cancer cells. Furthermore, γ-tocopherol inhibits DNA synthesis by 55% in DU-145 and Caco-2 cells, aligning with its impact on cell-cycle progression [85]. Another study explored the effects of α- and γ-tocopherol on the human androgen-independent PC-3 prostate cancer cell line, revealing that both forms of tocopherol inhibited cell growth, with γ-tocopherol showing significantly higher potency. Also, treatment with γ-tocopherol decreased levels of cyclins D1 and E and increased both the expression and activity of transglutaminase 2 (TG2), which is an enzyme associated with cell differentiation and apoptosis. Moreover, both tocopherols reduced the number of cells entering the S phase of the cell cycle, thereby slowing down cell division, with the effect of γ-tocopherol being more prominent [86]. Interestingly, while Sato et al. explored the combined effects of δ-tocotrienol (δ-T3) and γ-tocopherol on prostate cancer cell growth using the androgen-sensitive LNCaP cell line, they showed that γ-tocopherol-induced arrest at the G2/M phase, suggesting a possible correlation with alterations in the expression of cyclin B, a key regulator of this phase [87] (Figure 4).

5.3. Modulation of Cell Proliferation

In a study investigating the effect of a γ-TmT, containing 58% of γ-tocopherol, on estrogen (E2) -induced mammary hyperplasia in rats, Smolarek et al. showed that γ-TmT inhibits cell proliferation by downregulating estrogen receptor α (ERα) expression, reducing serum estrogen levels, and upregulating peroxisome proliferator-activated receptor (PPARγ) and Nrf2 [42]. Furthermore, Campbell et al. found that γ-tocopherol significantly increased both mRNA and protein expression of PPARγ in SW480 human colon cancer cells in a dose-dependent manner [88]. PPARγ is a ligand-activated transcription factor that regulates gene expression by binding to DNA. It plays a crucial role in carcinogenesis due to its involvement in the normal growth and differentiation of cells. Researchers believe that an agent's ability to modulate PPARγ could be significant, potentially leading to clinical implications for cancer chemoprevention, particularly in the colon [88]. When compared to α-tocopherol in vitro, γ-tocopherol showed a higher potency in inhibiting the growth and proliferation of MCF-7 breast cancer cells. Furthermore, γ-tocopherol significantly decreased ERα protein levels in these cells, which is a key finding since estrogen, acting through ERα, promotes the growth of many breast cancers. Indeed, by lowering ERα levels, γ-tocopherol disrupts this growth-promoting pathway. In vivo, when included in the diet of mice bearing MCF-7 xenografts, γ-tocopherol also markedly suppresses tumor growth [83] (Figure 4).

6. Clinical Trials and Therapeutic Potential

γ-tocopherol has been investigated in numerous clinical trials for its potential therapeutic benefits, particularly in managing inflammatory and oxidative stress-related conditions. Trials have demonstrated promising results, including increased serum levels of γ-tocopherol and its metabolite γ-CEHC, reduced nitrosative stress, and inhibition of NF-κB and JNK signaling pathways. γ-tocopherol supplementation also decreased pro-inflammatory cytokines such as IL-1β, IL-6, IL-8, and TNF-α, as well as neutrophil and eosinophil counts elevated by environmental exposures. Moreover, it prevented wood-smoke–induced airway inflammation and improved mucociliary clearance while reducing total sputum mucins. Additionally, γ-tocopherol showed vascular benefits, such as improved flow-mediated dilation, although some trials reported no significant changes in nitric oxide pathways or other oxidative stress markers. These findings highlight its potential role as an anti-inflammatory and antioxidant therapeutic agent. (Figure 5, Table 2).

Table 2. Clinical trials using γ-tocopherol found on https://clinicaltrials.gov (accessed on 25 December 2024).

Intervention/Treatment	Phase	Actual Enrollment	Identifier	Responsible Party	Results
3 doses of 1400 mg γ-tocopherol (2 capsules, each is 700 mg), at 12 h intervals	Not Applicable	10 subjects 18 to 50 years	NCT02610829	University of North Carolina, Chapel Hill	• Increased serum levels of γ-tocopherol and its metabolite γ-CEHC; • Reduced levels of nitrosative stress; • Inhibit NF-κB and JNK pathways; • Reduced cytokine secretion in PBMCs; • Reduced reactive nitrogen oxide species (RNOS) generation.
Oral doses 1200 mg of γ-tocopherol (2 capsules of the γ-tocopherol enriched vitamin E preparation)	Phase 1	25 subjects 18 to 50 years	NCT00836368	Chapel Hill, North Carolina, United States	No Results Posted
14 days of daily high dose (1200 mg) γ-tocopherol. Subjects will receive a 14 days supply (28 softgel capsules, approximately 600 g of γ-tocopherol each	Phase 1	8 subjects 18 to 50 years	NCT00466596	University of North Carolina, Chapel Hill	No Results Posted
γ-tocopherol 1400 mg, taken as 2700 mg capsules every 12 h for a total of 4 doses	Phase 2	18 subjects 18 to 45 years	NCT02911688	University of North Carolina, Chapel Hill	• Reduction in neutrophil and eosinophil counts elevated by ozone exposure; • Decreased levels of pro-inflammatory cytokines IL-1β, IL-6, IL-8, and TNF-α induced by ozone.
γ-tocopherol Maxi Gamma softgels 1200 mg	Phase 1	18 subjects 18 to 50 years	NCT00631085	University of North Carolina, Chapel Hill	• Decreased LPS-induced neutrophil and eosinophil percentage; • Decreased IL-1β levels.
1400 mg of γ-tocopherol -enriched supplement once daily for 7 days	Phase 2	16 subjects 18 to 45 years	NCT03444298	University of North Carolina, Chapel Hill	• Prevention of wood smoke–induced eosinophilic airway inflammation.
600 mg and 1200 mg of γ-tocopherol safety	Phase 1	16 subjects 18 to 50 years	NCT00386178	University of North Carolina, Chapel Hill	No Results Posted
1200 mg of γ-tocopherol daily for 14 days	Phase 1 Phase 2	23 subjects 18 to 50 years	NCT02104505	University of North Carolina, Chapel Hill	• γ-tocopherol treatment prevented the slowdown in mucociliary clearance observed 4 h post-challenge in the placebo group; • Total sputum mucins (excluding MUC5AC) were reduced 24 h post-challenge with γ-tocopherol treatment compared to placebo.
One administration of 500 mg γ-tocopherol	Phase 1 Phase 2	67 subjects 18 to 60 years	NCT01314443	University of Connecticut	• Improved flow-mediated dilation; • No changes in nitric oxide (NO•)-mediated pathways or other oxidative stress markers (e.g., oxLDL, MDA).

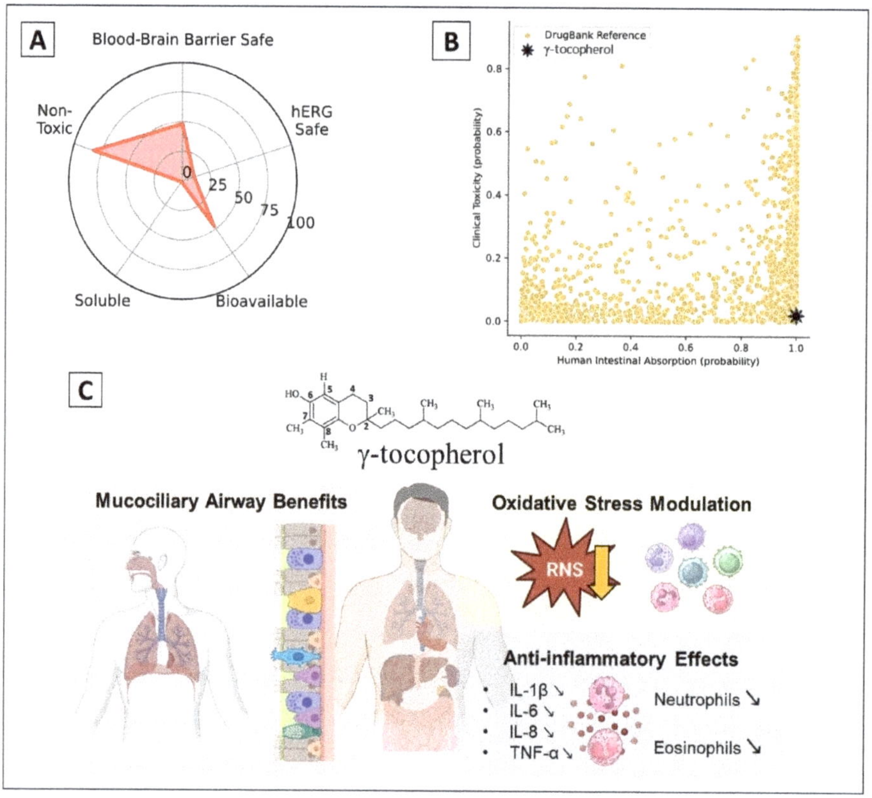

Figure 5. γ-tocopherol's properties and effects. (**A**) Favorable pharmacokinetic and safety profile (bioavailability, solubility, and non-toxicity)—information obtained from https://admet.ai.greenstonebio.com (accessed on 16 December 2024); (**B**) high absorption and low toxicity compared to other compounds—information obtained from https://admet.ai.greenstonebio.com (accessed on 16 December 2024); and (**C**) clinical trial's key biological effects.

7. Conclusions

γ-tocopherol, a significant isoform of vitamin E, emerges as a potent antioxidant with diverse protective effects against oxidative stress-induced damage. Its unique structural properties, including the chromanol ring and hydrophobic phytyl tail, enable its integration into lipid membranes, where it plays a pivotal role in neutralizing ROS and RNS. Furthermore, the anti-inflammatory activity of γ-tocopherol is also closely intertwined with its antioxidant properties. By neutralizing reactive oxygen and nitrogen species, γ-tocopherol not only reduces oxidative stress, a key driver of inflammation, but also modulates signaling pathways, such as NF-κB and Nrf2, thereby suppressing the production of pro-inflammatory cytokines and mediators. These combined effects underscore the critical role of its antioxidant capacity in contributing to its anti-inflammatory efficacy. Additionally, γ-tocopherol demonstrates remarkable efficacy in reducing lipid peroxidation markers, such as MDA and 8-isoprostane, across various biological systems, underscoring its potential in preventing cellular and tissue damage. Preclinical studies highlight its protective effects against DNA oxidative damage by reducing markers like 8-oxo-dG and γH2AX in multiple models, indicating its role in safeguarding genetic material. Nevertheless, further investigation into the precise molecular mechanisms and their influence on DNA structure and stability would be valuable. Additionally, γ-tocopherol's antioxidant effects extend to enhancing the body's defense systems, potentially offering therapeutic benefits for conditions linked to oxidative stress, including metabolic syndrome, environmental toxin exposure, and chronic inflammation.

Despite these promising findings, the complexity of its mechanisms and interactions with other antioxidants, such as α-tocopherol, necessitates further research. Future studies should focus on elucidating its precise molecular pathways and exploring its clinical applicability as a dietary supplement or therapeutic agent. Overall, γ-tocopherol stands out as a vital natural compound with immense potential to combat oxidative stress and related disorders, reinforcing the importance of diet-derived antioxidants in health promotion and disease prevention. Given the promising antioxidant and protective effects of γ-tocopherol demonstrated in various studies, the exploration of nanoformulation strategies becomes essential. These approaches could enhance its stability, bioavailability, and targeted delivery, addressing the limitations of conventional formulations. Since the existing research on nanoformulated γ-tocopherol is limited but encouraging, further investigations are warranted to optimize its therapeutic potential and broaden its application in oxidative stress-related disorders.

Author Contributions: Conceptualization, R.E.K., Y.L. (Youness Limami) and R.E.D.; data curation, B.E.-S., H.W., S.B., S.R. and Y.L. (Yassir Laazziouez); figures conception, H.W.; writing—original draft preparation, B.E.-S., H.W., S.B., S.R. and Y.L. (Yassir Laazziouez); and writing—review and editing, R.E.K., Y.L. (Youness Limami) and R.E.D. All authors have read and agreed to the published version of the manuscript.

Funding: This research was funded by the Ministry of Higher Education, Scientific Research and Innovation of Morocco (MESRSI), the National Agency for Medicinal and Aromatic Plants (ANPMA), and the National Center for Scientific and Technical Research (CNRST) within the framework of the fourth edition of the research project in the field of Valorization of Medicinal and Aromatic Plants VPMA4 (n°2022/17).

Institutional Review Board Statement: Not applicable.

Informed Consent Statement: Not applicable.

Data Availability Statement: Not applicable.

Acknowledgments: B.E.S gratefully acknowledges the CNRST (Centre National pour la Recherche Scientifique et Technique) in Morocco for the Phd Associate Scholarship PASS (2024–2027).

Conflicts of Interest: The authors declare no conflicts of interest.

References

1. Rizvi, S.; Raza, S.T.; Ahmed, F.; Ahmad, A.; Abbas, S.; Mahdi, F. The role of vitamin e in human health and some diseases. *Sultan Qaboos Univ. Med. J.* **2014**, *14*, e157–e165. [PubMed]
2. Jiang, Q.; Im, S.; Wagner, J.G.; Hernandez, M.L.; Peden, D.B. Gamma-tocopherol, a major form of vitamin E in diets: Insights into antioxidant and anti-inflammatory effects, mechanisms, and roles in disease management. *Free Radic. Biol. Med.* **2022**, *178*, 347–359. [CrossRef]
3. Pizzino, G.; Irrera, N.; Cucinotta, M.; Pallio, G.; Mannino, F.; Arcoraci, V.; Squadrito, F.; Altavilla, D.; Bitto, A. Oxidative Stress: Harms and Benefits for Human Health. *Oxid. Med. Cell Longev.* **2017**, *2017*, 8416763. [CrossRef] [PubMed]
4. Reddy, V.P. Oxidative Stress in Health and Disease. *Biomedicines* **2023**, *11*, 2925. [CrossRef] [PubMed]
5. Hba, S.; Ghaddar, S.; Wahnou, H.; Pinon, A.; El Kebbaj, R.; Pouget, C.; Sol, V.; Liagre, B.; Oudghiri, M.; Limami, Y. Natural Chalcones and Derivatives in Colon Cancer: Pre-Clinical Challenges and the Promise of Chalcone-Based Nanoparticles. *Pharmaceutics* **2023**, *15*, 2718. [CrossRef] [PubMed]
6. Zhao, H.; Wu, L.; Yan, G.; Chen, Y.; Zhou, M.; Wu, Y.; Li, Y. Inflammation and tumor progression: Signaling pathways and targeted intervention. *Signal Transduct. Target. Ther.* **2021**, *6*, 263. [CrossRef]
7. Reuter, S.; Gupta, S.C.; Chaturvedi, M.M.; Aggarwal, B.B. Oxidative stress, inflammation, and cancer: How are they linked? *Free Radic. Biol. Med.* **2010**, *49*, 1603–1616. [CrossRef]

8. Bindu, S.; Mazumder, S.; Bandyopadhyay, U. Non-steroidal anti-inflammatory drugs (NSAIDs) and organ damage: A current perspective. *Biochem. Pharmacol.* **2020**, *180*, 114147. [CrossRef]
9. McEvoy, L.; Carr, D.F.; Pirmohamed, M. Pharmacogenomics of NSAID-Induced Upper Gastrointestinal Toxicity. *Front. Pharmacol.* **2021**, *12*, 684162. [CrossRef]
10. Wahnou, H.; Ndayambaje, M.; Ouadghiri, Z.; Benayad, S.; Elattar, H.; Chgari, O.; Naya, A.; Zaid, Y.; Oudghiri, M. Artemisia herba-alba: Antioxidant capacity and efficacy in preventing chronic arthritis in vivo. *Inflammopharmacology* **2024**, *32*, 1855–1870. [CrossRef] [PubMed]
11. Pahrudin Arrozi, A.; Shukri, S.N.S.; Wan Ngah, W.Z.; Mohd Yusof, Y.A.; Ahmad Damanhuri, M.H.; Jaafar, F.; Makpol, S. Comparative Effects of Alpha- and Gamma-Tocopherol on Mitochondrial Functions in Alzheimer's Disease In Vitro Model. *Sci. Rep.* **2020**, *10*, 8962. [CrossRef] [PubMed]
12. Reiter, E.; Jiang, Q.; Christen, S. Anti-inflammatory properties of alpha- and gamma-tocopherol. *Mol. Asp. Med.* **2007**, *28*, 668–691. [CrossRef] [PubMed]
13. Devaraj, S.; Jialal, I. Failure of Vitamin E in Clinical Trials: Is Gamma-Tocopherol the Answer? *Nutr. Rev.* **2005**, *63*, 290–293. [CrossRef]
14. Kuo, F.; Subramanian, B.; Kotyla, T.; Wilson, T.A.; Yoganathan, S.; Nicolosi, R.J. Nanoemulsions of an anti-oxidant synergy formulation containing gamma tocopherol have enhanced bioavailability and anti-inflammatory properties. *Int. J. Pharm.* **2008**, *363*, 206–213. [CrossRef] [PubMed]
15. Das Gupta, S.; Suh, N. Tocopherols in cancer: An update. *Mol. Nutr. Food Res.* **2016**, *60*, 1354–1363. [CrossRef] [PubMed]
16. Colombo, M.L. An Update on Vitamin E, Tocopherol and Tocotrienol—Perspectives. *Molecules* **2010**, *15*, 2103–2113. [CrossRef] [PubMed]
17. Xu, Z.; Harvey, K.; Pavlina, T.; Zaloga, G.; Siddiqui, R. Tocopherol and Tocotrienol Homologs in Parenteral Lipid Emulsions. *Eur. J. Lipid Sci. Technol.* **2015**, *117*, 15–22. [CrossRef] [PubMed]
18. Charlton, N.C.; Mastyugin, M.; Török, B.; Török, M. Structural Features of Small Molecule Antioxidants and Strategic Modifications to Improve Potential Bioactivity. *Molecules* **2023**, *28*, 1057. [CrossRef] [PubMed]
19. Atkinson, J.; Harroun, T.; Wassall, S.R.; Stillwell, W.; Katsaras, J. The location and behavior of alpha-tocopherol in membranes. *Mol. Nutr. Food Res.* **2010**, *54*, 641–651. [CrossRef] [PubMed]
20. Shahidi, F.; de Camargo, A.C. Tocopherols and Tocotrienols in Common and Emerging Dietary Sources: Occurrence, Applications, and Health Benefits. *Int. J. Mol. Sci.* **2016**, *17*, 1745. [CrossRef] [PubMed]
21. Balakrishnan, G.; Schneider, R.G. Tocopherol degradation and lipid oxidation during storage of Chenopodium quinoa. *J. Food Compos. Anal.* **2023**, *123*, 105549. [CrossRef]
22. Bruscatto, M.H.; Zambiazi, R.C.; Sganzerla, M.; Pestana, V.R.; Otero, D.; Lima, R.; Paiva, F. Degradation of Tocopherols in Rice Bran Oil Submitted to Heating at Different Temperatures. *J. Chromatogr. Sci.* **2009**, *47*, 762–765. [CrossRef]
23. Aoyama, S.; Nishio, T.; Moriya, D.; Munemasa, S.; Murata, Y.; Nakamura, Y.; Nakamura, T. The Metabolite of γ-Tocopherol, 2,7,8-Trimethyl-2-(2′-Carboxyethyl)-6-Hydroxychroman, Exerts Intracellular Antioxidant Activity via Up-Regulation of Heme Oxygenase-1 in Hepatocytes. *Nutraceuticals* **2024**, *4*, 409–416. [CrossRef]
24. Thbayh, D.K.; Mentes, D.; Boros, Z.R.; Palusiak, M.; Farkas, L.; Viskolcz, B.; Fiser, B. α-Tocopherol and Trolox as Effective Natural Additives for Polyurethane Foams: A DFT and Experimental Study. *Molecules* **2024**, *29*, 6037. [CrossRef]
25. Amić, A.; Mastil'ák Cagardová, D. A DFT study of the antioxidant potency of α-tocopherol and its derivatives: PMHC, Trolox, and α-CEHC. *J. Mol. Liq.* **2024**, *403*, 124796. [CrossRef]
26. Benbrahim, N.; Zeddour-Brahim, K.; Zizi, Z.; Bengharez, Z. Stability and Reactivity of Tocopherols: Theoretical Study. *Chem. Proc.* **2023**, *14*, 20. [CrossRef]
27. Wagner, K.H.; Kamal-Eldin, A.; Elmadfa, I. Gamma-tocopherol—An underestimated vitamin? *Ann. Nutr. Metab.* **2004**, *48*, 169–188. [CrossRef] [PubMed]
28. Yamabe, S.; Tsuchida, N.; Yamazaki, S. A DFT Study on Transition States of Inhibition of Oxidation by α-Tocopherol. *ChemistrySelect* **2020**, *5*, 9184–9194. [CrossRef]
29. Abyar, F.; Farrokhpour, H. Ionization energies and photoelectron spectra of fat-soluble vitamins in the gas phase: A theoretical study. *RSC Adv.* **2014**, *4*, 35975–35987. [CrossRef]
30. Warren, J.J.; Mayer, J.M. Predicting organic hydrogen atom transfer rate constants using the Marcus cross relation. *Proc. Natl. Acad. Sci. USA* **2010**, *107*, 5282–5287. [CrossRef] [PubMed]
31. Povalishev, V.N.; Polozov, G.I.; Shadyro, O.I. Effects of α-tocopherol and related compounds on reactions involving various organic radicals. *Bioorganic Med. Chem. Lett.* **2006**, *16*, 1236–1239. [CrossRef] [PubMed]
32. Shin, H.; Eo, H.; Lim, Y. Similarities and differences between alpha-tocopherol and gamma-tocopherol in amelioration of inflammation, oxidative stress and pre-fibrosis in hyperglycemia induced acute kidney inflammation. *Nutr. Res. Pract.* **2016**, *10*, 33–41. [CrossRef]

33. Chung, M.Y.; Yeung, S.F.; Park, H.J.; Volek, J.S.; Bruno, R.S. Dietary α- and γ-tocopherol supplementation attenuates lipopolysaccharide-induced oxidative stress and inflammatory-related responses in an obese mouse model of nonalcoholic steatohepatitis. *J. Nutr. Biochem.* **2010**, *21*, 1200–1206. [CrossRef] [PubMed]
34. Yamamoto, Y.; Sousse, L.E.; Enkhbaatar, P.; Kraft, E.R.; Deyo, D.J.; Wright, C.L.; Taylor, A.; Traber, M.G.; Cox, R.A.; Hawkins, H.K.; et al. γ-tocopherol nebulization decreases oxidative stress, arginase activity, and collagen deposition after burn and smoke inhalation in the ovine model. *Shock* **2012**, *38*, 671–676. [CrossRef] [PubMed]
35. Das Gupta, S.; So, J.Y.; Wall, B.; Wahler, J.; Smolarek, A.K.; Sae-Tan, S.; Soewono, K.Y.; Yu, H.; Lee, M.J.; Thomas, P.E.; et al. Tocopherols inhibit oxidative and nitrosative stress in estrogen-induced early mammary hyperplasia in ACI rats. *Mol. Carcinog.* **2015**, *54*, 916–925. [CrossRef]
36. Devaraj, S.; Leonard, S.; Traber, M.G.; Jialal, I. Gamma-tocopherol supplementation alone and in combination with alpha-tocopherol alters biomarkers of oxidative stress and inflammation in subjects with metabolic syndrome. *Free Radic. Biol. Med.* **2008**, *44*, 1203–1208. [CrossRef]
37. Kowalska, M.; Piekut, T.; Prendecki, M.; Sodel, A.; Kozubski, W.; Dorszewska, J. Mitochondrial and Nuclear DNA Oxidative Damage in Physiological and Pathological Aging. *DNA Cell Biol.* **2020**, *39*, 1410–1420. [CrossRef] [PubMed]
38. Chen, J.X.; Liu, A.; Lee, M.J.; Wang, H.; Yu, S.; Chi, E.; Reuhl, K.; Suh, N.; Yang, C.S. δ- and γ-tocopherols inhibit phIP/DSS-induced colon carcinogenesis by protection against early cellular and DNA damages. *Mol. Carcinog.* **2017**, *56*, 172–183. [CrossRef] [PubMed]
39. Li, G.; Lee, M.J.; Liu, A.B.; Yang, Z.; Lin, Y.; Shih, W.J.; Yang, C.S. The antioxidant and anti-inflammatory activities of tocopherols are independent of Nrf2 in mice. *Free Radic. Biol. Med.* **2012**, *52*, 1151–1158. [CrossRef]
40. Wiser, J.; Alexis, N.E.; Jiang, Q.; Wu, W.; Robinette, C.; Roubey, R.; Peden, D.B. In vivo gamma-tocopherol supplementation decreases systemic oxidative stress and cytokine responses of human monocytes in normal and asthmatic subjects. *Free Radic. Biol. Med.* **2008**, *45*, 40–49. [CrossRef]
41. Dey, P.; Mah, E.; Li, J.; Jalili, T.; Symons, J.D.; Bruno, R.S. Improved hepatic γ-tocopherol status limits oxidative and inflammatory stress-mediated liver injury in db/db mice with nonalcoholic steatohepatitis. *J. Funct. Foods* **2018**, *40*, 670–678. [CrossRef]
42. Smolarek, A.K.; So, J.Y.; Thomas, P.E.; Lee, H.J.; Paul, S.; Dombrowski, A.; Wang, C.X.; Saw, C.L.; Khor, T.O.; Kong, A.N.; et al. Dietary tocopherols inhibit cell proliferation, regulate expression of ERα, PPARγ, and Nrf2, and decrease serum inflammatory markers during the development of mammary hyperplasia. *Mol. Carcinog.* **2013**, *52*, 514–525. [CrossRef] [PubMed]
43. Huang, Y.; Khor, T.O.; Shu, L.; Saw, C.L.; Wu, T.Y.; Suh, N.; Yang, C.S.; Kong, A.N. A γ-tocopherol-rich mixture of tocopherols maintains Nrf2 expression in prostate tumors of TRAMP mice via epigenetic inhibition of CpG methylation. *J. Nutr.* **2012**, *142*, 818–823. [CrossRef] [PubMed]
44. Gaschler, M.M.; Stockwell, B.R. Lipid peroxidation in cell death. *Biochem. Biophys. Res. Commun.* **2017**, *482*, 419–425. [CrossRef] [PubMed]
45. Luo, J.; Mills, K.; le Cessie, S.; Noordam, R.; van Heemst, D. Ageing, age-related diseases and oxidative stress: What to do next? *Ageing Res. Rev.* **2020**, *57*, 100982. [CrossRef]
46. Korošec, T.; Tomažin, U.; Horvat, S.; Keber, R.; Salobir, J. The diverse effects of α- and γ-tocopherol on chicken liver transcriptome. *Poult. Sci.* **2017**, *96*, 667–680. [CrossRef] [PubMed]
47. Kamal, R.; Kharbach, M.; Vander Heyden, Y.; Doukkali, Z.; Ghchime, R.; Bouklouze, A.; Cherrah, Y.; Alaoui, K. In vivo anti-inflammatory response and bioactive compounds' profile of polyphenolic extracts from edible Argan oil (*Argania spinosa* L.), obtained by two extraction methods. *J. Food Biochem.* **2019**, *43*, e13066. [CrossRef] [PubMed]
48. Bouchab, H.; Ishaq, A.; El Kebbaj, R.; Nasser, B.; Saretzki, G. Protective effect of argan oil on DNA damage in vivo and in vitro. *Biomarkers* **2021**, *26*, 425–433. [CrossRef] [PubMed]
49. Bouchab, H.; Essadek, S.; El Kamouni, S.; Moustaid, K.; Essamadi, A.; Andreoletti, P.; Cherkaoui-Malki, M.; El Kebbaj, R.; Nasser, B. Antioxidant Effects of Argan Oil and Olive Oil against Iron-Induced Oxidative Stress: In Vivo and In Vitro Approaches. *Molecules* **2023**, *28*, 5924. [CrossRef]
50. Haleng, J.; Pincemail, J.; Defraigne, J.O.; Charlier, C.; Chapelle, J.P. Oxidative stress. *Rev. Medicale Liege* **2007**, *62*, 628–638.
51. El Faqer, O.; Rais, S.; Ouadghiri, Z.; El Faqer, A.; Benchama, Z.; El Ouaddari, A.; Dakir, M.; El Amrani, A.; Mtairag, E.M. Physicochemical properties, GC–MS profiling, and antibacterial potential of Allium sativum essential oil: In vitro and in silico approaches. *Sci. Afr.* **2024**, *26*, e02484. [CrossRef]
52. El Faqer, O.; Elkoraichi, I.; Latif, M.; Debierre-Grockiego, F.; Ouadghiri, Z.; Rais, S.; Dimier-Poisson, I.; Mtairag, E.M. Pharmacological insights into Laurus nobilis: HPLC profiling and evaluation of its anti-Toxoplasma, antioxidant, and anti-hemolytic properties. *Biochem. Syst. Ecol.* **2024**, *117*, 104891. [CrossRef]
53. Saldeen, T.; Li, D.; Mehta, J.L. Differential effects of alpha- and gamma-tocopherol on low-density lipoprotein oxidation, superoxide activity, platelet aggregation and arterial thrombogenesis. *J. Am. Coll. Cardiol.* **1999**, *34*, 1208–1215. [CrossRef] [PubMed]

54. El Kebbaj, R.; Bouchab, H.; Tahri-Joutey, M.; Rabbaa, S.; Limami, Y.; Nasser, B.; Egbujor, M.C.; Tucci, P.; Andreoletti, P.; Saso, L.; et al. The Potential Role of Major Argan Oil Compounds as Nrf2 Regulators and Their Antioxidant Effects. *Antioxidants* **2024**, *13*, 344. [CrossRef] [PubMed]
55. Arulselvan, P.; Fard, M.T.; Tan, W.S.; Gothai, S.; Fakurazi, S.; Norhaizan, M.E.; Kumar, S.S. Role of Antioxidants and Natural Products in Inflammation. *Oxid. Med. Cell Longev.* **2016**, *2016*, 5276130. [CrossRef]
56. Wahnou, H.; Hmimid, F.; Errami, A.; Nait Irahal, I.; Limami, Y.; Oudghiri, M. Integrating ADMET, enrichment analysis, and molecular docking approach to elucidate the mechanism of Artemisia herba alba for the treatment of inflammatory bowel disease-associated arthritis. *J. Toxicol. Environ. Health Part A* **2024**, *87*, 836–854. [CrossRef]
57. Jiang, Q.; Ames, B.N. Gamma-tocopherol, but not alpha-tocopherol, decreases proinflammatory eicosanoids and inflammation damage in rats. *FASEB J.* **2003**, *17*, 816–822. [CrossRef]
58. Jiang, Q.; Moreland, M.; Ames, B.N.; Yin, X. A combination of aspirin and gamma-tocopherol is superior to that of aspirin and alpha-tocopherol in anti-inflammatory action and attenuation of aspirin-induced adverse effects. *J. Nutr. Biochem.* **2009**, *20*, 894–900. [CrossRef]
59. Elisia, I.; Kitts, D.D. Tocopherol isoforms (α-, γ-, and δ-) show distinct capacities to control Nrf-2 and NfκB signaling pathways that modulate inflammatory response in Caco-2 intestinal cells. *Mol. Cell. Biochem.* **2015**, *404*, 123–131. [CrossRef] [PubMed]
60. Jiang, Q.; Elson-Schwab, I.; Courtemanche, C.; Ames, B.N. gamma-tocopherol and its major metabolite, in contrast to alpha-tocopherol, inhibit cyclooxygenase activity in macrophages and epithelial cells. *Proc. Natl. Acad. Sci. USA* **2000**, *97*, 11494–11499. [CrossRef]
61. Wells, S.R.; Jennings, M.H.; Rome, C.; Hadjivassiliou, V.; Papas, K.A.; Alexander, J.S. Alpha-, gamma- and delta-tocopherols reduce inflammatory angiogenesis in human microvascular endothelial cells. *J. Nutr. Biochem.* **2010**, *21*, 589–597. [CrossRef] [PubMed]
62. Mah, E.; Pei, R.; Guo, Y.; Ballard, K.D.; Barker, T.; Rogers, V.E.; Parker, B.A.; Taylor, A.W.; Traber, M.G.; Volek, J.S.; et al. γ-Tocopherol-rich supplementation additively improves vascular endothelial function during smoking cessation. *Free Radic. Biol. Med.* **2013**, *65*, 1291–1299. [CrossRef]
63. Lee, H.; Lim, Y. Gamma-tocopherol ameliorates hyperglycemia-induced hepatic inflammation associated with NLRP3 inflammasome in alloxan-induced diabetic mice. *Nutr. Res. Pract.* **2019**, *13*, 377–383. [CrossRef] [PubMed]
64. Gans, M.D.; Gavrilova, T. Understanding the immunology of asthma: Pathophysiology, biomarkers, and treatments for asthma endotypes. *Paediatr. Respir. Rev.* **2020**, *36*, 118–127. [CrossRef] [PubMed]
65. Hernandez, M.L.; Wagner, J.G.; Kala, A.; Mills, K.; Wells, H.B.; Alexis, N.E.; Lay, J.C.; Jiang, Q.; Zhang, H.; Zhou, H.; et al. Vitamin E, γ-tocopherol, reduces airway neutrophil recruitment after inhaled endotoxin challenge in rats and in healthy volunteers. *Free Radic. Biol. Med.* **2013**, *60*, 56–62. [CrossRef]
66. Burbank, A.J.; Duran, C.G.; Pan, Y.; Burns, P.; Jones, S.; Jiang, Q.; Yang, C.; Jenkins, S.; Wells, H.; Alexis, N.; et al. Gamma tocopherol-enriched supplement reduces sputum eosinophilia and endotoxin-induced sputum neutrophilia in volunteers with asthma. *J. Allergy Clin. Immunol.* **2018**, *141*, 1231–1238.e1. [CrossRef] [PubMed]
67. Peden, D.B.; Almond, M.; Brooks, C.; Robinette, C.; Wells, H.; Burbank, A.; Hernandez, M.; Hinderliter, A.; Caughey, M.; Jiang, Q.; et al. A pilot randomized clinical trial of γ-tocopherol supplementation on wood smoke-induced neutrophilic and eosinophilic airway inflammation. *J. Allergy Clin. Immunol. Glob.* **2023**, *2*, 100177. [CrossRef] [PubMed]
68. Wahnou, H.; Liagre, B.; Sol, V.; El Attar, H.; Attar, R.; Oudghiri, M.; Duval, R.E.; Limami, Y. Polyphenol-Based Nanoparticles: A Promising Frontier for Enhanced Colorectal Cancer Treatment. *Cancers* **2023**, *15*, 3826. [CrossRef]
69. Patra, J.K.; Das, G.; Fraceto, L.F.; Campos, E.V.R.; del Rodriguez-Torres, M.P.; Acosta-Torres, L.S.; Diaz-Torres, L.A.; Grillo, R.; Swamy, M.K.; Sharma, S.; et al. Nano based drug delivery systems: Recent developments and future prospects. *J. Nanobiotechnol.* **2018**, *16*, 71. [CrossRef] [PubMed]
70. Nikoletopoulou, V.; Markaki, M.; Palikaras, K.; Tavernarakis, N. Crosstalk between apoptosis, necrosis and autophagy. *Biochim. Biophys. Acta—Mol. Cell Res.* **2013**, *1833*, 3448–3459. [CrossRef] [PubMed]
71. Chgari, O.; Wahnou, H.; Ndayambaje, M.; Moukhfi, F.; Benkhnigue, O.; Marnissi, F.; Limami, Y.; Oudghiri, M. *Orbea variegata* (L.) Haw in skin carcinogenesis: Insights from an in vivo male Swiss mouse model study. *J. Toxicol. Environ. Health Part A* **2024**, *87*, 630–645. [CrossRef] [PubMed]
72. Towers, C.G.; Wodetzki, D.; Thorburn, A. Autophagy and cancer: Modulation of cell death pathways and cancer cell adaptations. *J. Cell Biol.* **2020**, *219*, e201909033. [CrossRef] [PubMed]
73. Fan, J.; Jiang, T.; He, D. Emerging insights into the role of ferroptosis in the pathogenesis of autoimmune diseases. *Front. Immunol.* **2023**, *14*, 1120519. [CrossRef] [PubMed]
74. Benayad, S.; Wahnou, H.; El Kebbaj, R.; Liagre, B.; Sol, V.; Oudghiri, M.; Saad, E.M.; Duval, R.E.; Limami, Y. The Promise of Piperine in Cancer Chemoprevention. *Cancers* **2023**, *15*, 5488. [CrossRef] [PubMed]

75. Stockwell, B.R.; Friedmann Angeli, J.P.; Bayir, H.; Bush, A.I.; Conrad, M.; Dixon, S.J.; Fulda, S.; Gascón, S.; Hatzios, S.K.; Kagan, V.E.; et al. Ferroptosis: A Regulated Cell Death Nexus Linking Metabolism, Redox Biology, and Disease. *Cell* **2017**, *171*, 273–285. [CrossRef] [PubMed]
76. Di Pardo, A.; Basit, A.; Armirotti, A.; Amico, E.; Castaldo, S.; Pepe, G.; Marracino, F.; Buttari, F.; Digilio, A.F.; Maglione, V. De novo Synthesis of Sphingolipids Is Defective in Experimental Models of Huntington's Disease. *Front. Neurosci.* **2017**, *11*, 698. [CrossRef] [PubMed]
77. Jiang, Q.; Rao, X.; Kim, C.Y.; Freiser, H.; Zhang, Q.; Jiang, Z.; Li, G. Gamma-tocotrienol induces apoptosis and autophagy in prostate cancer cells by increasing intracellular dihydrosphingosine and dihydroceramide. *Int. J. Cancer* **2012**, *130*, 685–693. [CrossRef] [PubMed]
78. Gopalan, A.; Yu, W.; Jiang, Q.; Jang, Y.; Sanders, B.G.; Kline, K. Involvement of de novo ceramide synthesis in gamma-tocopherol and gamma-tocotrienol-induced apoptosis in human breast cancer cells. *Mol. Nutr. Food Res.* **2012**, *56*, 1803–1811. [CrossRef]
79. Jiang, Q.; Wong, J.; Fyrst, H.; Saba, J.D.; Ames, B.N. γ-Tocopherol or combinations of vitamin E forms induce cell death in human prostate cancer cells by interrupting sphingolipid synthesis. *Proc. Natl. Acad. Sci. USA* **2004**, *101*, 17825–17830. [CrossRef]
80. Jang, Y.; Park, N.-Y.; Rostgaard-Hansen, A.L.; Huang, J.; Jiang, Q. Vitamin E metabolite 13′-carboxychromanols inhibit pro-inflammatory enzymes, induce apoptosis and autophagy in human cancer cells by modulating sphingolipids and suppress colon tumor development in mice. *Free Radic. Biol. Med.* **2016**, *95*, 190–199. [CrossRef]
81. Yan, R.; Lin, B.; Jin, W.; Tang, L.; Hu, S.; Cai, R. NRF2, a Superstar of Ferroptosis. *Antioxidants* **2023**, *12*, 1739. [CrossRef] [PubMed]
82. Plati, J.; Bucur, O.; Khosravi-Far, R. Apoptotic cell signaling in cancer progression and therapy. *Integr. Biol.* **2011**, *3*, 279–296. [CrossRef] [PubMed]
83. Bazzaz, R.; Bijanpour, H.; Pirouzpanah, S.M.B.; Yaghmaei, P.; Rashtchizadeh, N. Adjuvant therapy with γ-tocopherol-induce apoptosis in HT-29 colon cancer via cyclin-dependent cell cycle arrest mechanism. *J. Biochem. Mol. Toxicol.* **2019**, *33*, e22399. [CrossRef] [PubMed]
84. Rezaei, M.; Zeidooni, L.; Hashemitabar, M.; Razzazzadeh, S.; Mahdavinia, M.; Ghasemi, K. Gamma-tocopherol enhances apoptotic effects of lovastatin in human colorectal carcinoma cell line (HT29). *Nutr. Cancer* **2014**, *66*, 1386–1393. [CrossRef] [PubMed]
85. Gysin, R.; Azzi, A.; Visarius, T. Gamma-tocopherol inhibits human cancer cell cycle progression and cell proliferation by down-regulation of cyclins. *FASEB J.* **2002**, *16*, 1952–1954. [CrossRef] [PubMed]
86. Torricelli, P.; Caraglia, M.; Abbruzzese, A.; Beninati, S. γ-Tocopherol inhibits human prostate cancer cell proliferation by up-regulation of transglutaminase 2 and down-regulation of cyclins. *Amino Acids* **2013**, *44*, 45–51. [CrossRef]
87. Sato, C.; Kaneko, S.; Sato, A.; Virgona, N.; Namiki, K.; Yano, T. Combination Effect of δ-Tocotrienol and γ-Tocopherol on Prostate Cancer Cell Growth. *J. Nutr. Sci. Vitaminol.* **2017**, *63*, 349–354. [CrossRef] [PubMed]
88. Campbell, S.E.; Stone, W.L.; Whaley, S.G.; Qui, M.; Krishnan, K. Gamma (γ) tocopherol upregulates peroxisome proliferator activated receptor (PPAR) gamma (γ) expression in SW 480 human colon cancer cell lines. *BMC Cancer* **2003**, *3*, 25. [CrossRef] [PubMed]

molecules

MDPI

Review

The Extraction, Determination, and Bioactivity of Curcumenol: A Comprehensive Review

Jie Li [1,2,3], Yitian Sun [1,2], Guohua Li [1,2], Chunsong Cheng [4], Xinbing Sui [2] and Qibiao Wu [1,5,6,*]

1 State Key Laboratory of Quality Research in Chinese Medicines, Faculty of Chinese Medicine, Macau University of Science and Technology, Macau 999078, China; 18256960482@sina.cn (J.L.)
2 College of Pharmacy, Hangzhou Normal University, Hangzhou 311121, China
3 School of Environmental and Chemical Engineering, Zhaoqing University, Zhaoqing 526061, China
4 Lushan Botanical Garden, Chinese Academy of Sciences, Jiujiang 332900, China
5 Zhuhai M.U.S.T. Science and Technology Research Institute, Zhuhai 519031, China
6 Guangdong-Hong Kong-Macao Joint Laboratory for Contaminants Exposure and Health, Guangzhou 510006, China
* Correspondence: qbwu@must.edu.mo

Abstract: *Curcuma wenyujin* is a member of the *Curcuma zedoaria* (zedoary, Zingiberaceae) family, which has a long history in traditional Chinese medicine (TCM) due to its abundant biologically active constituents. Curcumenol, a component of *Curcuma wenyujin*, has several biological activities. At present, despite different pharmacological activities being reported, the clinical usage of curcumenol remains under investigation. To further determine the characteristics of curcumenol, the extraction, determination, and bioactivity of the compound are summarized in this review. Existing research has reported that curcumenol exerts different pharmacological effects in regard to a variety of diseases, including anti-inflammatory, anti-oxidant, anti-bactericidal, anti-diabetic, and anti-cancer activity, and also ameliorates osteoporosis. This review of curcumenol provides a theoretical basis for further research and clinical applications.

Keywords: *Curcuma wenyujin*; curcumenol; biological activity; TCM

1. Introduction

Curcuma wenyujin Y. H. Chen et C. Ling (*Curcuma wenyujin*) is a member of the *Curcuma zedoaria* (zedoary, Zingiberaceae) family, which is primarily distributed in Zhejiang Province, China [1]. It has a long history of extensive utilization in TCM. It is one of the members of the well-known TCM group named "Zhebawei", which includes eight authentic Chinese herbal medicines from Zhejiang Province (*Atractylodes, paeony, Zhejiang fritillaria, Chrysanthemum morifolium, Corydalis glaucescens, Scrophularia ningpoensis Hemsl., Ophiopogon japonicus (Thunb.) Ker-Gawl*, and *Wen Yujin*) [2] (Figure 1). *Curcuma wenyujin* can be incorporated into three different Chinese medicines, "Acruginous Turmeric Rhizome", "Wenyujin", and "Wenyujin Rhizoma Concisum", and exhibits anti-inflammatory [3], anti-tumor [4], anti-oxidant [5], anti-bacterial, anti-viral, and hepatoprotective properties. The pharmacological value of *Curcuma wenyujin* is derived from its constituents, which have been reported to comprise a total of 169 compounds, thus far. These include monoterpenoids, sesquiternoids, diterpenoids, curcuminoids, etc. [6]. Curcumin, an active ingredient of *Curcuma wenyujin*, has been widely investigated and applied [7]. Curcumenol is one of the other sesquiterpenoid active ingredients (following the Chinese Pharmacopoeia), which has been extracted from the edible rhizome of *Curcuma zedoaria* since the 1960s [8]; it has been reported as one of the primary constituents in *Curcuma* plant essential oils, as well as an important component of food and traditional remedies [9–12]. A growing number of studies have suggested that it exhibits useful biological effects, including anti-cancer [13–16], anti-inflammatory [17–19], anti-bactericidal [20,21], and liver protection activities [22], among others. *Curcumina longa*, another important member of the *Curcuma zedoaria* (zedoary, Zingiberaceae) family, has also attracted research interest. The differences in the components and physiological effects of the two herbs are shown in Table 1.

Figure 1. Picture of *Curcuma wenyujin* and the structure of curcumenol.

Despite the huge pharmacological potential of curcumenol, its further application has been hindered by unsatisfactory physical properties (e.g., low water solubility). In this research, relevant studies and reports were reviewed to summarize the methods related to the extraction, determination, and pharmacology of curcumenol, with the aim of promoting the investigation and utilization of curcumenol in human healthcare.

Table 1. Differences between the components and physiological effects of *Curcuma wenyujin* and *Curcumina longa*.

Curcuma wenyujin		*Curcumina longa*
The Content of Unique Chemical Markers to Distinguish the Two Herbs [23]		
Curcumenol	>	Curcumenol
Curcumenone	>	Curcumenone
Neocurdione	>	Neocurdione
Curdione	>	Curdione
Curcumin	<	Curcumin
Physiological effects		
Anti-inflammatory [24]		Anti-inflammatory [25]
Anti-oxidant [26]		Anti-oxidant [27]
Antimicrobial [28] and antiviral activity [29]		Antimicrobial [30] and antiviral activity [31]

Table 1. *Cont.*

Curcuma wenyujin	*Curcumina longa*
Anti-diabetic [32]	Anti-diabetic [33]
Hepatoprotective [34]	Hepatoprotective [35]
Effects on the cardiovascular system [36]	Effects on the cardiovascular system [37]
Analgesic effect [38]	Analgesic effect [39]
Effect on the nervous system: Alzheimer's disease [40]	Effect on the nervous system: improve memory impairment [41]
Anti-cancer: glioma [42], breast cancer [43], liver cancer [44], gastric cancer [45], colon cancer [46], leukemia [47], cervical cancer [48], lung cancer [49]	Anti-cancer: liver cancer [50], colon cancer [51], cervical cancer [52], human thyroid cancer [53], human nasopharyngeal carcinoma cells [54], lung cancer [55]
	Effects on the respiratory system [56]
	Effect on the digestive system [57]
	Anticoagulant effect [58]

2. Traditional Medicinal Uses of Curcumenol-Related Herbs

Curcuma is an important herb in Chinese medicine, with a long history of use for about 1500 years. It has significance in healthcare due to its various medicinal effects. It has been reported that *Curcuma* can promote blood circulation, promote the flow of qi, unblock meridians, and relieve pain [1]. In-depth investigations have also been conducted to elucidate which exact compounds exert the abovementioned traditional curative effects, based on current medical theories. A series of compounds related to these treatment effects have been identified, including curcumenol. Curcumenol exhibits anti-inflammatory, anti-tumor, anti-virus, anti-oxidation, and hepatoprotective effects, as listed in Table 2. The above-described pharmacologica activity of curcumenol provides further evidence for the use of *Curcuma* in traditional medicines.

Table 2. Traditional uses of curcumenol-related herbal medicines.

Scientific Name	Traditional Name	Plant Part Used	Effective Compounds and Ref.	Traditional Medicine Use	Monotherapy or an Adjunct and Ref.
Curcuma zedoaria	Zedoary, Zingiberaceae	Rhizome	Curcumenol [17]	Osteoarthritis	An adjunct to allopathic medication [59]
Curcuma zedoaria	White turmeric	Rhizome	Curcumenol [18]	Anti-inflammatory	An adjunct to allopathic medication [60]
Curcuma zedoaria	White turmeric	Rhizome	Curcumenol and six other compounds [19]	Psoriasis	An adjunct to allopathic medication [61]
Curcuma	*Curcuma phaeocaulis*, *Curcuma kwangsiensis*, *Curcuma wenyujin*	Rhizome	Curcumenol, Curcumol, β-elemene, curdione [21]	Anti-fungal	An adjunct to allopathic medication [62]

Table 2. *Cont.*

Scientific Name	Traditional Name	Plant Part Used	Effective Compounds and Ref.	Traditional Medicine Use	Monotherapy or an Adjunct and Ref.
Curcuma zedoaria	Zingiberaceae	Rhizome	Curcumenol and some principal sesquiterpenes [22]	Hepatoprotective	An adjunct to allopathic medication [63]
Curcuma zedoaria	Zingiberaceae	Rhizome	Curcumenol, dihydrocurdione [64,65]	Analgesic	An adjunct to allopathic medication [66]
Curcuma	Zingiberaceae	*Curcuma aromatica* Salisb. rhizome	Curzerene, isoprocurcumenol, and (+)-curcumenol [67]	Coronary heart disease	An adjunct to allopathic medication [68]
Curcuma zedoaria	Zingiberaceae	Rhizome	Curcumenol [69]	Osteoporosis	An adjunct to allopathic medication [70]
Curcuma zedoary		Rhizome	Curcumenol [71]	Liver cancer	An adjunct to allopathic medication [72]
Curcuma		Rhizome	Curcumenol [73]	Cervical cancer	An adjunct to allopathic medication [74]
Curcuma wenyujin	Wenyujin	Rhizome	Curcumenol [13]	Lung cancer	An adjunct to allopathic medication [75]
Curcuma zedoaria rhizome	Zingiberaceae	*C. zedoaria* rhizomes	Curcumenol, 4,8-dioxo-6β-methoxy-7α,11-epoxycarabrane, and zedoarofuran [15]	Gastric cancer	An adjunct to allopathic medication [76]
Curcuma zedoaria	Temu putih	Rhizome	Curcumenone, Curcumenol [16]	Breast cancer	An adjunct to allopathic medication [77]

3. Sources of Curcumenol

Curcumenol was extracted from *Curcuma zedoaria*'s edible rhizome for the first time in the mid-20th century. In tropical forests, Zingerberaceae covers 52 genera, with 1500 species in total. Most of them are planted and reported in South East Asia [78]. Zingerberaceae, also known as "temu putih", acts as a kind of additive; it is widely used in food preparation for nursing women after childbirth and as a spice in native dishes in Malaysia [79,80]. Curcumenol has also been identified in several other plants (e.g., the flower tea *Chrysanthemum indicum* [81], *Neolitsea pallens* [82], and *Torilis japonica* [83]). Zhu Shunying et al. [81] reported essential oils from samples of *Chrysanthemum indicum*, including monoterpene hydrocarbons (0.42–6.72%), sesquiterpenes (9.18–39.29%), and oxygenated monoterpenes (43.29–78.92%); curcumenol was derived from the oxygenated monoterpenes. G.C. Kharkwal et al. [83] investigated *Torilis japonica* (Houtt.) DC., which grows in the central Himalayan region. As revealed by their research, the oils are abundant in sesquiterpine hydrocarbons (approximately 60%). In total, 43 compounds with curcumenol accounted for 94.28% of the oils.

Rajendra C. Padalia et al. [82] investigated the composition of volatile constituents in the leaf, stem, and bark of *N. pallens* through nuclear magnetic resonance (NMR) spectroscopy, gas chromatography–mass spectrometry (GC–MS), and gas chromatography (GC) methods. Curcumenol (5.3%) was reported to occur in the oxygenated sesquiterpenoid, making up the major proportion of the bark oil. Curcumenol has also been extracted from bacterium and animal samples. Noor Akbar et al. [20] purified docosanedioic acid, Di-rhamnolipids, L-homotyrosine, N-acyl-homoserine lactone, and curcumenol from the gut bacteria of animals (e.g., cockroaches, water monitor lizards, and turtles).

4. Physicochemical Properties of Curcumenol

$C_{15}H_{22}O_2$, 5β-Guaia-7 (11), 9-dien-8α-ol, 5, 8-epoxy-, 234.33, and 19431–84-6 are the molecular formula, chemical name, molecular weight, and chemical abstract service (CAS) number of curcumenol, respectively. Curcumenol has a white needle-like crystal structure at ambient temperatures, with a melting point of 113–115 °C, and it is insoluble in water, but soluble in methanol, ethanol, dimethyl sulfoxide, and some organic solvents. As indicated by the results achieved using spectroscopic and chemical methods, curcumenol exists in solution as a pair of hemiacetal–ketone tautomers [84]. More insights have been gained into the ultraviolet (UV) and electron-capture dissociation spectroscopic properties of curcumenol through theoretical time-dependent density functional theory calculations and modeling. Co-existing curcumenotone has a large tensional and unstable structure, contributing notably to the anti-oxidant process [84].

5. Extraction Methods

The extraction method for the components is of critical significance in the qualitative and quantitative process of herbal drug research [85]. Appropriate extraction methods should be developed to deal with different samples. Extraction methods for the different samples (e.g., biological samples, crude herb samples, and preparations) are detailed in Table 3.

A few investigations have reported on the extraction of curcumenol from biological samples. Salwa Mansur Ali et al. identified curcumenol and 11 other compounds from tissues and hemolymphs, which was extracted from the red-headed centipede, through the addition of protease inhibitors (serine/cysteine/metalloproteases) [86]. Subsequently, the samples were treated at 4 °C; then, ten cycles of freezing and thawing were performed on the gut and muscle tissues and, before analysis, the samples were subjected to sonication for further extraction [86].

Table 3. Sample preparation methods employed to extract curcumenol.

Matrix	Pretreatment/Extraction Approach	Procedure	Recovery (%)	Ref.
		Biological Samples		
Crude hemolymph extract	Organ lysis	The exposed hemolymph was taken in an aseptic manner and suspended in a quantity of sterile distilled water. Ten cycles of freezing and thawing were applied to the gut and muscle tissues; then, they were homogenized, sonicated, centrifuged, and analyzed.	–	[86]

Table 3. *Cont.*

Matrix	Pretreatment/Extraction Approach	Procedure	Recovery (%)	Ref.
		Plant		
Crude turmeric samples	Pressurized liquid water extraction	*Curcuma* Radix was dried at 60 °C and sieved with a 60-mesh sieve. The dried powder and diatomaceous earth were added to a stainless steel extraction cell. The extraction of the sample was obtained using the following parameters: 40% of the flush volume, methanol at 100 °C, static extraction for 5 min, 1000 psi.	_	[23]
Powdered rhizome	Ultrasonication in methanol	The powdered rhizome was extracted with methanol at 40 kHz, 200 W for 1 h using ultrasound, then filtered through 0.45 μm pores before analysis.	103.62	[87]
Powdered rhizome	Steam distillation	Powdered rhizome was exposed to vapor at 90 °C, followed by extraction for 30 min.	_	[88]
Powdered rhizome	Extraction with n-hexane	n-Hexane was shaken with samples for 30 min at 40 °C, and centrifuged at 4000 rpm.	_	[89]
Powdered rhizome	Pressurized liquid extraction	The extraction of the sample was achieved with the following conditions: static cycle, 1, filtering through a 0.45 μm icon filter for 40% of the flush volume; pressure, 1000 psi; static extraction period, 5 min; particle size, 0.2–0.3 mm; temperature, 100 °C, with methanol.	100.5	[90]
Different parts of the plant, powdered	Extraction with dichloromethane	Dichloromethane was added and the sample was macerated for 7 days at 15–20 °C.	_	[64]
Powdered rhizome	Cold immersion with methanol	A 0.5 g sample of powder was placed in a 100 mL triangle bottle, then 50 mL methanol was added, before cold soaking for 12 h, shaking every several minutes, until the supernatant was poured out.	99.6	[91]
Four sources of *Curcuma* Radix decoction pieces	Heating reflux with water	Decoction pieces were added in 10-fold amounts (m L/g) of water, soaked for 1 h, heated and reflowed twice, filtered, and concentrated.	116.8	[92]
Powdered rhizome	Ultrasonication in ethanol	Ultrasound with 70% ethanol for 45 min.	95.8–104.1	[93]

Table 3. *Cont.*

Matrix	Pretreatment/Extraction Approach	Procedure	Recovery (%)	Ref.
Powdered rhizome	Pressurized liquid extraction	Methanol; 120 °C; particle size, 0.2–0.3 mm; static extraction time, 5 min; pressure, 1500 psi; static cycle, 1; and 60% flush volume.	–	[94]
		Formulation		
Decoction	Immersed in water and ethanol	Immersed in 50% ethanol, 70% ethanol, 100% ethanol, and water for 1 h each, then 2 h reflux.	–	[95]

In general, curcumenol is extracted from crude herb samples using various conditions and solvents (e.g., ultrasonication, pressurized liquid, and steam distillation) (Table 2). Zang Yuanfang et al. [96] developed an ultrasonic extraction method (power 600 W, frequency 40 kHz) with 70% methanol for 30 min, with a recovery rate of 99.8%. Yang F.Q. et al. indicated that a pressurized liquid extraction method could yield 90% recovery of curcumenol under methanol at 100 °C, with the particle size ranging from 0.2 to 0.3 mm, a static extraction time of 5 min, a pressure of 1000 psi, a static cycle, and a flush volume of 40% [97].

6. Methods for Qualitative and Quantitative Analysis

During the extraction of target compounds from different matrices, specific and effective analysis methods are also important. Curcumenol is a component of essential oils in many plant samples, and the accurate determination of volatile components in these samples is necessary. Thin-layer chromatography (TLC), colorimetry, HPLC, and GC with flame ionization detection or mass detection, have been used to perform quality control of essential oils in herbal medicines [98,99]. Nevertheless, the low separation efficiency of TLC and poor selectivity of colorimetry have hampered both their further application and development. A more powerful analysis tool was needed for the quality assessment of the abovementioned essential oils. Therefore, different qualitative and quantitative analysis methods have been established by researchers. Chromatography is a powerful and common tool used in research on curcumenol, involving different samples, covering GC and liquid chromatography (LC) (Table 4). For instance, Komatsu K. et al. [88] established a GC–MS method for the identification and detection of curcumenol in rhizomes of *C. kwangsiensis* from Guangdong Province, and *C. wenyujin* from Zhejiang Province, China, and *C. kwangsiensis* from the Guangxi Zhuangzu Autonomous Region (Guangxi A. R.). LC techniques have been widely applied for quantification in different fields, and are usually combined with other detectors (e.g., fluorescence, UV, and MS) [100–102]. F.Q. Yang et al. [90] developed a detection approach utilizing HPLC in combination with a diode array, with the aim of conducting rapid and simultaneous quantitative analysis of curcumenol and other compounds in the three species of *Curcuma* rhizomes. The measurement of the 11 above-described components in three different species of *Curcuma* rhizomes using the abovementioned method showed high repeatability, with inter-day and intra-day variances lower than 1.98% and 1.57%, respectively. It was effective in quantifying target components in samples of different *Curcuma* species. Gang Yin et al. [95] established UPLC-QQQ-MS to determine curcumenol and other bioactive components for the quality assessment of an AR–CR herb pair. The recovery (90.20–107.60%), repeatability (relative standard deviation (RSD) 5.98%), stability (RSD 4.29%), intra- and inter-day precision ((RSD) 3.64%, RSD 5.68%), limit of quantification (0.81–2.54 ng/mL), and limit of detection (0.33–10.78 ng/mL), were all used to optimize and validate the proposed method. Therefore, the primary bioactive ingredients in the AR–CR herb combination and the individual herbs were successfully compared using the established methodology.

Table 4. Spectroscopic and chromatographic methods used to assay the samples.

Analytical Method	Research Objectives/Title	Matrix and Sample Preparation Method	Result	Ref.
High-performance liquid chromatography (HPLC)	Determination of curcumenol in Yujin.	Powdered rhizome	Recovery value of curcumenol 103.62%.	[87]
HPLC	Simultaneous determination of 11 characteristic components in three species of *Curcuma* rhizomes using pressurized liquid extraction and HPLC.	Powdered rhizome	With intra- and inter-day fluctuations of less than 1.57% and 1.98%, respectively, this approach demonstrated good repeatability for the quantification of the abovementioned 11 components in three species of *Curcuma* rhizomes.	[90]
HPLC	Quantitation of Curcumenol extracted from water.	Four sources of *Curcuma* Radix decoction pieces	For *C. kwangsiensis*, *C. wenyujin*, *C. longa*, and *C. phaeocaulis*, the contents of curcumenol in the water extract were 0.066, 0.271, 0.058, and 0.310 mg/g, respectively.	[92]
Ultra-performance liquid chromatography (UPLC)	Determination of the curcumenol content in *Curcuma kwangsiensis*, vinegar-boiled *Curcuma kwangsiensis*, and water extract residues using UPLC.	Powdered rhizome and water residues	The established analysis method is accurate, stable, and sensitive; it can be used for the quality assay of *Curcuma kwangsiensis*, and it provides a scientific basis for the resource utilization of water extract residues.	[103]
High resolution gas chromatography (HRGC)	Quantitation of curcumenol.	Different parts of this plant, powdered	Standard samples of curcumenol within the concentration range of 0.03–0.93 mg/mL.	[64]
Liquid chromatography tandem–mass spectrometry (LC–MS)	Fingerprint of *Curcuma phaeocaulis* using LC–MS.	Powdered rhizome	Peaks 4–6 and 9 were identified as curcumin, demethoxycurcumin, curcumenol, and curcumone, respectively, through the contrast of the retention time and the online UV spectra and the molecular weight of the chemical standards.	[104]
Ultra-high-performance liquid chromatography coupled with the triple quadrupole tandem mass spectrometry method (UPLC-QQQ-MS)	Identification of curcumenol.	Decoction, dispensing granule decoction	It was the first time that 17 chemicals in the *Astragali* Radix–*Curcumae* Rhizoma (AR–CR) herb pair were simultaneously analyzed using UPLC-QQQ-MS, offering a workable approach in terms of overall quality control.	[95]
Ultra-high-performance liquid chromatography–quadrupole time-of-flight mass spectrometry	Identification of curcumenol.	Crude turmeric samples, pressurized liquid extractions	This new method led to the discovery of curcumenol and some other components as unique chemical markers for identification.	[23]

Table 4. *Cont.*

Analytical Method	Research Objectives/Title	Matrix and Sample Preparation Method	Result	Ref.
Gas chromatography–mass spectrometry (GC/GC–MS)	Identification of curcumenol.	Powdered rhizome	Identified curcumenol.	[88]
GC–MS	Identification of curcumenol.	Powdered rhizome	Two isomeric forms of curcumenol comprised the GC chromatogram's major components, taking up 28.68 ± 0.91% and 17.96 ± 0.69% of the control plants' total volatiles, respectively.	[89]
GC–MS	Identification and quantitation of curcumenol in *Curcuma* rhizomes.	Powdered rhizome	Curcumenol and four other compounds were optimized as markers for the quality control of *Ezhu*.	[94]
GC–MS	Determination of curcumenol in *Curcuma* rhizomes.	Powdered rhizome	Nine sesquiterpenoids were effectively quantified using the validated method in 18 samples of three *Curcuma* species, utilized as *Ezhu*.	[97]
GC–MS	Determination of curcumenol in the fractionation of volatile constituents originating from the *Curcuma* rhizome using GC.	Powdered rhizome	The structures of the compound were identified as curcumenol using a mass spectrometer (MS) and NMR spectra, respectively.	[105]

7. Other Factors Influencing the Content of Curcumenol

The contents of the components in plants are affected by their variety, place of origin, growth environment, processing methods, etc. [106–109]. Some factors relevant to the content of curcumenol (e.g., the constituents of minerals in the soil and the harvest season of herbs) have been illustrated. Rabia F. El-Hawaz et al. [89] examined the mineral concentration effects on the content of volatile constituents in rhizomes. The interaction of Ca^{2+} with KNO_3 affected the contents of β-elemenone, isocurcumenol, germacrone, and curcumenol isomers I and II. The abovementioned findings demonstrated that minerals in the in vitro bioreactor medium during rhizome growth had an impact on the production of volatile components in turmeric when transferred to the greenhouse after six months. Christiane Regina Pamplona et al. [64] revealed that dihydrocurdione (2) and curcumenol (1) are two active terpenoids from different portions of *Curcuma zedoaria* cultivated in Brazil and that they displayed seasonal variations. As revealed by the results, the production of 1 and 2 was nearly three times greater in the mother rhizome in the autumn (Tables 5 and 6). There are an insufficient number of studies focusing on the influencing factors of the content of curcumenol in *Curcuma wenyujin*. However, as a component of the volatile oil in *Curcuma wenyujin*, the content of volatile oil was affected by the areas of production [110], is shown in Table 7.

Table 5. Content of curcumenol in various seasons and in a wide range of parts of *C. zedoaria* (mg/100 g dried plant) [64]. Reproduced with permission from Christiane Regina Pamplona, et al., Zeitschrift fur Naturforschung. C, Journal of biosciences; published by Verl. d. Zeitschrift für Naturforschung, 2006.

Part	Autumn	Winter	Spring	Summer
Roots	15.70 ± 0.14	8.90 ± 0.15	8.70 ± 0.12	1.5 ± 0.10
Mother rhizome	33.10 ± 0.12	9.10 ± 0.08	5.90 ± 0.05	6.0 ± 0.01
Rugous rhizome	10.40 ± 0.03	3.10 ± 0.03	2.00 ± 0.04	2.9 ± 0.03

Table 6. Content of dihydrocurdione in various seasons and in a wide range of parts of *C. zedoaria* (mg/100 g dried plant) [64]. Reproduced with permission from Christiane Regina Pamplona, et al., Zeitschrift fur Naturforschung. C, Journal of biosciences; published by Verl. d. Zeitschrift für Naturforschung, 2006.

Part	Autumn	Winter	Spring	Summer
Roots	8.50 ± 0.27	3.10 ± 0.15	4.40 ± 0.06	1.60 ± 0.10
Mother rhizome	25.00 ± 0.25	9.40 ± 0.17	6.10 ± 0.15	7.40 ± 0.14
Rugous rhizome	6.70 ± 0.03	2.90 ± 0.08	1.50 ± 0.06	1.50 ± 0.03

Table 7. Comparison of the content of volatile oil in *wenyujin* from different producing areas [110].

Producing Areas	Content of Volatile Oils
Taoshan Town, Ruian City, Zhejiang Province, China	2.0%
Mayu Town, Ruian City, Zhejiang Province, China	1.8%
Siqian Town, Wenzhou City, Zhejiang Province, China	1.5%
Yongjia County, Wenzhou City, Zhejiang Province, China	3.1%
Leqing City, Zhejiang Province, China	2.0%

8. Bioactivity

Existing research has confirmed that curcumenol has numerous biological effects, including anti-tumor [13–16], anti-inflammatory [17,18], anti-virus [20,21], hepatoprotection [32], and neuroprotection activities [5]. Curcumin is another major bioactive substance contained in the *Curcuma* species; these two compounds have aroused increasing attention from chemists and biologists, and are expected to be investigated in-depth in the future. A comparison of the physiological effects of the two compounds is shown in Table 8.

Table 8. The physiological effects of curcumenol and curcumin.

	Curcumenol				Curcumin		
Disease/ Curcumenol Activity	**Model**	**Treatment Doses**	**Mechanism**	**Ref.**	**Treatment Doses**	**Model**	**Ref.**
Anti-inflammatory activity	BV-2 microglial cells	5–20 μM	Inhibiting Akt-dependent nuclear factor kappa-B (NF-κB) activation and the downregulation of Akt and p38 mitogen-activated protein kinase (MAPK) signaling	[18]	5–20 μM	BV-2 microglial cells	[111]
Anti-inflammatory activity	HaCaT cells	200 μg/mL	Inhibits the overexpression of inflammatory factors	[19]	20 μM	HaCaT cells	[112]
Analgesic effects	Mice	10 mg/kg	-	[64]	1000 mg/kg to 2000 mg/kg	Mice	[113]
Analgesic effects	Mice	12, 22, 29 μmol/kg	No involvement with the opioid system	[65]	-	-	-
Anti-oxidant ability and neuroprotective activity	NG108-15 cells	4 μM	-	[5]	25–100 μM	NG108-15 cells	[114]
Against coronary heart disease	Coronary heart disease rats	-	Improving blood lipid level, blood stasis, and myocardial infarction, and controlling the signaling pathway of PI3K/AKT/mTOR	[67]	-	-	-
Anti-diabetic activity	Human hepatocellular carcinoma (HepG2) cells	10 μM	-	[32]	-	Human	[115]
Ameliorating osteoporosis	Mice	100 μM	Impairs the stability of TRAF6 enhanced by IPMK and suppresses excessive osteoclast activity	[69]	100 mg/kg	Mice	[116]
SARS-CoV-2 infection	-	-	High-affinity interaction with proteins involved in coronavirus infection	[117]	10 μg/mL	DG614 strain and Delta variant	[118]
Anti-bacteriostatic effects	Gram-negative and Gram-positive bacteria	50 μg/mL	-	[20]	-	-	-

Table 8. *Cont.*

	Curcumenol				Curcumin		
Disease/ Curcumenol Activity	**Model**	**Treatment Doses**	**Mechanism**	**Ref.**	**Treatment Doses**	**Model**	**Ref.**
Anti-bacteriostatic effects	*P. capsici*	20 μg/mL	Damages the cell membrane	[21]	-	-	-
Breast cancer	MCF-7 cells	9.3 ± 0.3 μg/mL	Anti-proliferative activity and induces apoptotic cell death	[16]	20 μM	MCF-7 cells	[119]
Gastric cancer	AGS cells	263.34 ± 2.97 μM	Inhibition of proliferation	[15]	32 μM	AGS cells	[120]
Lung cancer	H1299 and H460 cells	100–400 μg/mL	Via the lncRNA H19/miR-19b-3p/FTH1 axis for lung cancer cell ferroptosis	[13]	5–30 μM	H460 cells	[121]
Cervical cancer	HeLa and C33A Cells	-	Through the tyrosine 3-monooxygenase/ tryptophan 5-monooxygenase activation protein gamma polypeptide (YWHAG) pathway, facilitating cervical cancer cell apoptosis	[73]	20 μM	HeLa and C33A Cells	[122]
Inhibitory to cytochrome P450 (CYP) enzymes	Cytochrome P450 3A4 (CYP3A4)	12.6 ± 1.3 μM	-	[123]	11.93 ± 3.49 μM	CYP3A4	[124]
Binding to human serum albumin (HSA)	HSA	60 μM	-	[125]	-	HSA	[126]

8.1. Anti-Inflammatory Activity

Inflammation is the defense response of the immune system to tissue damage arising from trauma, infection, chemical exposure, etc. [127–129]. It is helpful in the repair of damaged tissues and removes harmful substances, but it is also destructive and disease may be triggered under excessive inflammatory conditions [130,131]. The production of inflammatory cytokines, followed by a decline in the biomechanics and structural integrity of the knee joint, results in osteoarthritis (OA) [132]. Xiao Yang et al. suggested that curcumenol reduced inflammation in ATDC5 chondrocytes and primary mouse chondrocytes, and ameliorated OA in a mouse model of medial meniscus-induced destabilization [17]. They employed tumor necrosis factor alpha (TNF-α) and Interleukin-1β (IL-1β) to treat ATDC5 chondrocyte and primary chondrocyte cells to induce inflammation. Curcumenol inhibited the progression of inflammation through downregulation of the expression levels of matrix metalloproteinase-3 (MMP3), inactivation of the MAPK and NF-κB signaling pathways, and the amelioration of catalytic changes and degradation of the extracellular matrix. Curcumenol can, therefore, be adopted to treat OA and inflammatory-mediated neurodegenerative diseases. Jia Ye Lo et al. [18] investigated the regulatory effect of curcumenol on the lipopolysaccharide-induced inflammation of BV-2 microglia, and reported that it could downregulate the Akt and p38 MAPK signaling pathways by inhibiting Akt-dependent NF-κB activation. Thus, the expression of proinflammatory mediators and regulatory genes is downregulated. Wang Jiafeng [19] suggested that 10 μg/mL curcumenol and other active components in *Curcuma* could inhibit the excessive secretion of four inflammatory cytokines (IL-1β, interleukin-6, interleukin-8, and interleukin 33) induced by TNF-α to different degrees in the treatment of psoriasis. Moreover, IκB-α and p-p65 expression in the NF-κB pathway could be suppressed. As revealed by the above results, the possible mechanism in the treatment of psoriasis by curcumenol and other active components in *Curcuma* is the inhibition of the TNF-α pathway and the inducement of the NF-κB pathway; subsequently, the inflammation and excessive proliferation of epidermal cells in the affected area can be alleviated.

8.2. Analgesic Effects

Curcuma rhizomes have long served as analgesics throughout history [6]. Some researchers have conducted phytochemical and analgesic activity analyses and attempted to determine the antinociceptive active compounds of *Curcuma zedoaria* R. Br. over the past few years [64,65]. Navarro et al. [65] reported that curcumenol exhibited manifold more potent and dose-associated analgesic activities in mice when evaluated in several models as compared with reference drugs. However, the mechanism of curcumenol does not cover the participation of the opioid system, as revealed by insufficient effects in the hot plate test.

8.3. Anti-Oxidant Ability and Neuroprotective Activity

Oxidative stress has been reported to arise from mitochondrial dysfunction, and it is also considered one of the critical contributors to neurodegeneration [133,134]. Compounds extracted from *Curcuma*, ginger, *Ginkgo biloba*, cinnamon, and related plants are characterized by their preventive and therapeutic roles in neurodegenerative disorders [135–139]. Omer Abdalla Ahmed Hamdi et al. [5] investigated the effect of components from *Curcuma zedoaria* (Christm.) Rosc. on hydrogen peroxide-induced oxidative stress in NG108-15 cells. The results showed that when NG108-15 cells were exposed to H_2O_2 (400 μM) for 24 h, the cell viability decreased to 67.6%. If pretreated with curcumenol, the protective effect is 100% at 4 μM and 97.7% when the curcumenol concentration is increased to 30 μM. Therefore, curcumenol has neuroprotective effects. Curcumenol also showed anti-oxidant activity in the oxygen radical anti-oxidant capacity (ORAC) assay [5].

8.4. Coronary Heart Disease (CHD)

CHD is the most common cause of death globally [140]. The *Curcuma aromatica Salisb. rhizome* (CASR) has often been employed to treat CHD arising from blood stasis syndrome in clinics, exhibiting multi-component and multi-pathway effects [67]. Chenghao Fei et al. [67] suggested that CASR is capable of significantly improving blood stasis, myocardial infarction, and blood lipid levels, and regulating the PI3K/AKT/mTOR signaling pathway, in CHD rats. Lipidomics has further shown that CASR can control the metabolism of glycerophospholipids, aberrant sphingolipids, and glycerolipids. Its effects may be attributed to (+)-curcumenol, curzerene, and isoprocurcumenol, which are reportedly the major active CASR compounds and are capable of mitigating vascular endothelial damage, upregulating blood lipid levels, and reducing blood viscosity [67]. These components could become promising drugs in CHD treatment.

8.5. Anti-Diabetic Activity

Over the past 20 years, obesity has spread over the world like wildfire, endangering lives by impacting nearly every organ system. It is currently a major public health concern and one of the most prevalent non-communicable diseases (NCDs) [141–143]. Changxin Zhou et al. [32] acquired extract compounds from *Curcuma wenyujin*, and adopted a glucose consumption model of HepG2 cells to evaluate their anti-diabetic effect. As the results showed, curcumenol and eight other compounds exhibited promising activity, with an over 45% increase in glucose consumption at 10 μM.

8.6. Ameliorating Osteoporosis

Osteoporosis is characterized by reduced bone mass and damaged bone microstructure; it is a systemic bone disease, which can result in bone brittleness and easy fracturing [144]. Despite substantial research into the pharmaceutical management of osteoporosis, better and more efficient treatment approaches are still needed [145]. Shiyu Wang et al. [69] revealed for the first time that curcumenol inhibits osteoclast differentiation in vitro and ameliorates osteoporosis in ovariectomy mice. Curcumenol can inhibit the protective effect of inositol polyphosphate multikinase on osteoclast differentiation, while hindering key downstream pathway activation when the receptor activator of nuclear factor-κB ligand induces osteoclastogenesis. Given the novel involvement of inositol polyphosphate multikinase in osteoclastogenesis and the possible therapeutic effect of curcumenol in regulating osteoporosis, inhibitors targeting inositol polyphosphate multikinase are likely to represent a novel direction for developing anti-osteoporosis medications.

8.7. Anti-Viral and Anti-Bacteriostatic Effects

With the rapid spread of the novel coronavirus disease over the past few years, further effective treatments are still required [146]. Gaurav S. Dave et al. selected curcumenol and three other potent candidates from natural compounds to treat and prevent coronavirus infection. Comparing the open state of severe acute respiratory syndrome coronavirus 2 to the closed state, curcumenol achieved an impressive docking score. This finding provides evidence in favor of the preventative ability of curcumenol to reduce the binding capacity of severe acute respiratory syndrome coronavirus in healthy humans [117].

As revealed by Noor Akbar et al. [20], the gut bacteria of animals that live in polluted environments serve as a promising source of anti-bacterial compounds. Using purified chemicals (e.g., curcumenol) extracted from the gut bacteria of such animals, these researchers determined the anti-bacterial and cytotoxic effects of these chemicals; cell viability tests were performed using lactate dehydrogenase release and methylthiazolyldiphenyl-tetrazolium bromide (MTT) assay, respectively. The findings showed that purified curcumenol and other compounds demonstrated significant anti-bacterial activities against several Gram-positive and Gram-negative bacteria, with effective minimal inhibitory concentrations of MIC50 and MIC90 at μg levels. Additionally, these concentration levels only exert minor effects on human cells, as indicated by the lactate dehydrogenase and MTT research [20].

As indicated by Wang Bi's team, natural plant-derived zedoary turmeric oil (ZTO) (e.g., curcumenol) and several other components (curcumol, β-elemene, and curdione) serve as the main effective chemicals, exhibiting superb anti-fungal activity against *P. capsici* both in vitro and in vivo. It has been speculated that it exerts its effects by disrupting cell membrane integrity [21]. ZTO is a promising natural anti-fungal compound for the treatment of phytophthora blight arising from P. capsici.

8.8. Diversity of Anti-Cancer Effects

Cancer is a large category of diseases that can occur in nearly every organ or tissue in the body, and is characterized by uncontrollable abnormal cell growth, invading adjacent parts of the body outside of the usual boundaries, and/or spreading to other organs. Chinese herbal medicines have long served as anti-cancer treatments, since they contain ample anti-cancer chemicals, exert promising cytotoxicity effects, regulate the tumor microenvironment and cancer immunity, and result in improved chemotherapy effects [147]. Existing research has suggested that curcumenol extracted from *Curcuma wenyujin* exhibits appreciable anti-tumor efficacy (e.g., suppressing breast tumor cells [148,149], digestive tumor cells [150], liver cancer cells [71], and lung cancer cells [13]), as shown in Figure 2.

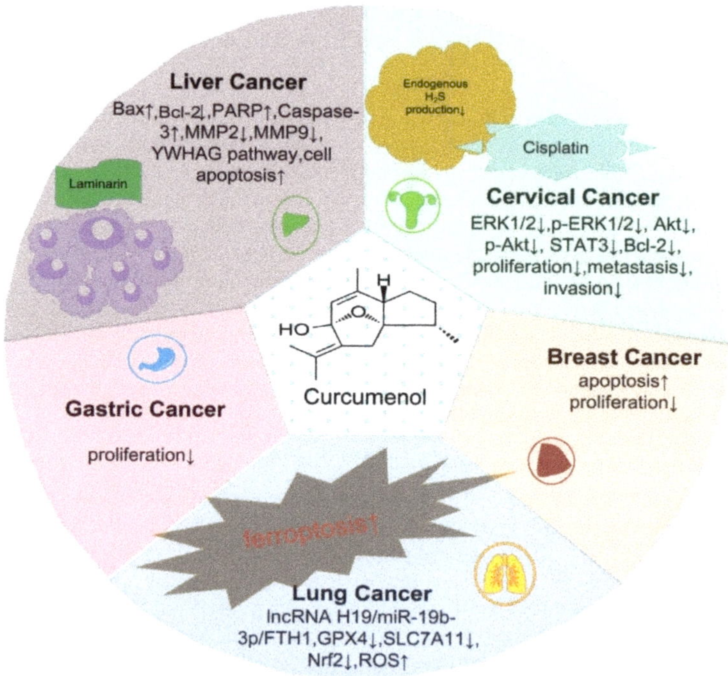

Figure 2. Diversity of anti-cancer effects. In this figure, we can conclude that curcumenol exerts five anti-cancer effects: suppressing breast tumor cells, digestive tumor cells, liver cancer cells, cervical cancer cells, and lung cancer cells. In cervical cancer, curcumenol can inhibit YWHAG expression in cervical cancer and reduce cervical cancer cell proliferation, invasion, and MMP2 and MMP9 expression when combined with cisplatin; the abovementioned combination of drugs leads to increased apoptosis. In liver cancer, the combination of curcumenol and laminarin can inhibit the proliferation, migration, and invasion of human hepatoma HepG2 cells. In lung cancer, curcumenol can suppress cell proliferation and induce cell death in lung cancer cells. Ferroptosis is the mechanism by which curcumenol induces lung cancer cell death; the lncRNA H19/miR-19b-3p/FTH1 axis was found to be crucial for this process. In gastric cancer, curcumenol has demonstrated the dose-dependent suppression of gastric cancer cell proliferation. In breast cancer, curcumenol showed remarkable anti-proliferative action against MCF-7 cells, which might cause apoptotic cell death. The increasing and decreasing expression of proteins were signed as ↑ and ↓, respectively.

8.8.1. Effects on Breast Cancer

Breast cancer ranks highly among women with cancer-related mortality and is a major cause of death for women in their 40s [151,152]. Many researchers have identified effective compounds to treat breast cancer [153]. The chemical components in *Curcuma zedoaria*'s hexane and dichloromethane fractions were studied by Omer Abdalla Ahmed Hamdi et al. [16]. In total, 19 substances were examined using an MTT assay against Ca Ski, MCF-7, the human prostatic carcinoma cell line (PC-3), and HT-29 cancer cell lines, to determine whether they could inhibit cell proliferation. The two key ingredients of *Curcuma zedoaria*, curcumenone and curcumenol, both showed remarkable anti-proliferative action against MCF-7 cells, with the half-maximal inhibitory concentrations (IC50 values) reaching 8.3 ± 1.0 and 9.3 ± 0.3 μg/mL, respectively. As revealed by the results achieved based on the Hoechst 33342/PI double-staining experiment and phase contrast, they might cause apoptotic cell death. These findings underpin the ethnomedical use of *Curcuma zedoaria* as a breast cancer treatment.

8.8.2. Effects on Gastric Cancer

Gastric adenocarcinoma ranks third in cancer-related deaths, with approximately 800,000 fatalities worldwide, and is the fifth most frequent malignancy in terms of incidence [154]. Although many chemical therapies exist, a new and effective candidate compound is still needed. Curcumenol has demonstrated the dose-dependent suppression of gastric cancer cell proliferation [15]. Curcumenol, 11-epoxycarabrane, 4,8-dioxo-6-methoxy-7, and zedoarofuran, exert lethal effects against stomach cancer AGS cells, with IC50 values ranging from 212 to 392 μM, as found by Tae Kyoung Lee's research team. As the primary cytotoxic components in *C. zedoaria* rhizomes, curcumenol and three other compounds offer additional experimental support for the traditional usage of *C. zedoaria* rhizomes in treating gastric cancer.

8.8.3. Effects on Lung Cancer

Lung cancer tumors have the highest mortality rate in the world; however, even though there have been many improvements in the therapy of lung cancer [155], developing new effective drugs is still urgently required. Zhang Ruonan et al. [13] reported that curcumenol could suppress cell proliferation and induce cell death in lung cancer cells. Their research showed that ferroptosis was the mechanism by which curcumenol induced lung cancer cell death, and that the lncRNA H19/miR-19b-3p/FTH1 axis was crucial for this process. In light of this, curcumenol could be used to treat lung cancer patients.

8.8.4. Synergistic Anti-Cancer Effects

Drug combinations have been demonstrated to alleviate toxic side effects and improve efficacy in the treatment of cancer [156–159]. In addition to directly acting on different cancer cells, it has been reported that curcumenol can increase anti-tumor effects in combination with other anti-cancer compounds, such as cisplatin and laminarin [71,73]. Zhijie Mao et al. [73] determined that curcumenol extracted from *Curcuma* could inhibit YWHAG expression in cervical cancer and reduce cervical cancer cell proliferation, invasion, and MMP2 and MMP9 expression when combined with cisplatin, compared with cisplatin alone. The abovementioned combination of drugs led to increased apoptosis, downregulated B cell lymphoma 2 (Bcl-2) expression, and the upregulated expression of Bcl-2 antagonist X, polyadenosine diphosphate-ribose polymerase, and caspase-3. Accordingly, curcumenol is capable of increasing the anti-tumor effect of cisplatin against cancer cell proliferation, migration, invasion, and apoptosis. Curcumenol also exhibits enhanced anti-cancer effects against human hepatoma HepG2 cells when combined with laminarin. Huanxiao Han et al. [71] suggested that the combination of curcumenol and laminarin could inhibit the proliferation, migration, and invasion of human hepatoma HepG2 cells; the levels of pSTAT3 and Bcl-2 decreased in cystathionine beta synthase knockdown HepG2 cells. Furthermore, it can notably contribute to the use of kelp and *Curcuma zedoary* in liver cancer treatment. The mechanism of synergistic anti-cancer effects is shown in Figure 3.

8.9. Pharmacokinetics Study

Curcumenol is the key component of extensively employed natural products, requiring the determination of the pharmacokinetic parameters in order to guarantee safety and effectiveness. However, few studies have been conducted on its pharmacokinetic characteristics, as depicted in the following paragraphs (e.g., the interactions with human liver cytochrome P450 enzymes and the binding effect on human serum albumin).

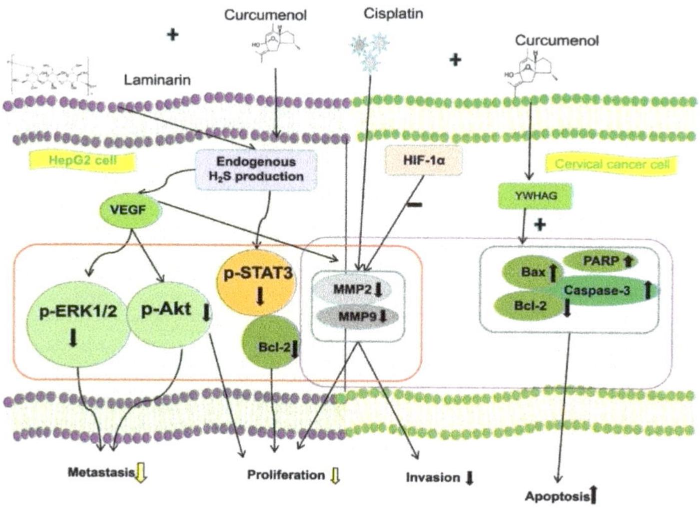

Figure 3. The mechanism of synergistic anti-cancer effects [71,73]. The combination of curcumenol and laminarin could inhibit the proliferation, migration, and invasion of HepG2 cells, and it may inhibit the angiogenesis of hepatoma via H2S and the VEGF pathway, consequently inhibiting the metastasis of HepG2 cells. In cervical cancer, curcumenol could inhibit YWHAG expression and reduce cervical cancer cell proliferation, invasion, and MMP2 and MMP9 expression when combined with cisplatin. The abovementioned combination of drugs led to increased apoptosis, downregulated Bcl-2 expression, and the upregulated expression of Bax, PARP, and caspase-3. The increasing and decreasing trend were signed as ↑ and ↓, respectively.

8.9.1. Inhibitory Effect on Human Liver Cytochrome P450 Enzymes

Curcumenol has been extensively used to inhibit cancer and inflammation, such that it is likely to be employed in combination with different drugs in many conditions to enhance treatment efficacy. It is imperative to evaluate the pharmacokinetics of drug–drug interactions induced by curcumenol. Dong-Xue Sun et al. [123] confirmed that CYP3A4 could be strongly inhibited (IC50 = 12.6 ± 1.3 μM) by curcumenol in the investigation of inhibitory effects on seven CYP isoforms. Curcumenol does not serve as a mechanism-based inhibitor, according to research into time- and nicotinamide adenine dinucleotide phosphate-dependent inhibition. However, given the limited pharmacokinetic information available, the promising clinical effects of curcumenol on patients are unlikely to have been adequately evaluated. Accordingly, more research should be conducted to determine the extent of the drug–drug interactions potentially arising due to curcumenol.

8.9.2. Binding to Human Serum Albumin (HSA) In Vitro

The interaction of a compound with serum albumin can influence its pharmacokinetic characteristics, such as distribution, metabolism, elimination, and bioavailability, in bodily processes [160,161]. Based on this theory, Omer Abdalla Ahmed Hamdi et al. [125] investigated the binding of curcumenol and curcumenone to HSA through fluorescence quench titration. Molecular docking was performed to gain thorough insights into the interactions with HSA. Analysis of the fluorescence data indicated that there was a moderate binding affinity between the ligands and HSA. This showed that the binding constants of curcumenone and curcumenol were $2.46 \times 105\ M^{-1}$ and $1.97 \times 104\ M^{-1}$, respectively. The binding properties of curcumenol to HSA offer useful parameters for the detection of pharmacokinetic profiles based on the fluorescence spectroscopy and molecular docking findings. Nevertheless, since there are scant in vivo pharmacokinetic data, the pharmacokinetic parameters of curcumol, an analogue of curcumenol, in beagle dogs [150] can serve as a reference if required. However, specific pharmacokinetic data, in vitro and in vivo, remain to be obtained more substantially in the future.

8.10. Other Effects

In addition to the abovementioned bioactivities, further pharmacology effects have been identified. Curcumenol and some principal sesquiterpenes isolated from Zedoariae Rhizoma can exert prominent protective effects against liver injury mice [22]. Curcumenol may also be an effective candidate agent for treating structural and functional problems involving the skeletal muscle [84]. Zhang et al. [84] investigated the effect of curcumenol on myogenic differentiation and mitochondrial function. After increasing the mitochondria mass and function in myotubes, it was reported to promote the initiation of myogenesis and the formation of functional myotubes. Guanghui Zhong found that curcumenol could improve renal function in 5/6 nephrectomy rats with chronic renal failure [162].

Overall, curcumenol has immense value in the health field, as shown in Figure 4, and further in-depth investigations are warranted.

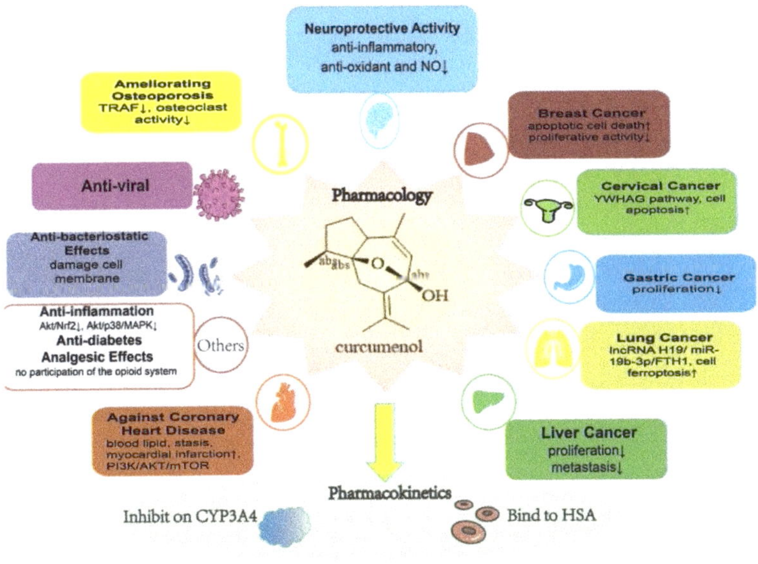

Figure 4. Biological activities of curcumenol. In this figure, we can conclude that the biological activities of curcumenol are very abundant, ranging from pharmacology to pharmacokinetics; in addition to anti-cancer effects, curcumenol exerts anti-inflammatory activity, analgesic effects, anti-oxidant ability, neuroprotective activity against coronary heart disease, anti-diabetic activity, osteoporosis-ameliorating, anti-viral activity, and anti-bacteriostatic effects. The anti-cancer mechanism is detailed in Figure 2. Neuroprotective effects, involving anti-inflammatory and anti-oxidant effects, and NO levels, were decreased. The amelioration of osteoporosis is associated with decreased TRAF expression and decreased osteoclast activity. Curcumenol has also exhibited anti-viral effects in various studies. The anti-bacteriostatic effects mechanism of curcumenol involves damaging the cell membrane. The anti-inflammation effect of curcumenol involves decreases in Akt, Nrf2, p38, and MAPK expression. It has also shown anti-diabetes effects and analgesic effects, but its mechanism does not involve the participation of the opioid system. The mechanism of action against coronary heart disease involves the PI3K/AKT/mTOR pathway. The increasing and decreasing trend were signed as ↑ and ↓, respectively.

Although there are scarce pharmacokinetics data concerning curcumenol, it has shown an inhibitory effect on CYP3A, and binding effects to HSA, which are two key factors in the pharmacokinetics process in vivo.

9. Conclusions and Prospects

Wenyujin, a member of the *Curcuma zedoaria* family, has been reported to comprise a total of 169 compounds, thus far. It has a long history and has been extensively employed in TCM due to its abundant components, including curcuminoids, monoterpenoids, sesquiterpenoids, diterpenoids, etc. To elucidate the exact compounds that exert the abovementioned traditional curative effects using current medical theories and to determine the targets of its biological effects, so as to change the dosage form, improve the effect of treatment, and reduce the waste of TCM resources, many researchers have conducted studies and found a series of compounds related to those treatment effects, including curcumenol. To more effectively extract curcumenol from different plant samples, researchers have developed several methods (e.g., pressurized liquid water extraction, steam distillation, ultrasonication in methanol or ethanol, extraction with n-hexane, extraction with dichloromethane, cold immersion with methanol, and heating reflux with water). Different detection methods have also been established (e.g., HPLC, UPLC, HRGC, LC-MS/MS, and GC–MS methods) to precisely measure and identify traces of curcumenol. Curcumenol also exhibits extensive bioactive effects (e.g., anti-cancer, anti-inflammatory, anti-bactericidal, and anti-diabetic activities, and the amelioration of osteoporosis).

In traditional applications of *Curcuma*, it is often used as an adjuvant of pre-existing therapies with other Chinese medicine to treat diseases. Moreover, curcumenol, as an important active ingredient, can be combined with the main chemical components of other Chinese medicine compounds in preparations of clinical applications to achieve better clinical effects. In anti-cancer treatments, it should be combined with chemotherapy drugs to reduce the toxicity of these medications to normal cells, while enhancing the sensitivity of tumor cells to chemotherapy drugs, reducing the drug resistance of tumor cells, and reducing side effects. However, it can dissolve in organic solvents but is insoluble in water, which has seriously limited its pharmacological effects in basic research and clinical treatment. To address this, new preparations should be developed to increase the water solubility and biological target effects of curcumenol. It is also possible to modify the structure of curcumenol to increase its solubility and biological effects.

More pharmacokinetic parameters of curcumenol should be determined to establish its safety and efficacy, such that further references can be provided for applications in clinical practice. Furthermore, it could facilitate the development of TCM and safeguard consumers' health and safety.

Author Contributions: Conceptualization, J.L.; writing—original draft, J.L.; writing—review and editing, Y.S. and G.L.; supervision, C.C., X.S. and Q.W. All authors have read and agreed to the published version of the manuscript.

Funding: This work was financially funded by the Science and Technology Development Fund, Macau SAR (No: 0098/2021/A2 and 0048/2023/AFJ), Jiangxi Province Double Thousand Talents-Leader of Natural Science Talents Project, and the Jiangxi Province urgently needed overseas talent project (20223BCJ25027), Macau University of Science and Technology's Faculty Research Grant (No: FRG-23-003-FC).

Institutional Review Board Statement: Not applicable.

Informed Consent Statement: Not applicable.

Data Availability Statement: Data are contained within the article.

Conflicts of Interest: The authors declare no conflicts of interest.

Abbreviations

AGS cells	human gastric adenocarcinoma cells
AR	*Astragali* Radix
AR–CR	*Astragali* Radix–*Curcuma* Rhizoma
ATDC5 chondrocytes	mouse embryonic tumor cells
CAS	chemical abstracts service
CaSki	human cervical cancer
CASR	*Curcuma aromatica* salisb. rhizome
CHD	coronary heart disease
CR	*Curcuma* Rhizoma
CYP	cytochrome P450
CYP3A4	cytochrome P450 3A4
GC	gas chromatography
GC–MS	gas chromatography–mass spectrometry
MS	mass spectrometry
H460 cells	human large cell lung cancer cells
HaCaT cells	human immortalized keratinocytes
HeLa	human cervical carcinoma cell line
HPLC	high-performance liquid chromatography
HRGC	high-resolution gas chromatography
HSA	human serum albumin
HT-29	human colorectal adenocarcinoma cells
IL-1β	Interleukin-1β
LC-MS	liquid chromatography tandem mass spectrometry
MAPK	mitogen-activated protein kinases
MCF-7 cells	human breast cancer cell line
MIC	minimal inhibitory concentration
MMP	matrix metalloproteinase
MTT	methylthiazolyldiphenyl-tetrazolium bromide
NF-κB	nuclear factor kappa-B
NMR	nuclear magnetic resonance
OA	osteoarthritis
pSTAT3	signal transducer and activator of transcription 3 protein phosphorylation
PC-3	human prostate cancer cell
RSD	relative standard deviation
TCM	traditional Chinese medicine
TLC	thin-layer chromatography
TNF-α	tumor necrosis factor alpha
UPLC	ultra-performance liquid chromatography;
UPLC-QQQ-MS	ultra-high-performance liquid chromatography coupled to triple quadrupole tandem mass spectrometry method
UV	ultraviolet
YWHAG	tyrosine 3-monooxygenase/tryptophan 5-monooxygenase activation protein gamma polypeptide
ZTO	zedoary turmeric oil

References

1. Fang, F.; Cheng, Z.H.; Guo, Y.L.; Cai, Y.B. Comparative Analysis of the Volatile Components in the Fresh Roots and Rhizomes of *Curcuma wenyujin* by Static Headspace Gas Chromatography Mass Spectrometry. *Chin. J. Chem.* **2006**, *24*, 1346–1351. [CrossRef]
2. Yin, M.; Chu, S.; Zhao, Y.; Zheng, X.; HS, P. Materia medica illustrations related to the regional names in Zhejiang province in Ben Cao Tu Jing. *Zhonghua Yi Shi Za Zhi* **2022**, *52*, 131–139. [CrossRef] [PubMed]
3. Cai, Y.; Li, W.; Tu, H.; Chen, N.; Zhong, Z.; Yan, P.; Dong, J. Curcumolide reduces diabetic retinal vascular leukostasis and leakage partly via inhibition of the p38MAPK/NF-κ B signaling. *Bioorg. Med. Chem. Lett.* **2017**, *27*, 1835–1839. [CrossRef]
4. Ma, J.; Chen, J.; Zhao, B.; Jiang, Z.; Feng, L.; Jia, X. Advance in research on anticancer drug β-elemene and its derivatives. *Chin. Tradit. Herb. Drug.* **2018**, *49*, 1184–1191. [CrossRef]

5. Hamdi, O.A.A.; Ye, L.J.; Kamarudin, M.N.A.; Hazni, H.; Paydar, M.; Looi, C.Y.; Shilpi, J.A.; Kadir, H.A.; Awang, K. Neuroprotective and Antioxidant Constituents from *Curcuma zedoaria* Rhizomes. *Rec. Nat. Prod.* **2015**, *9*, 349–355.
6. Li, Y.; Wu, Y.; Li, Y.; Guo, F. Review of the traditional uses, phytochemistry, and pharmacology of *Curcuma wenyujin* Y. H. Chen et C. Ling. *J. Ethnopharmacol.* **2021**, *269*, 113689. [CrossRef] [PubMed]
7. Kotha, R.R.; Luthria, D.L. Curcumin: Biological, Pharmaceutical, Nutraceutical, and Analytical Aspects. *Molecules* **2019**, *24*, 2930. [CrossRef] [PubMed]
8. Hikino, H.; Sakurai, Y.; Numabe, S.; Takemoto, T. Structure of curcumenol. *Chem. Pharm. Bull.* **1968**, *16*, 39–42. [CrossRef] [PubMed]
9. Xu, J.; Ji, F.; Kang, J.; Wang, H.; Li, S.; Jin, D.-Q.; Zhang, Q.; Sun, H.; Guo, Y. Absolute Configurations and NO Inhibitory Activities of Terpenoids from *Curcuma longa*. *J. Agric. Food. Chem.* **2015**, *63*, 5805–5812. [CrossRef] [PubMed]
10. Asem, S.D.; Laitonjam, W.S. A new guaianolide sesquiterpene lactone from *Curcuma leucorrhiza* Roxb. *Nat. Prod. Res.* **2014**, *28*, 477–482. [CrossRef]
11. Qiu, G.; Yan, P.; Shao, W.; Zhou, J.; Lin, W.; Fang, L.; Zhao, X.; Dong, J. Two new sesquiterpenoids including a sesquiterpenoid lactam from *Curcuma wenyujin*. *Chem. Pharm. Bull.* **2013**, *61*, 983–986. [CrossRef] [PubMed]
12. Aspollah Sukari, M.; Wah, T.S.; Saad, S.M.; Rashid, N.Y.; Rahmani, M.; Lajis, N.H.; Hin, T.-Y.Y. Bioactive sesquiterpenes from *Curcuma ochrorhiza* and *Curcuma heyneana*. *Nat. Prod. Res.* **2010**, *24*, 838–845. [CrossRef] [PubMed]
13. Zhang, R.; Pan, T.; Xiang, Y.; Zhang, M.; Xie, H.; Liang, Z.; Chen, B.; Xu, C.; Wang, J.; Huang, X.; et al. Curcumenol triggered ferroptosis in lung cancer cells via lncRNA H19/miR-19b-3p/FTH1 axis. *Bioact. Mater.* **2022**, *13*, 23–36. [CrossRef]
14. Peng, B.; Zhou, X.; Shi, J.; Li, Z. Effects of volatile oil and three main components from *Curcuma phaeocaulis* Valeton on liver cancer and endometrial carcinoma cell lines. *West China J. Pharm. Sci.* **2007**, *2007*, 312–313. [CrossRef]
15. Lee, T.K.; Lee, D.; Lee, S.R.; Ko, Y.-J.; Sung Kang, K.; Chung, S.J.; Kim, K.H. Sesquiterpenes from *Curcuma zedoaria* rhizomes and their cytotoxicity against human gastric cancer AGS cells. *Bioorg. Chem.* **2019**, *87*, 117–122. [CrossRef] [PubMed]
16. Ahmed Hamdi, O.A.; Syed Abdul Rahman, S.N.; Awang, K.; Abdul Wahab, N.; Looi, C.Y.; Thomas, N.F.; Abd Malek, S.N. Cytotoxic Constituents from the Rhizomes of *Curcuma zedoaria*. *Sci. World J.* **2014**, *2014*, 321943. [CrossRef]
17. Yang, X.; Zhou, Y.; Chen, Z.; Chen, C.; Han, C.; Li, X.; Tian, H.; Cheng, X.; Zhang, K.; Zhou, T.; et al. Curcumenol mitigates chondrocyte inflammation by inhibiting the NF-κB and MAPK pathways, and ameliorates DMM-induced OA in mice. *Int. J. Mol. Med.* **2021**, *48*, 192. [CrossRef]
18. Lo, J.Y.; Kamarudin, M.N.A.; Hamdi, O.A.A.; Awang, K.; Kadir, H.A. Curcumenol isolated from *Curcuma zedoaria* suppresses Akt-mediated NF-κB activation and p38 MAPK signaling pathway in LPS-stimulated BV-2 microglial cells. *Food. Funct.* **2015**, *6*, 3550–3559. [CrossRef]
19. Wang, J. Inhibition of HaCaT Cell Proliferation by Active Constituents of Rhizoma Curcumae and Its Mechanism of Action Based on NF-κB Signaling Pathway. Master's Thesis, University of Electronic Science and Technology of China, Chengdu, China, 2019.
20. Akbar, N.; Siddiqui, R.; Iqbal, M.; Khan, N.A. Antibacterial Activities of Selected Pure Compounds Isolated from Gut Bacteria of Animals Living in Polluted Environments. *Antibiotics* **2020**, *9*, 190. [CrossRef] [PubMed]
21. Wang, B.; Liu, F.; Li, Q.; Xu, S.; Zhao, X.; Xue, P.; Feng, X. Antifungal activity of zedoary turmeric oil against Phytophthora capsici through damaging cell membrane. *Pestic. Biochem. Physiol.* **2019**, *159*, 59–67. [CrossRef] [PubMed]
22. Matsuda, H.; Ninomiya, K.; Morikawa, T.; Yoshikawa, M. Inhibitory effect and action mechanism of sesquiterpenes from *Zedoariae* Rhizoma on D-galactosamine/lipopolysaccharide-induced liver injury. *Bioorg. Med. Chem. Lett.* **1998**, *8*, 339–344. [CrossRef] [PubMed]
23. Liu, F.; Bai, X.; Yang, F.Q.; Zhang, X.J.; Hu, Y.; Li, P.; Wan, J.B. Discriminating from species of *Curcumae* Radix (*Yujin*) by a UHPLC/Q-TOFMS-based metabolomics approach. *Chin. Med.* **2016**, *11*, 21. [CrossRef] [PubMed]
24. Hu, X.; Wang, L.; Zhang, L.; Zhang, T. β-Elemene inhibits 7,12-dimethylbenz(a)anthracene/12-O-tetradecanoylphorbol-13-acetate-induced skin tumorigenesis through suppression of NF-κB-associated signaling events in the mouse skin model. *J. Biochem. Mol. Toxicol.* **2020**, *34*, e22550. [CrossRef] [PubMed]
25. Gouthamchandra, K.; Sudeep, H.V.; Chandrappa, S.; Raj, A.; Naveen, P.; Shyamaprasad, K. Efficacy of a Standardized Turmeric Extract Comprised of 70% Bisdemothoxy-Curcumin (REVERC3) against LPS-Induced Inflammation in RAW264.7 Cells and Carrageenan-Induced Paw Edema. *J. Inflamm. Res.* **2021**, *14*, 859–868. [CrossRef]
26. Lee, H.Y.; Kim, S.W.; Lee, G.H.; Choi, M.K.; Chung, H.W.; Lee, Y.C.; Kim, H.R.; Kwon, H.J.; Chae, H.J. Curcumin and *Curcuma longa* L. extract ameliorate lipid accumulation through the regulation of the endoplasmic reticulum redox and ER stress. *Sci. Rep.* **2017**, *7*, 6513. [CrossRef] [PubMed]
27. Mhillaj, E.; Tarozzi, A.; Pruccoli, L.; Cuomo, V.; Trabace, L.; Mancuso, C. Curcumin and Heme Oxygenase: Neuroprotection and Beyond. *Int. J. Mol. Sci.* **2019**, *20*, 2419. [CrossRef] [PubMed]
28. Zhang, L.; Yang, Z.; Chen, D.; Huang, Z.; Li, Y.; Lan, X.; Su, P.; Pan, W.; Zhou, W.; Zheng, X.; et al. Variation on Composition and Bioactivity of Essential Oils of Four Common *Curcuma* Herbs. *Chem. Biodivers.* **2017**, *14*, e1700280. [CrossRef]
29. Qin, Y.; Zhao, Q.; Zhao, Y.; Ren, X.; Bao, K.; Li, X.; Jiang, C. Exploring mechanism of *Curcuma wenyujin* against COVID-19. *Chin. Tradit. Herb. Drugs* **2020**, *51*, 1977–1983. [CrossRef]
30. Salehi, B.; Rodrigues, C.F.; Peron, G.; Dall'Acqua, S.; Sharifi-Rad, J.; Azmi, L.; Shukla, I.; Singh Baghel, U.; Prakash Mishra, A.; Elissawy, A.M.; et al. Curcumin nanoformulations for antimicrobial and wound healing purposes. *Phytother. Res.* **2021**, *35*, 2487–2499. [CrossRef]

31. Hu, Y.; Zhang, J.; Kong, W.; Zhao, G.; Yang, M. Mechanisms of antifungal and anti-aflatoxigenic properties of essential oil derived from turmeric (*Curcuma longa* L.) on *Aspergillus flavus*. *Food Chem.* **2017**, *220*, 1–8. [CrossRef]
32. Zhou, C.; Zhang, L.; Chen, F.; Wu, H.; Mo, J.; Gan, L. Terpenoids from *Curcuma wenyujin* increased glucose consumption on HepG2 cells. *Fitoterapia* **2017**, *121*, 141–145. [CrossRef] [PubMed]
33. Lekshmi, P.C.; Arimboor, R.; Indulekha, P.S.; Menon, A.N. Turmeric (*Curcuma longa* L.) volatile oil inhibits key enzymes linked to type 2 diabetes. *Int. J. Food Sci. Nutr.* **2012**, *63*, 832–834. [CrossRef] [PubMed]
34. Xie, H.; Su, D.; Zhang, J.; Ji, D.; Mao, J.; Hao, M.; Wang, Q.; Yu, M.; Mao, C.; Lu, T. Raw and vinegar processed *Curcuma wenyujin* regulates hepatic fibrosis via bloking TGF-β/Smad signaling pathways and up-regulation of MMP-2/TIMP-1 ratio. *J. Ethnopharmacol.* **2020**, *246*, 111768. [CrossRef] [PubMed]
35. Pivari, F.; Mingione, A.; Brasacchio, C.; Soldati, L. Curcumin and Type 2 Diabetes Mellitus: Prevention and Treatment. *Nutrients* **2019**, *11*, 1837. [CrossRef] [PubMed]
36. Lin, W.; Tu, H.; Zhu, Y.; Guan, Y.; Liu, H.; Ling, W.; Yan, P.; Dong, J. Curcumolide, a unique sesquiterpenoid from *Curcuma wenyujin* displays anti-angiogenic activity and attenuates ischemia-induced retinal neovascularization. *Phytomedicine* **2019**, *64*, 152923. [CrossRef] [PubMed]
37. Abolfazli, S.; Mortazavi, P.; Kheirandish, A.; Butler, A.E.; Jamialahmadi, T.; Sahebkar, A. Regulatory effects of curcumin on nitric oxide signaling in the cardiovascular system. *Nitric Oxide* **2023**, *143*, 16–28. [CrossRef] [PubMed]
38. Ma, L.T.; Bai, Y.; Cao, P.; Ren, K.X.; Chen, J.; Zhang, T.; Fan, B.Y.; Qiao, Y.; Yan, H.Y.; Wang, J.J.; et al. The analgesic effects of β-elemene in rats with neuropathic pain by inhibition of spinal astrocytic ERK activation. *Mol. Pain.* **2022**, *18*, 17448069221121562. [CrossRef]
39. Limcharoen, T.; Dasuni Wasana, P.W.; Hasriadi Muangnoi, C.; Vajragupta, O.; Rojsitthisak, P.; Towiwat, P. Curcumin Diglutaric Acid, a Prodrug of Curcumin Reduces Pain Hypersensitivity in Chronic Constriction Injury of Sciatic Nerve Induced-Neuropathy in Mice. *Pharmaceuticals* **2020**, *13*, 212. [CrossRef]
40. Qi, Y.; Qin, W.; Kang, K.; Jiang, H.; Li, Z.; Wang, Y.; Jia, D. Effects of *Wenyujin* Essential Oil on tau Protein Phosphorylation in Mice with Aβ-induced Alzheimer Disease through PI3k/Akt Pathway. *Chin. J. Inf. Tradit. Chin. Med.* **2017**, *24*, 45–48. [CrossRef]
41. Eun, C.S.; Lim, J.S.; Lee, J.; Lee, S.P.; Yang, S.A. The protective effect of fermented *Curcuma longa* L. on memory dysfunction in oxidative stress-induced C6 gliomal cells, proinflammatory-activated BV2 microglial cells, and scopolamine-induced amnesia model in mice. *BMC Complement. Altern. Med.* **2017**, *17*, 367. [CrossRef] [PubMed]
42. Li, J.; Zeng, H.; You, Y.; Wang, R.; Tan, T.; Wang, W.; Yin, L.; Zeng, Z.; Zeng, Y.; Xie, T. Active targeting of orthotopic glioma using biomimetic liposomes co-loaded elemene and cabazitaxel modified by transferritin. *J. Nanobiotechnol.* **2021**, *19*, 289. [CrossRef] [PubMed]
43. Zhong, Z.F.; Tan, W.; Tian, K.; Yu, H.; Qiang, W.A.; Wang, Y.T. Combined effects of furanodiene and doxorubicin on the migration and invasion of MDA-MB-231 breast cancer cells in vitro. *Oncol. Rep.* **2017**, *37*, 2016–2024. [CrossRef] [PubMed]
44. Wu, J.; Tang, X.; Shi, Y.; Ma, C.; Zhang, H.; Zhang, J.; Lu, Y.; Wei, J.; Li, L.; Han, L. Crosstalk of LncRNA HOTAIR and SP1-mediated repression of PDK1 contributes to β-Elemene-inhibited proliferation of hepatocellular carcinoma cells. *J. Ethnopharmacol.* **2022**, *283*, 114456. [CrossRef] [PubMed]
45. Deng, M.; Liu, B.; Song, H.; Yu, R.; Zou, D.; Chen, Y.; Ma, Y.; Lv, F.; Xu, L.; Zhang, Z.; et al. β-Elemene inhibits the metastasis of multidrug-resistant gastric cancer cells through miR-1323/Cbl-b/EGFR pathway. *Phytomedicine* **2020**, *69*, 153184. [CrossRef] [PubMed]
46. Sun, R.; Gu, J.; Chang, X.; Liu, F.; Liang, Y.; Yang, X.; Liang, L.; Tang, D. Metabonomics study on orthotopic transplantion mice model of colon cancer treated with Astragalus membranaceus-*Curcuma wenyujin* in different proportions via UPLC-Q-TOF/MS. *J. Pharm. Biomed. Anal.* **2021**, *193*, 113708. [CrossRef] [PubMed]
47. Ying, J.; Yang, W.; Xie, C.Y.; Ni, Q.C.; Pan, X.D.; Dong, J.H.; Liu, Z.M.; Wang, X.S. Induction of caspase-3-dependent apoptosis in human leukemia HL-60 cells by δ-elemene. *Yakugaku Zasshi* **2011**, *131*, 1383–1394. [CrossRef] [PubMed]
48. Lim, C.B.; Ky, N.; Ng, H.M.; Hamza, M.S.; Zhao, Y. *Curcuma wenyujin* extract induces apoptosis and inhibits proliferation of human cervical cancer cells in vitro and in vivo. *Integr. Cancer Ther.* **2010**, *9*, 36–49. [CrossRef]
49. Song, G.; Lu, H.; Chen, F.; Wang, Y.; Fan, W.; Shao, W.; Lu, H.; Lin, B. Tetrahydrocurcumin-induced autophagy via suppression of PI3K/Akt/mTOR in non-small cell lung carcinoma cells. *Mol. Med. Rep.* **2018**, *17*, 5964–5969. [CrossRef]
50. Abdel-Lateef, E.; Mahmoud, F.; Hammam, O.; El-Ahwany, E.; El-Wakil, E.; Kandil, S.; Abu Taleb, H.; El-Sayed, M.; Hassenein, H. Bioactive chemical constituents of *Curcuma longa* L. rhizomes extract inhibit the growth of human hepatoma cell line (HepG2). *Acta Pharm.* **2016**, *66*, 387–398. [CrossRef]
51. Lee, Y.H.; Song, N.Y.; Suh, J.; Kim, D.H.; Kim, W.; Ann, J.; Lee, J.; Baek, J.H.; Na, H.K.; Surh, Y.J. Curcumin suppresses oncogenicity of human colon cancer cells by covalently modifying the cysteine 67 residue of SIRT1. *Cancer Lett.* **2018**, *431*, 219–229. [CrossRef]
52. Jiang, J.L.; Jin, X.L.; Zhang, H.; Su, X.; Qiao, B.; Yuan, Y.J. Identification of antitumor constituents in curcuminoids from *Curcuma longa* L. based on the composition-activity relationship. *J. Pharm. Biomed. Anal.* **2012**, *70*, 664–670. [CrossRef]
53. Zhang, L.; Cheng, X.; Gao, Y.; Zhang, C.; Bao, J.; Guan, H.; Yu, H.; Lu, R.; Xu, Q.; Sun, Y. Curcumin inhibits metastasis in human papillary thyroid carcinoma BCPAP cells via down-regulation of the TGF-β/Smad2/3 signaling pathway. *Exp. Cell Res.* **2016**, *341*, 157–165. [CrossRef]

54. Kuo, C.L.; Wu, S.Y.; Ip, S.W.; Wu, P.P.; Yu, C.S.; Yang, J.S.; Chen, P.Y.; Wu, S.H.; Chung, J.G. Apoptotic death in curcumin-treated NPC-TW 076 human nasopharyngeal carcinoma cells is mediated through the ROS, mitochondrial depolarization and caspase-3-dependent signaling responses. *Int. J. Oncol.* **2011**, *39*, 319–328. [CrossRef]
55. Lin, C.Y.; Hung, C.C.; Wang, C.C.N.; Lin, H.Y.; Huang, S.H.; Sheu, M.J. Demethoxycurcumin sensitizes the response of non-small cell lung cancer to cisplatin through downregulation of TP and ERCC1-related pathways. *Phytomedicine* **2019**, *53*, 28–36. [CrossRef]
56. Emami, B.; Shakeri, F.; Ghorani, V.; Boskabady, M.H. Relaxant effect of *Curcuma longa* on rat tracheal smooth muscle and its possible mechanisms. *Pharm. Biol.* **2017**, *55*, 2248–2258. [CrossRef]
57. Bundy, R.; Walker, A.F.; Middleton, R.W.; Booth, J. Turmeric extract may improve irritable bowel syndrome symptomology in otherwise healthy adults: A pilot study. *J. Altern. Complement. Med.* **2004**, *10*, 1015–1018. [CrossRef] [PubMed]
58. Kim, D.C.; Ku, S.K.; Bae, J.S. Anticoagulant activities of curcumin and its derivative. *BMB Rep.* **2012**, *45*, 221–226. [CrossRef]
59. Feng, M.; Liao, T.; Yang, N.; Fang, Y.; Wang, P.; Liu, Z. LC-MS Combined with Network Pharmacology to Explore the Mechanism of "*Wenyujin* Rhizoma *Concisum-Angelicae Sinensis* Radix" Essential Oil in the Treatment of Knee Osteoarthritis. *Chin. J. Mod. Appl. Pharm.* **2023**, *40*, 2403–2413. [CrossRef]
60. Wang, K.; Shi, X.; Ke, J.; Ma, Y.; Ye, M. Mechanism of action of Rhizoma *Acori graminei-Curcuma aromatic* in treatment of Alzheimer's disease: A study based on network pharmacology. *Hum. J. Tradit. Chin. Med.* **2023**, *39*, 163–170. [CrossRef]
61. Yu, J.; Li, Z.; He, Z.; Lu, C. Meta-analysis of Efficacy and Safety of Yinxieling Tablet and Its Optimized Formula in the Treatment of Psoriasis Vulgaris. *Tradit. Chin. Drug Res. Clin. Pharmacol.* **2021**, *32*, 1048–1054. [CrossRef]
62. Liu, C.; Dong, Y. Clinical study on the vaginal flora of acrobic vaginitis and the effects of the treatments. *Prog. Obstet. Gynecol.* **2009**, *18*, 832–835. [CrossRef]
63. Zhang, S.; Peng, Q.; Zhao, X.; Xie, M.; Guo, P. Combined Sanleng *Ezhu* Decoction with Western medicine in treating liver fibrosis in chronic hepatitis B of liver spleen deficiency and blood stasis syndrome. *Hebei J. TCM.* **2022**, *44*, 1834–1838. [CrossRef]
64. Pamplona, C.R.; Souza, M.M.d.; Machado, M.d.S.; Filhoa, V.C.; Navarro, D.; Yunes, R.A.; Monache, F.D.; Niero, R. Seasonal Variation and Analgesic Properties of Different Parts from *Curcuma zedoaria* Roscoe (Zingiberaceae) Grown in Brazil. *Z. Naturforsch. C. J. Biosci.* **2006**, *61*, 6–10. [CrossRef]
65. De Fátima Navarro, D.; De Souza, M.M.; Neto, R.A.; Golin, V.; Niero, R.; Yunes, R.A.; Monache, F.D.; Cechinel-Filho, V. Phytochemical analysis and analgesic properties of *Curcuma zedoaria* grown in Brazil. *Phytomedicine* **2002**, *9*, 427–432. [CrossRef] [PubMed]
66. Qin, L.; Bai, M. Preparation and application of Yu Jin powder. *J. Chin. Med. Mater.* **1994**, *7*, 46. [CrossRef]
67. Fei, C.; Ji, D.; Tong, H.; Li, Y.; Su, L.; Qin, Y.; Bian, Z.; Zhang, W.; Mao, C.; Li, L.; et al. Therapeutic mechanism of *Curcuma aromatica* Salisb. rhizome against coronary heart disease based on integrated network pharmacology, pharmacological evaluation and lipidomics. *Front. Pharmacol.* **2022**, *13*, 950749. [CrossRef]
68. Li, J. Clinical observation on stable angina of coronary heart disease treated with Sanleng *Ezhu* decoction. *Tianjin J. Tradit. Chin. Med.* **2007**, *24*, 470–471.
69. Wang, S.; Ma, Q.; Xie, Z.; Shen, Y.; Zheng, B.; Jiang, C.; Yuan, P.; An, Q.; Fan, S.; Jie, Z. An Antioxidant Sesquiterpene Inhibits Osteoclastogenesis Via Blocking IPMK/TRAF6 and Counteracts OVX-Induced Osteoporosis in Mice. *J. Bone. Miner. Res.* **2021**, *36*, 1850–1865. [CrossRef]
70. Lu, J.; Li, Z.; Du, G.; Zhang, J.; Pan, F.; Zhang, Z.; Wang, Y.; Shi, J.; Lian, Y.; Zhan, H. Systematic review and Meta-analysis of Gusongbao preparation in treatment of primary osteoporosis. *China J. Chin. Mater. Medica* **2023**, *48*, 3086–3096. [CrossRef]
71. Han, H.; Wang, L.; Liu, Y.; Shi, X.; Zhang, X.; Li, M.; Wang, T. Combination of *Curcuma zedoary* and kelp inhibits growth and metastasis of liver cancerin vivoandin vitro via reducing endogenous H_2S levels. *Food. Funct.* **2019**, *10*, 224–234. [CrossRef]
72. Wang, G. Curative Effect and Clinical Evaluation of the Anticancer of Compound Zedoary Turmeric Oil Hepatic Artery Embolism in Treatment of Liver Cancer. *China Foreign Med. Treat.* **2016**, *35*, 110–111. [CrossRef]
73. Mao, Z.; Zhong, L.; Zhuang, X.; Liu, H.; Peng, Y.; Zhang, H. Curcumenol Targeting YWHAG Inhibits the Pentose Phosphate Pathway and Enhances Antitumor Effects of Cisplatin. *Evid. Based. Complement. Alternat. Med.* **2022**, *2022*, 3988916. [CrossRef] [PubMed]
74. Ruan, Q.; Zhong, G. Effect analysis of *Curcuma zedoary* oil combined with conventional chemotherapy on ovarian cancer. *J. Med. Theory Pract.* **2018**, *31*, 554–555. [CrossRef]
75. Li, P.; Ke, H. Effect Analysis of Zedoary Turmeric Oil Combined with Cisplatin and Docetaxel in Treatment of Advanced Non-Small Cell Lung Cancer. *Chin. Community Dr.* **2023**, *39*, 77–79.
76. Gao, D.; Zhang, J.; Li, H. Effects of *Chanpi Ezhu* Decoction on Adjuvant Chemotherapy in the Treatment of Gastric Cancer. *World J. Integr. Tradit. West. Med.* **2021**, *16*, 2329–2333. [CrossRef]
77. Zeng, X.; Cui, J.; Zhou, Q.; Li, N.; Liu, J.; Li, J.; Huang, D.; Cai, P.; Zhou, R.; Yan, J.; et al. Research status of Jinrong granule against breast hyperplasia and breast cancer and its clinical application. *Chin. J. Clin. Pharmacol.* **2023**, *39*, 295–299. [CrossRef]
78. Lobo, R.; Prabhu, K.S.; Shirwaikar, A.; Shirwaikar, A. *Curcuma zedoaria* Rosc. (white turmeric): A review of its chemical, pharmacological and ethnomedicinal properties. *J. Pharm. Pharmacol.* **2009**, *61*, 13–21. [CrossRef]
79. Etoh, H.; Kondoh, T.; Yoshioka, N.; Sugiyama, K.; Ishikawa, H.; Tanaka, H. 9-Oxo-neoprocurcumenol from *Curcuma aromatica* (Zingiberaceae) as an Attachment Inhibitor against the Blue Mussel, Mytilus edulis galloprovincialis. *Biosci. Biotechnol. Biochem.* **2003**, *67*, 911–913. [CrossRef]

80. Yoshioka, T.; Fujii, E.; Endo, M.; Wada, K.; Tokunaga, Y.; Shiba, N.; Hohsho, H.; Shibuya, H.; Muraki, T. Antiinflammatory potency of dehydrocurdione, a zedoary-derived sesquiterpene. *Inflamm. Res.* **1998**, *47*, 476–481. [CrossRef] [PubMed]
81. Shunying, Z.; Yang, Y.; Huaidong, Y.; Yue, Y.; Guolin, Z. Chemical composition and antimicrobial activity of the essential oils of Chrysanthemum indicum. *J. Ethnopharmacol.* **2005**, *96*, 151–158. [CrossRef] [PubMed]
82. Padalia, R.C.; Chanotiya, C.S.; Thakuri, B.C.; Mathela, C.S. Germacranolide Rich Essential Oil from *Neolitsea pallens*. *Nat. Prod. Commun.* **2006**, *2*, 291–293. [CrossRef]
83. Kharkwala, G.C.; Pandea, C.; Tewari, G.; Panwarb, A.; Pandeb, V. Terpenoid composition and antimicrobial activity of essential oil from *Torilis japonica* (Houtt.) DC. *J. India. Chem. Soc.* **2017**, *94*, 191–194.
84. Zhang, L.; Shen, S.; Gao, Y.; Shi, S.; Zhou, C.; Mo, J.; Xu, Y.; Lin, L.; Gan, L. Tautomerism and bioactivities of curcumenol, a common sesquiterpenoid widely existing in edible plants. *Food. Funct.* **2019**, *10*, 1288–1294. [CrossRef] [PubMed]
85. Indrayanto, G. The importance of method validation in herbal drug research. *J. Pharm. Biomed. Anal.* **2022**, *214*, 114735. [CrossRef]
86. Ali, S.M.; Khan, N.A.; Sagathevan, K.; Anwar, A.; Siddiqui, R. Biologically active metabolite(s) from haemolymph of red-headed centipede Scolopendra subspinipes possess broad spectrum antibacterial activity. *AMB. Express.* **2019**, *9*, 95. [CrossRef] [PubMed]
87. Wang, L.; Wang, X.; Liu, X.; Wang, Y.; Ren, X.; Dong, Y.; Song, R.; Ma, J.; Fan, Q.; Wei, J.; et al. Fast discrimination and quantification analysis of *Curcumae* Radix from four botanical origins using NIR spectroscopy coupled with chemometrics tools. *Spectrochim. Acta. A Mol. Biomol. Spectrosc.* **2021**, *254*, 119626. [CrossRef] [PubMed]
88. Komatsu, K.; Sasaki, Y.; Tanaka, K.; Kuba, Y.; Fushimi, H.; Cai, S.-Q. Morphological, genetic, and chemical polymorphism of *Curcuma kwangsiensis*. *J. Nat. Med.* **2008**, *62*, 413–422. [CrossRef]
89. El-Hawaz, R.F.; Grace, M.H.; Janbey, A.; Lila, M.A.; Adelberg, J.W. In vitro mineral nutrition of *Curcuma longa* L. affects production of volatile compounds in rhizomes after transfer to the greenhouse. *BMC. Plant. Biol.* **2018**, *18*, 122. [CrossRef]
90. Yang, F.Q.; Wang, Y.T.; Li, S.P. Simultaneous determination of 11 characteristic components in three species of *Curcuma* rhizomes using pressurized liquid extraction and high-performance liquid chromatography. *J. Chromatogr. A* **2006**, *1134*, 226–231. [CrossRef]
91. Chen, S. Determination of Three Effective Constituents in Rhizoma *Curcumae* from Different Area by HPLC. *Chin. Pharm. J.* **2009**, *44*, 1742–1744.
92. Shi, D.; Su, B.; Zhang, J.; Dai, Y. Study on Analgesic Effect of 4 Sources of *Curcumae* Radix Decoction Pieces and Comparison of Curcuminol Content in Its Water Extracts. *Chin. Pharm.* **2020**, *31*, 2209–2213. [CrossRef]
93. Wang, M.; Yu, M.; Peng, M.; Yin, Z.; Mao, C.; Su, L.; Ji, D.; Lu, T. Quality evaluation of *Curcumae* Radix from different origins based on UPLC characteristic chromatogram, multicomponent content, and chemometrics. *China J. Chin. Mater. Med.* **2022**, *47*, 2964–2974. [CrossRef]
94. Yang, F.Q.; Li, S.P.; Chen, Y.; Lao, S.C.; Wang, Y.T.; Dong, T.T.X.; Tsim, K.W.K. Identification and quantitation of eleven sesquiterpenes in three species of *Curcuma* rhizomes by pressurized liquid extraction and gas chromatography–mass spectrometry. *J. Pharm. Biomed. Anal.* **2005**, *39*, 552–558. [CrossRef]
95. Yin, G.; Cheng, X.; Tao, W.; Dong, Y.; Bian, Y.; Zang, W.; Tang, D. Comparative analysis of multiple representative components in the herb pair *Astragali* Radix-*Curcumae* Rhizoma and its single herbs by UPLC-QQQ-MS. *J. Pharm. Biomed. Anal.* **2018**, *148*, 224–229. [CrossRef] [PubMed]
96. Zang, Y.; Xu, L.; Liu, H.; Luan, Y.; Xu, X.; Yu, Y.; Lin, Y. Determination of 8 Main Active Compounds in *Curcumae* Rhizoma by HPLC Wavelength Switching Method. *Chin. J. Mod. Appl. Pharm.* **2021**, *38*, 2227–2233. [CrossRef]
97. Yang, F.Q.; Li, S.P.; Zhao, J.; Lao, S.C.; Wang, Y.T. Optimization of GC–MS conditions based on resolution and stability of analytes for simultaneous determination of nine sesquiterpenoids in three species of *Curcuma* rhizomes. *J. Pharm. Biomed. Anal.* **2007**, *43*, 73–82. [CrossRef] [PubMed]
98. Shiea, J.; Lin, H.; Bhat, S.M.; Lee, C.; Huang, M.; Ponnusamy, V.K.; Cheng, S. Thin layer chromatography/desorption flame-induced atmospheric pressure chemical ionization/mass spectrometry for the analysis of volatile and semi-volatile mixtures. *Rapid. Commun. Mass. Spectrom.* **2022**, *36*, e9409. [CrossRef] [PubMed]
99. Ayoub, I.M.; Abdel-Aziz, M.M.; Elhady, S.S.; Bagalagel, A.A.; Malatani, R.T.; Elkady, W.M. Valorization of *Pimenta racemosa* Essential Oils and Extracts: GC-MS and LC-MS Phytochemical Profiling and Evaluation of *Helicobacter pylori* Inhibitory Activity. *Molecules* **2022**, *27*, 7965. [CrossRef] [PubMed]
100. Depmeier, T.; Lange, T.; Hanekamp, W.; Strünker, T.; Lehr, M. HPLC fluorescence assay for measuring the activity of diacylglycerol lipases and the action of inhibitors thereof. *Anal. Biochem.* **2022**, *657*, 114889. [CrossRef] [PubMed]
101. Sammut Bartolo, N.; Vella Szijj, J.; Ferrito, V.; Serracino-Inglott, A. HPLC-UV Method Development and Validation to Monitor Difluprednate Synthesis. *J. Chromatogr. Sci.* **2023**, *61*, 322–328. [CrossRef]
102. Gajula, S.N.R.; Khairnar, A.S.; Jock, P.; Kumari, N.; Pratima, K.; Munjal, V.; Kalan, P.; Sonti, R. LC-MS/MS: A sensitive and selective analytical technique to detect COVID-19 protein biomarkers in the early disease stage. *Expert. Rev. Proteom.* **2023**, *20*, 5–18. [CrossRef]
103. Liang, Z.; Liang, L.; Zeng, Z.; Wei, M.; Lei, Y.; Lin, W. Determination of curcumenol content in *Curcuma kwangsiensis*, Vingar-Boiled *Curcuma kwangsiensis*, and Water Residues by UPLC. *Chem. Bio Eng.* **2020**, *37*, 62–65. [CrossRef]
104. Zhou, X.; Chen, G.; Zhao, C.; Gong, X.; Li, M. Fingerprint of *Curcuma phaeocaulis* by LC-MS. *China J. Chin. Mater. Med.* **2008**, *33*, 2218–2221.
105. Yang, F.Q.; Wang, H.K.; Chen, H.; Chen, J.D.; Xia, Z.N. Fractionation of Volatile Constituents from *Curcuma Rhizome* by Preparative Gas Chromatography. *J. Autom. Methods Manag. Chem.* **2011**, *2011*, 942467. [CrossRef] [PubMed]

106. Chen, L.; Liu, Y.; Huang, X.; Zhu, Y.; Li, J.; Miao, Y.; Du, H.; Liu, D. Comparison of Chemical Constituents and Pharmacological Effects of Different Varieties of *Chrysanthemum Flos* in China. *Chem. Biodivers.* **2021**, *18*, e2100206. [CrossRef] [PubMed]
107. Çakmakçı, R.; Mosber, G.; Milton, A.H.; Alatürk, F.; Ali, B. The Effect of Auxin and Auxin-Producing Bacteria on the Growth, Essential Oil Yield, and Composition in Medicinal and Aromatic Plants. *Curr. Microbiol.* **2020**, *77*, 564–577. [CrossRef] [PubMed]
108. Saffariha, M.; Azarnivand, H.; Zare Chahouki, M.A.; Tavili, A.; Nejad Ebrahimi, S.; Jahani, R.; Potter, D. Changes in the essential oil content and composition of *Salvia limbata* C.A. Mey at different growth stages and altitudes. *Biomed. Chromatogr.* **2021**, *35*, e5127. [CrossRef] [PubMed]
109. Gu, S.; Li, L.; Huang, H.; Wang, B.; Zhang, T. Antitumor, Antiviral, and Anti-Inflammatory Efficacy of Essential Oils from *Atractylodes macrocephala* Koidz. Produced with Different Processing Methods. *Molecules* **2019**, *24*, 2956. [CrossRef]
110. Zhen, F.; Sun, X.; Jin, X. Phenological period and quality comparison of different producing areas of *Wenyujin* in southern Zhejiang. *J. Zhejiang Agr. Sci.* **2019**, *60*, 823–824. [CrossRef]
111. Yu, Y.; Shen, Q.; Lai, Y.; Park, S.Y.; Ou, X.; Lin, D.; Jin, M.; Zhang, W. Anti-inflammatory Effects of Curcumin in Microglial Cells. *Front. Pharmacol.* **2018**, *9*, 386. [CrossRef]
112. Cho, J.W.; Lee, K.S.; Kim, C.W. Curcumin attenuates the expression of IL-1beta, IL-6, and TNF-alpha as well as cyclin E in TNF-alpha-treated HaCaT cells; NF-kappaB and MAPKs as potential upstream targets. *Int. J. Mol. Med.* **2007**, *19*, 469–474.
113. Sengupta, S.; Tripathi, A. Evaluation of analgesic and prophylactic activity of curcumin against chikungunya-infected acute/chronic arthralgic mice. *J. Med. Virol.* **2023**, *95*, e28661. [CrossRef]
114. Mahakunakorn, P.; Tohda, M.; Murakami, Y.; Matsumoto, K.; Watanabe, H.; Vajaragupta, O. Cytoprotective and cytotoxic effects of curcumin: Dual action on H_2O_2-induced oxidative cell damage in NG108-15 cells. *Biol. Pharm. Bull.* **2003**, *26*, 725–728. [CrossRef]
115. Marton, L.T.; Pescinini, E.S.L.M.; Camargo, M.E.C.; Barbalho, S.M.; Haber, J.; Sinatora, R.V.; Detregiachi, C.R.P.; Girio, R.J.S.; Buchaim, D.V.; Cincotto Dos Santos Bueno, P. The Effects of Curcumin on Diabetes Mellitus: A Systematic Review. *Front. Endocrinol.* **2021**, *12*, 669448. [CrossRef]
116. Fan, D.; Lu, J.; Yu, N.; Xie, Y.; Zhen, L. Curcumin Prevents Diabetic Osteoporosis through Promoting Osteogenesis and Angiogenesis Coupling via NF-κB Signaling. *Evid. Based Complement. Altern. Med.* **2022**, *2022*, 4974343. [CrossRef]
117. Dave, G.S.; Rakholiya, K.D.; Kaneria, M.J.; Galvadiya, B.P.; Vyas, S.R.; Kanbi, V.H.; Patel, M.P. High affinity interaction of Solanum tuberosum and *Brassica juncea* residue smoke water compounds with proteins involved in coronavirus infection. *Phytother. Res.* **2020**, *34*, 3400–3410. [CrossRef] [PubMed]
118. Marín-Palma, D.; Tabares-Guevara, J.H.; Zapata-Cardona, M.I.; Flórez-Álvarez, L.; Yepes, L.M.; Rugeles, M.T.; Zapata-Builes, W.; Hernandez, J.C.; Taborda, N.A. Curcumin Inhibits In Vitro SARS-CoV-2 Infection In Vero E6 Cells through Multiple Antiviral Mechanisms. *Molecules* **2021**, *26*, 6900. [CrossRef] [PubMed]
119. Liu, J.L.; Pan, Y.Y.; Chen, O.; Luan, Y.; Xue, X.; Zhao, J.J.; Liu, L.; Jia, H.Y. Curcumin inhibits MCF-7 cells by modulating the NF-κB signaling pathway. *Oncol. Lett.* **2017**, *14*, 5581–5584. [CrossRef] [PubMed]
120. Zheng, R.; Deng, Q.; Liu, Y.; Zhao, P. Curcumin Inhibits Gastric Carcinoma Cell Growth and Induces Apoptosis by Suppressing the Wnt/β-Catenin Signaling Pathway. *Med. Sci. Monit.* **2017**, *23*, 163–171. [CrossRef]
121. Ye, M.; Zhang, J.; Zhang, J.; Miao, Q.; Yao, L.; Zhang, J. Curcumin promotes apoptosis by activating the p53-miR-192-5p/215-XIAP pathway in non-small cell lung cancer. *Cancer Lett.* **2015**, *357*, 196–205. [CrossRef]
122. Kim, B.; Kim, H.S.; Jung, E.J.; Lee, J.Y.; Tsang, B.; Lim, J.M.; Song, Y.S. Curcumin induces ER stress-mediated apoptosis through selective generation of reactive oxygen species in cervical cancer cells. *Mol. Carcinog.* **2016**, *55*, 918–928. [CrossRef] [PubMed]
123. Sun, D.; Fang, Z.; Zhang, Y.; Cao, Y.; Yang, L.; Yin, J. Inhibitory effects of curcumenol on human liver cytochrome P450 enzymes. *Phytother. Res.* **2010**, *24*, 1213–1216. [CrossRef]
124. Shamsi, S.; Tran, H.; Tan, R.S.; Tan, Z.J.; Lim, L.Y. Curcumin, Piperine, and Capsaicin: A Comparative Study of Spice-Mediated Inhibition of Human Cytochrome P450 Isozyme Activities. *Drug Metab. Dispos.* **2017**, *45*, 49–55. [CrossRef]
125. Hamdi, O.; Feroz, S.; Shilpi, J.; Anouar, E.; Mukarram, A.; Mohamad, S.; Tayyab, S.; Awang, K. Spectrofluorometric and Molecular Docking Studies on the Binding of Curcumenol and Curcumenone to Human Serum Albumin. *Int. J. Mol. Sci.* **2015**, *16*, 5180–5193. [CrossRef]
126. Dezhampanah, H.; Shabanzade, Z. Investigation of binding interaction between human serum albumin with zirconium complex of curcumin and curcumin. *J. Biomol. Struct. Dyn.* **2022**, *40*, 722–732. [CrossRef] [PubMed]
127. Roe, K. An inflammation classification system using cytokine parameters. *Scand. J. Immunol.* **2020**, *93*, e12970. [CrossRef]
128. Krstanović, F.; Britt, W.J.; Jonjić, S.; Brizić, I. Cytomegalovirus Infection and Inflammation in Developing Brain. *Viruses* **2021**, *13*, 1078. [CrossRef] [PubMed]
129. Pollard, K.M.; Cauvi, D.M.; Toomey, C.B.; Hultman, P.; Kono, D.H. Mercury-induced inflammation and autoimmunity. *Biochim. Biophys. Acta. Gen. Subj.* **2019**, *1863*, 129299. [CrossRef]
130. Hong, Y.-K.; Chang, Y.-H.; Lin, Y.-C.; Chen, B.; Guevara, B.E.K.; Hsu, C.-K. Inflammation in Wound Healing and Pathological Scarring. *Adv. Wound. Care* **2023**, *12*, 288–300. [CrossRef]
131. Nedunchezhiyan, U.; Varughese, I.; Sun, A.R.; Wu, X.; Crawford, R.; Prasadam, I. Obesity, Inflammation, and Immune System in Osteoarthritis. *Front. Immunol.* **2022**, *13*, 907750. [CrossRef]
132. Motta, F.; Barone, E.; Sica, A.; Selmi, C. Inflammaging and Osteoarthritis. *Clin. Rev. Allergy. Immunol.* **2022**, *64*, 222–238. [CrossRef]

133. Peoples, J.N.; Saraf, A.; Ghazal, N.; Pham, T.T.; Kwong, J.Q. Mitochondrial dysfunction and oxidative stress in heart disease. *Exp. Mol. Med.* **2019**, *51*, 1–13. [CrossRef] [PubMed]
134. Dionísio, P.A.; Amaral, J.D.; Rodrigues, C.M.P. Oxidative stress and regulated cell death in Parkinson's disease. *Ageing. Res. Rev.* **2021**, *67*, 101263. [CrossRef] [PubMed]
135. Nebrisi, E.E. Neuroprotective Activities of Curcumin in Parkinson's Disease: A Review of the Literature. *Int. J. Mol. Sci.* **2021**, *22*, 11248. [CrossRef] [PubMed]
136. El-Akabawy, G.; El-Kholy, W. Neuroprotective effect of ginger in the brain of streptozotocin-induced diabetic rats. *Ann. Anat.* **2014**, *196*, 119–128. [CrossRef] [PubMed]
137. Singh, S.K.; Srivastav, S.; Castellani, R.J.; Plascencia-Villa, G.; Perry, G. Neuroprotective and Antioxidant Effect of *Ginkgo biloba* Extract Against AD and Other Neurological Disorders. *Neurotherapeutics* **2019**, *16*, 666–674. [CrossRef]
138. Yulug, B.; Kilic, E.; Altunay, S.; Ersavas, C.; Orhan, C.; Dalay, A.; Tuzcu, M.; Sahin, N.; Juturu, V.; Sahin, K. Cinnamon Polyphenol Extract Exerts Neuroprotective Activity in Traumatic Brain Injury in Male Mice. *CNS. Neurol. Disord. Drug Targets* **2018**, *17*, 439–447. [CrossRef] [PubMed]
139. Bian, Y.; Zhao, C.; Lee, S.M.-Y. Neuroprotective Potency of Saffron Against Neuropsychiatric Diseases, Neurodegenerative Diseases, and Other Brain Disorders: From Bench to Bedside. *Front. Pharmacol.* **2020**, *11*, 579052. [CrossRef]
140. World Health Organization. The Top 10 Causes of Death. Available online: https://www.who.int/newsroom/fact-sheets/detail/the-top-10-causes-of-death (accessed on 1 December 2023).
141. Tsai, A.G.; Bessesen, D.H. Obesity. *Ann. Intern. Med.* **2019**, *170*, ITC33–ITC48. [CrossRef]
142. Shiriki Kumanyika, W.H.D. Solving Population-wide Obesity-Progress and Future Prospects. *N. Engl. J. Med.* **2020**, *383*, 2197–2200. [CrossRef]
143. Blüher, M. Obesity: Global epidemiology and pathogenesis. *Nat. Rev. Endocrinol.* **2019**, *15*, 288–298. [CrossRef]
144. Aspray, T.J.; Hill, T.R. Osteoporosis and the Ageing Skeleton. In *Biochemistry and Cell Biology of Ageing: Part II Clinical Science*; Subcellular Biochemistry; Springer: New York, NY, USA, 2019; pp. 453–476.
145. Arceo-Mendoza, R.M.; Camacho, P.M. Postmenopausal Osteoporosis. *Endocrinol. Metab. Clin. N. Am.* **2021**, *50*, 167–178. [CrossRef] [PubMed]
146. Hu, B.; Guo, H.; Zhou, P.; Shi, Z.L. Characteristics of SARS-CoV-2 and COVID-19. *Nat. Rev. Microbiol.* **2021**, *19*, 141–154. [CrossRef] [PubMed]
147. Han, X.H.; Ye, Y.Y.; Guo, B.F.; Liu, S. Effects of platycodin D in combination with different active ingredients of Chinese herbs on proliferation and invasion of 1T1 and MDA MB 231 breast cancer cell lines. *Chin. J. Integr. Med.* **2012**, *10*, 67–75. [CrossRef] [PubMed]
148. Al-Amin, M.; Eltayeb, N.M.; Khairuddean, M.; Salhimi, S.M. Bioactive chemical constituents from *Curcuma caesia* Roxb. rhizomes and inhibitory effect of curcuzederone on the migration of triple-negative breast cancer cell line MDA-MB-231. *Nat. Prod. Res.* **2019**, *35*, 3166–3170. [CrossRef]
149. Jung, E.B.; Trinh, T.A.; Lee, T.K.; Yamabe, N.; Kang, K.S.; Song, J.H.; Choi, S.; Lee, S.; Jang, T.S.; Kim, K.H.; et al. Curcuzedoalide contributes to the cytotoxicity of *Curcuma zedoaria* rhizomes against human gastric cancer AGS cells through induction of apoptosis. *J. Ethnopharmacol.* **2018**, *213*, 48–55. [CrossRef]
150. Zhang, R.; Wang, B.; Zhao, H.; Li, X.; Wei, C.; Wei, R.G. Determ ination of curcumol in plasma by HPLC-MS/MS method and its pharmacokinetics in Beagle dogs. *Acta Pharm. Sin. B.* **2007**, *42*, 973–977. [CrossRef]
151. Center for Disease Control. Breast Cancer Statistics. Available online: https://www.cdc.gov/cancer/breast/statistics/index.htm (accessed on 1 December 2023).
152. American Cancer Society. Key Statistics for Breast Cancer. Available online: https://www.cancer.org/cancer/breast-cancer/about/how-common-is-breast-cancer.html (accessed on 1 December 2023).
153. Siddiqui, J.A.; Singh, A.; Chagtoo, M.; Singh, N.; Godbole, M.M.; Chakravarti, B. Phytochemicals for breast cancer therapy: Current status and future implications. *Curr. Cancer. Drug Targets* **2015**, *15*, 116–135. [CrossRef]
154. Smyth, E.C.; Nilsson, M.; Grabsch, H.I.; van Grieken, N.C.; Lordick, F. Gastric cancer. *Lancet* **2020**, *396*, 635–648. [CrossRef]
155. Sung, H.; Ferlay, J.; Siegel, R.L.; Laversanne, M.; Soerjomataram, I.; Jemal, A.; Bray, F. Global Cancer Statistics 2020: GLOBOCAN Estimates of Incidence and Mortality Worldwide for 36 Cancers in 185 Countries. *CA Cancer J. Clin.* **2021**, *71*, 209–249. [CrossRef]
156. Prager, G.W.; Taieb, J.; Fakih, M.; Ciardiello, F.; Van Cutsem, E.; Elez, E.; Cruz, F.M.; Wyrwicz, L.; Stroyakovskiy, D.; Pápai, Z.; et al. Trifluridine-Tipiracil and Bevacizumab in Refractory Metastatic Colorectal Cancer. *N. Engl. J. Med.* **2023**, *388*, 1657–1667. [CrossRef] [PubMed]
157. Duarte, D.; Rêma, A.; Amorim, I.; Vale, N. Drug Combinations: A New Strategy to Extend Drug Repurposing and Epithelial-Mesenchymal Transition in Breast and Colon Cancer Cells. *Biomolecules* **2022**, *12*, 190. [CrossRef] [PubMed]
158. Ducrey, E.; Castrogiovanni, C.; Meraldi, P.; Nowak-Sliwinska, P. Forcing dividing cancer cells to die; low-dose drug combinations to prevent spindle pole clustering. *Apoptosis* **2021**, *26*, 248–252. [CrossRef] [PubMed]
159. Nair, N.U.; Greninger, P.; Zhang, X.; Friedman, A.A.; Amzallag, A.; Cortez, E.; Sahu, A.D.; Lee, J.S.; Dastur, A.; Egan, R.K.; et al. A landscape of response to drug combinations in non-small cell lung cancer. *Nat. Commun.* **2023**, *14*, 3830. [CrossRef]
160. Yang, F.; Zhang, Y.; Liang, H. Interactive Association of Drugs Binding to Human Serum Albumin. *Int. J. Mol. Sci.* **2014**, *15*, 3580–3595. [CrossRef]

161. Ashraf, G.M.; Gupta, D.D.; Alam, M.Z.; Baeesa, S.S.; Alghamdi, B.S.; Anwar, F.; Alqurashi, T.M.A.; Al Abdulmonem, W.; Alyousef, M.A.; Alhumaydhi, F.A.; et al. Unravelling Binding of Human Serum Albumin with Galantamine: Spectroscopic, Calorimetric, and Computational Approaches. *ACS. Omega* **2022**, *7*, 34370–34377. [CrossRef]
162. Zhong, G.; Cai, X.; Wei, R.; Wei, S.; Cao, X. Curcumenol improves renal function in 5/6 nephrectomy-induced chronic renal failure rats via the SIRT1/NF-κB pathway. *Anat. Rec.* **2023**, *306*, 3189–3198. [CrossRef]

MDPI AG
Grosspeteranlage 5
4052 Basel
Switzerland
Tel.: +41 61 683 77 34

Molecules Editorial Office
E-mail: molecules@mdpi.com
www.mdpi.com/journal/molecules

www.ingramcontent.com/pod-product-compliance
Lightning Source LLC
Chambersburg PA
CBHW061628210926
51850CB00009B/2476
* 9 7 8 3 7 2 5 8 6 4 5 0 8 *